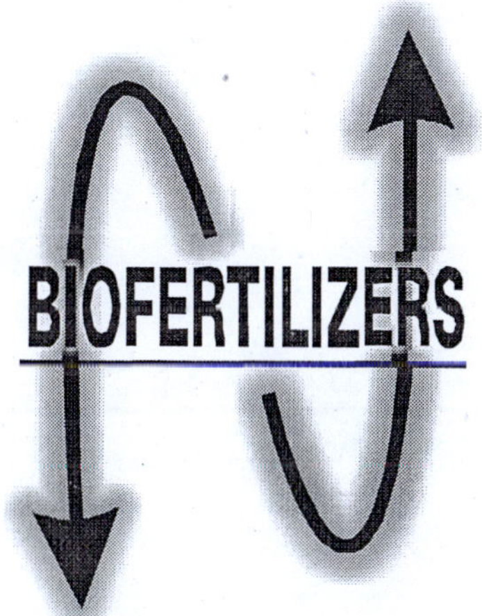

BIOFERTILIZERS

for Sustainable Agriculture

BIOFERTILIZERS

for Sustainable Agriculture

Arun K. Sharma
Scientist (Agronomy)
Central Arid Zone Research Institute
Jodhpur, India

Agrobios (India)

Published by:
Agrobios (India)
Agro House, Behind Nasrani Cinema
Chopasani Road, Jodhpur 342 003
Phone: 91-0291-2643993, 2642319
E. mail: agrobiosinia@gmail.com

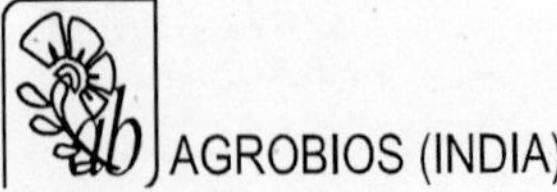

Disclaimer: The views expressed in this book are those of the author and not necessarily of the Institute he belongs to.

ISBN: 978-81-7754-118-2

Published by: Dr. Updesh Purohit for Agrobios (India), Jodhpur
Lasertypeset at: Yashee Computers, Jodhpur
Cover Design by: Shyam
Printed at: Agrobios Digitals, Jodhpur

Preface

Biofertilizers are the nutrient availability systems in which biological processes are involved. These biological systems are the essential part of life of several microbes found in soils and provide nutrients like nitrogen, phosphorus and several beneficial compounds for crop. Presence of these microbes makes soil alive and capable for sustainable support to the life of crop, the another life.

During the last fifty years due to the high use of synthetic fertilizers not only the overall soil fertility is decreased but also these biological systems adversely affected, visible in the decreasing trend of production even after using higher doses of fertilizers. Besides, the use of synthetic fertilizers is causing several environmental problems.

Experiencing the adverse effects of synthetic input dependent agriculture the concept of organic /sustainable/natural farming is gaining momentum. In these systems the production is based in synergism with nature, which makes systems of unending life i.e. sustainable

To revive the soil health and living which support to a sustainable production system the soil environment need to be made congenial for living of the useful microbial population, responsible for continuous availability of nutrients form natural sources.

In this book, the concept of making soil live for sustainable production is discussed. In chapter One, The organic matter which is the food and provide basic living environment for life in soil, is discussed with sources and dynamics. In chapter Two, Earthworm are not only good composter but also a indicator of health of life in the soil and even a farmer can easily know about the health of soil in his crop field. Considering these important roles of earthworm the information on vermiculture is given in chapter Three.

After making the living environment the scope, function and field applications and several other aspects of soil microbes responsible for Nitrogen fixation and Phosphorus mobilization are discussed from chapter Four to Seven. Chapter Eight deals with the holistic approach for making an autonomous cyclic mechanism of resources specially for nutrient for long term sustainability. In the chapter Nine, several aspects related to promotion the use of biofertilizer have been discussed.

Hopefully this book will provide all information for reviving of soil life/health with biofertilizer use to make long term sustainable production systems.

Arun K. Sharma

CHAPTER 1

The Living Soil

Long-term sustainability of life can only be possible in a living system and it is very difficult in the system having shortage of living elements e.g.survival of flora and fauna is very difficult in a desertified land. This fact is being realizing after using the synthetic fertilizers imprudently in the intensive agriculture for several decades that the life of soil or living system in the soil which supports the production of crop, the another living system, has been damaged to a great extent.

There are indications that the highly productive fertilizer and seed technologies introduced over the past several decades may be reaching a point of diminishing returns (Bouis 1993; Cassman et al. 1995; Flinn and De Datta 1984). Prospects for expanding low-cost irrigation, one of the driving forces behind yield increases, are also becoming more limited (Rosegrand and Svendsen 1993; Rosegrant 1997; Carruthers, Rosegrant, and Seckler 1997), as are the prospects for converting marginal lands into productive arable land (Backman et al. 1990; Crosson and Anderson 1992). Furthermore, new technologies such as genetically engineered, yield-increasing plants are not expected to be major factors in food production increases in developing countries during the next two decades (Hazell 1995; Peng, Khush, and Cassman 1994). Consequently, keeping pace with population growth and increasing land scarcity will be more difficult than in the recent past.

It is also established that many of the great gains in production made in the green revolution by use of high-yielding varieties with high inputs of inorganic fertilizer cannot be maintained indefinitely. Among the causes attributed to yield declines under long-term cultivation are changes in soil fertility associated with loss of organic matter and the accompanying decline in soil physical and chemical properties.

Concerns also are growing about the long-term sustainability of agriculture. Both the over- and under application of fertilizer and the poor management of resources have damaged the environment. In developed countries, for example, over-application of inorganic and organic fertilizer has led to environmental contamination of water supplies and soils (Canway and Pretty 1991; Bumb and Baanante 1996; NRC 1989). In developing countries, harsh climatic conditions, population pressure, land constraints, and the decline of traditional soil management practices have often reduced soil fertility (Stoorvogel and Smaling 1990; Tandon 1998; Henao and Baanante 1999; Bumb and Baanante 1996). Because agriculture is a soil-based industry that extracts nutrients from the soil, effective and efficient approaches to slowing that removal and returning nutrients to the sol will be required in order to maintain and increase crop productivity and sustain agriculture for the long term.

CURRENT ISSUES IN PLANT NUTRIENT USE AND SOIL FERTILITY

Slowing Yield Growth

Despite the continued development of new and improved modern varieties and greater use of chemical fertilizer, yield growth began to slow in the latter part of the 20th century. The world's annual cereal yield growth rate has declined from an average of 2.2 per cent in the 1970s to 1.1 per cent in the 1990s. Wheat yields in Asia grew at an average annual rate of 4.3 per cent during the 1970s. But during 1990-1997, wheat yields dropped to the far slower growth rate of 0.7 per cent per year. After rapid growth of almost 2.4 per cent per year during the 1980s, Asian rice yield growth fell to 1.5 per cent per year in the 1990s. This global slowdown has raised concerns that yield growth may have reached a plateau or begun to decline in many of the world's most fertile areas. These declines are partly attributable to poor soil management, which in turn has been accentuated by a number of other factors, including inappropriate policies, insufficient commitment to investment in agricultural research, falling agricultural prices, demograhic pressures, land availability constraints, and ill-defined property rights. The cumulative effect of all these factors has led to increased soil mining.

Slowdown in Investment

During the 1970s, governments and inter-governmental organizations gave investment in agricultural research a relatively high priority. In Asia, for example, real expenditure on public agricultural research and development grew at an average annual rate of 8.7 per cent. During the 1980s, however, real expenditure growth in Asia slowed to 6.2 per cent per year. Spending in Sub-Saharan Africa slowed even further, from 2.5 per cent during the 1980s. for developed countries, investment growth slowed as well from 2.7 per cent per year in the 1970s to 1.7 per cent in the 1980s (Alston, Pardey, and Smith 1998). Without continued investment in research on the development of yield enhancing technologies and sustainable agricultural management practices, crop yield growth could eventually stagnate and soil fertility could degrade to irrecoverable levels.

Investment in new irrigation infrastructure has also declined from its peak during the late 1970s and early 1980s. Both average annual public expenditures and annual lending and assistance for irrigation systems from international development agencies have fallen. Numerous factors have contributed to the reduction in irrigation investment, including the poor performance of some post investments, fewer low-cost irrigation sites for potential development, increased real capital costs for construction of new irrigation systems, and environmental concerns such as the spread of salinization (Rosegrant and Pingali 1994; Rosegrant and Svendsen 1993).

TABLE 1.1: Annual cereal crop yield growth rates, 1970s-1990s

Crop	Region	1970s	1980s (per cent)	1990s
Wheat	Asia	4.33	3.71	0.72
	Latin America	0.60	3.40	2.36
	Sub-Saharan Africa	3.54	0.92	-0.81
	World	2.10	2.78	0.42
Rice	Asia	1.61	2.42	1.55
	Latin America	0.70	2.97	3.71
	Sub-Saharan Africa	0.02	2.51	-0.56
	World	1.49	2.37	1.54
Maize	Asia	3.43	2.75	1.55
	Latin America	1.49	0.61	3.82
	Sub-Saharan Africa	2.26	1.72	2.09
	World	3.19	0.60	1.76
Cereals	Asia	2.90	2.79	1.46
	Latin America	1.69	1.28	3.12
	Sub-Saharan Africa	1.90	0.56	0.66
	World	2.18	1.79	1.12

Note: The 1990s refer to the period 1990-97.

Declining commodity prices during the 1980s also reduced governmental and intergovernmental incentives to make agriculture-related investments. Prices for coffee and cocoa during the 1990-92 period fell to 39 per cent of their nominal price in 1980-82. Similarly, the prices of wheat, maize, and rice in 1990-92 declined to 60, 61, and 50 per cent of the prices, respectively, in 1980-82 (OECD 1993). In addition to reducing investment incentives, declining prices reduced farmer's incomes, often forcing them to mine soils more intensively. If crop yields are to increase and if agriculture is to be sustainable over the long term, a renewed commitment to agricultural research and infrastructure will be necessary.

Nutrient Over-Application and Environment Contamination

Concern has also grown in recent years that the use of fertilizers, particularly inorganic fertilizers, can lead to serious environmental consequences. Environmental contamination of this type, however, is largely a problem in the developed world and a few regions of the developing world. As fertilizers make up a small share of the total production costs in many developed countries, farmers often apply fertilizer in excess of recommended levels in order to ensure high yields. Over-application of inorganic and organic fertilizers is estimated to have boosted nutrient capacity in the soil by about 2,000 kilograms of nitrogen, 700 kilograms of phosphorus, and 1,000 kilograms of potassium per hectare of arable land in Europe and North America during the past 30 years (World Bank 1996). Such oversupply of nutrients can lead to environmental contamination, which often has negative consequences for humans and animals. Over-application of nitrogen, for example, allows the nutrient to be carried away in ground-water and to contaminate surface water and underground aquifers. Ingestion of nitrate can be toxic to humans and animals when it is transformed within the body into nitrite, which affects the oxygen-carrying ability of red blood cells. Evidence also suggests that nitrite and the carcinogenic compounds it can create may also lead to goiter, birth defects, heart disease, and stomach, liver, and esophagus cancers (Conway and Pretty 1991).

Leaching and run-off of surplus nitrogen and phosphorus into rivers, lakes, and inlets can cause eutrophication-an excess accumulation of nutrients in water that promotes the overproduction of algae. Excess surface algae deprive underwater plants of sunlight, which in turn alters the aquatic food cycle. The decomposition of dead algae by bacteria reduces the amount of oxygen in the water available for fish.

Nitrogen also escapes into the atmosphere in the form of nitrogen gas and various nitrous oxides. In the upper atmosphere, nitrous oxides react to form acid rain, which can harm crops, acidify soil and water, and damage property. Cumulative application of acidifying ammonia-based fertilizers, together with acid rain, also contributes to soil acidification.

Evidence is mounting that excessive fertilization can damage the environment. In the United Kingdom, some 1.6 million people get water with nitrate levels that exceed guidelines, and Danish, Dutch, and German coastal regions show signs of eutrophication (Bockman et al. 1990. Furthermore, USDA (U.S. Department of Agriculture) estimated that agriculture causes nearly two-thirds of the pollution in U.S. rivers and that runoff from excess plant nutrient application causes close to 60 per cent of the pollution in lakes.

The Solution

The overall strategy for increasing crop yields and sustaining them at a high level must include an integrated approach to the management of soil nutrients, along with other complementary measures. An integrated approach recognizes that soils are the storehouse of most of the plant nutrients essential for plant growth and that the way in which nutrients are managed will have a major impact on plant growth, soil fertility, and agricultural sustainability. Farmers, researchers, institutions, and government all have an important role to play in sustaining agricultural productivity.

The use of inorganic fertilizers by farmers in developing countries, in contrast with those of the industrialized world, still have only limited access to inorganic fertilizers (FAO, 1993). This skewed distribution, and the high cost of fertilizer in most parts of the world, emphasizes the need for increasing the efficiency of their use, and many researchers look forward to the combination of inorganic with organic sources of nutrient as the best route for this purpose.

BIOLOGICAL MANAGEMENT OF SOIL FERTILITY

The practice of biological management of soil fertility is the old concept. It is the use of biological resources of the ecosystem, particularly those of the soil itself, for the manipulation of soil fertility. *It should be clear at the outset that this is not the same as organic farming* (NRC, 1989). It does not eschew the use of inorganic inputs, but rather focuses on increasing the efficiency of their use by biological means. Emergence from the circumstances of nutritional deficiency and poverty that characterize the lives of many of the small-scale farmers of the world can only be achieved where there are sufficient resources to raise food productivity. Biological and physicochemical management are essentially based on integrated approach to soil fertility management.

According to Sanchez (1994) "we should rely on biological processes by adapting germplasm to adverse soil conditions, enhancing soil biological activity and optimizing nutrient cycling to minimize external inputs and maximize the efficiency of their use".

It is an ecological approach to soil fertility management which has favored balanced farming systems. Swift et al. (1979), Tinsley and Darbyshire (1994), Pankhurst et al. (1994) and Woomer and Swift (1994) suggested in their review works that increased interest in a biological approach to soil fertility management will help to restore soil fertility and will solve many problems related to soil management.

The biological soil fertility management is an ecological approach for sustainable agriculture and is mainly concerned with the maintenance of yield, closely associated with desires to conserve natural resources, including a greater value accorded to maintenance of biodiversity. An increased efficient use of resource, including the need to utilize all available resources within economic limits will be realized in the long term as well as profitable in the short term too (Lynam and Herdt, 1989; Spencer and Swift, 1992).

SUSTAINABLE AGRICULTURE: NEED OF THE HOUR

There is an urgent need to develop farming techniques, which are sustainable from environmental, production, and socioeconomic points of view. The means to guarantee sufficient food production in the next decades and beyond is critical because modern agriculture production throughout the world does not appear to be sustainable in the long-term. The agricultural community is thus setting it hopes on *sustainable agriculture,* which will maintain the cycles of input-output and ecosystem balance. Definition form FAO (1988), seems appropriate in this context

Sustainable rural development is the management and conservation of the natural resource base, and the orientation of technological and institutional change in such a manner as to assure the attainment and continued satisfaction of human needs for the present and future generations. Such sustainable development, in the agriculture, forestry and fishers sectors, conserves land, water, plant and animal genetic resources, is environmentally non-degrading, technically appropriate, economically viable and socially acceptable.

Evolution of Sustainable Agriculture

In the 1960s and 1970s, a growing environmental agriculture movement evolved in response to increasing soil erosion, pesticide use, and groundwater contamination. Simultaneously, economic conditions for farmers were becoming more stressful and the number of family farms declined.

In 1980s, Wes Jackson of The Land Institute in Salina, K. S., began using the term *sustainable agriculture* to describe an alternative system of agriculture based upon resource conservation and quality of rural life.

While sustainable agriculture has become the umbrella under which, many of the above-mentioned alternatives farming systems fall, it is important to note that sustainable agriculture is really a long-term goal, not a specific set of farming practices. In temperate zones sustainable agriculture was defined as such:

> *Sustainable agriculture is a philosophy based on human goals and on understanding the long-term impact of our activities on the environment and on other species. Use of this philosophy guides our application of prior experience and the latest scientific advances to create integrated, resource-conserving, equitable farming systems. These systems reduce environmental degradation, maintain agricultural productivity, promote economic viability in both the short and long term and maintain stable rural communities and quality of life.*

In this context, sustainable agriculture embraces all agricultural systems, striving to meet these criteria. Many aspects of modern conventional agriculture are included in sustainable agriculture, just as are many aspects of alternative farming systems.

One aspect of modern agriculture receiving a lot of attention in the sustainable agriculture discussion is the use of chemical inputs to supply fertility and pest control. White agriculture chemicals will continue to play an important role in American agriculture. Many farmers are looking at alternatives due to environmental, economical, or regulatory reasons. In a transition to farming systems more reliant on biological methods of production, low-input farming serves as in intermediary step.

Sustainable agriculture emphasizes the conservation of its own resources. For a farm to be sustainable, it must produce adequate amounts of high-quality foods, be environmentally safe, and where appropriate, be profitable. Sustainable farms minimize their purchased inputs (fertilizers, energy and equipment) and rely, as much as possible on the renewable resources of the farm itself. This is especially important in the 90 per cent of farms that exist in the third world, where these inputs are often not available or affordable.

Agricultural ecosystems, unlike the natural ecosystems, are human manipulated ecosystems and regular and sometimes intense disturbances are a major part of economically viable agriculture management systems.

Sustainable agriculture is a complex issue associated with producing food while maintaining our biophysical resources including soil, water and biota with no adverse impacts on the wider environment. It should:

1. Maintain or improve the production of 'clean' food
2. Maintain or improve the quality of landscapes, which includes soils, water, biota and aesthetics.
3. Have minimal impact on the wide environment.
4. Be economically viable
5. Be acceptable to society.

There are different concepts of sustainable agriculture, but none is generally accepted. It embraces several forms of non-conventional agriculture that are often called organic, alternative, ecological or low input. However, from economic and ecological perspectives, two basic criteria must be met if agriculture is to be sustainable in the long term. These are:

Sustainable farming uses some form of integrated pest management for pest control, and this can include the use of chemical pesticides that are not used by organic farmers. Thus, sustainable agriculture does not mean a return to the farming methods of the late 1800's. Rather, it combines traditional techniques that stress conservation with modern technologies, such as improved seed, modern equipment for low-tillage practices, integrated pest management that relies heavily on biological control principles, and weed control that depends on crop rotations. Sustainable farms try to use wind or solar energy instead of purchased energy, and use organic animal manure and nitrogen-fixing legumes as green manure to maintain soil fertility, as much as possible, thereby minimizing the need to purchase inputs from outside the farm. The use of genetically engineered crop strains is certainly not excluded by sustainable farming. The emphasis is on maintaining the environment, not on rules about what can or cannot be done. Profits from sustainable farms can exceed those of conventional farms.

Innovative farmers have developed many alternative farming methods and systems. These systems consist of a wide variety of integrated practices and methods suited to the specific needs, limitations, resource bases and economic conditions of different farms. To make wider adoption possible, however, farmers need to receive information and technical assistance in developing new management skills.

SUSTAINABLE LIVELIHOOD

Scientists, the world over, have developed new technologies for application in the field. Over the years they have come to realize that these technologies must be integrated into existing farming systems, and have been going out into the fields to try to understand the problems and to see how the solutions are working. The large-scale adoption of technology-based agriculture can lead to a loss of knowledge of traditional farming methods, which were developed over many centuries and are adapted to the physical, biological and social environments of the people. The knowledge may be useful in devising strategies for sustaining agriculture in the developing regions. Traditional agricultural methods may be applicable for designing sustainable systems in the developed countries. In other words, the rich and sophisticated can learn a lot from the poor and unsophisticated.

During the past one-decade, there have been important changes in the thinking of agricultural scientists and policy makers concerned with agricultural research in developing countries. It is clear that the type of research that led to the Green Revolution has benefited the resource-rich farmers, whereas resource-poor families have been left behind.

It is increasingly recognized that who produces food and where it is produced may actually be as or more important than how much is produced?

If resource-rich farmers produce food that subsistence farmers cannot purchase, there is no improvement in the food situation. Population projections show that many rural areas with fragile ecologies will have to continue supporting large populations.

Food can be produced in a more sustainable way that protects the resource base. The techniques are known, available and used by some farmers. Although more researches are needed, however, these techniques could find more widespread use given the proper incentives. Currently, the incentives (price supports, tax policies and export subsidies) favour industrial agriculture. Society needs to replace them with an incentive that favours sustainable agriculture. Because sustainable agriculture uses fewer inputs, this may cause some dislocation in the industrial sector and will be strongly resisted. In the Third World, much research on indigenous farming methods is needed, and the top-down approach to research needs to be replaced and/or supplemented by on-farm research.

According to Dr. M.S. Swaminathan, *the greening of agriculture requires the greening of both technology and public policy*. Producing more food and agricultural commodities from less land, water and energy is a task that will call for the integration of the best in modern technology, with the ecological strengths of traditional farming practices.

MAKING SUSTAINABLE FARMING SYSTEMS

The shift from chemical to sustainable or ecological agriculture should, however, be gradual. A sudden switchover can spell disaster and discourage farmers from taking to this course. At least two to three seasons will be needed for the transition, and during the interim years the farmers could build up a sufficient organic base to fertilize the fields and improve the productivity of the soil. Once a state of ecological balance has been attained and there are a good number of beneficial organisms to check the explosion of pests and pathogens, crop yields of a high order can be ensured. The cost of cultivation can be brought down substantially as many farm-grown inputs can be integrated efficiently in the farming system.

Sustainable farming can be ecofriendly and profitable too. It is a self-reliant method and has to be an integrated system. Since it is a knowledge-intensive practice, one has to keep pace with the dynamics of nature to increase the biological productivity of the soil.

Since eco-farming uses several farm grown inputs, and less dependent on market purchased inputs, it is economically attractive to the growers. Case studies show that when a chemical far incurred about Rs. 11,250 towards the cost of cultivation per hectare of rice, an organic farm spent Rs. 10,590. Alternatively, if the biofertilizers and organic nutrient supplements such as neem cake are subsidized to extent the same monetary advantage as chemical fertilizers, the cost of ecological farming will come down significantly.

It has been observed that in chemical farming, fertilizers and plant protection chemicals accounted for nearly 30 per cent of the total cost. The cost of market purchased inputs in ecological farming was far less, and bulk of it went towards the purchase of neem cake.

In both the cases, the yield remained almost the same at 5,625 kg paddy per hectare. The straw yield was eight tonnes. The net returns from the ecological farming system at the current cost of Rs. 3. 34 per kg paddy is Rs. 8,197. 50. The chemical farming, the net profit is Rs. 7,500. If one gets a premium price for the poison-free, organically grown rice, the economic returns from the ecological returns from the ecological farming system will be highly encouraging. While the modern system of farming expended nearly 31,000 MJ of energy, the ecological farming method accounted for only 23,400 MJ. High energy inputs such as fertilizer, plant protection chemicals, herbicides and high speed diesel are substituted with biological inputs and low-energy human and cattle power. The output/input ratio in the case of modern farming is 3. 76, while in the case of ecological farming it is more than 4. 95.

Sustainable farming is more labour intensive, and several of its components such as vermicompositing, vermiwash, and generation of botanical pesticides and rearing beneficial insects such as *Trichogramma* parasites, ladybird beetles, and spiders for release in the field as bio-control agents offer scope for unemployed and landless youth.

Group and community endeavour holds the key for successful implementation of ecological farming over larger. Community participation in pest control and rodent eradication will be rewarding. Even in the case of developing biofertilizers such as blue green algae and *Azolla* group action will pay rich dividends.

The above practices are not unique to biological management but common to farming practice the world over. The use of biological inputs such as nitrogen fixing bacteria, mycorrhiza, or soil fauna as a means of enhancing the endemic biological activities are the means of biological soil management. Direct management is also achieved by the use of organic matter inputs – in effect a means of selectively feeding the heterotrophic biological populations of soil – a practice of very ancient origin but at times eschewed in modern agriculture. Equally direct but usually unintentional effects are also achieved by the use of pesticides, which may kill particular groups of soil organisms that are involved in processes of significance to soil fertility. Management techniques such as tillage and fertilization also influence the activity of the biota indirectly by altering the physical and chemical environment of the soil.

The success of biological management practices depends on following two preconditions, which must be satisfied:

1. The availability of a management practice that is practically and economically acceptable to the farmer.
2. The demonstration by the scientist that the practice leads to enhanced soil fertility.

MANAGEMENT OF ORGANIC INPUTS

Regulation of Nutrient Dynamics by Resource Quality Control

Biological management of soil fertility depends on the manipulation of organic inputs to the soil more than on any other practice. The organic matter supplies energy and nutrients to the soil biota to stimulate their activities and therefore promote soil fertility.

Organic inputs are processed by a complex web of soil organisms, however, biological management is based on the assumption that the outcomes are relatively consistent. These outcomes are multiple and include the generation of inorganic nutrients such as NH_4, NO_3, PO_4, and SO_4 from organic compounds of these elements; the synthesis of soil organic matter (SOM) from precursor compounds in the organic inputs; and modification of the physical structure of the soil as a result of stimulation of biological activities. Quantitative variation in these outcomes occurs as result of following variety of potential aspects:

1. Organic inputs
2. The diversity of organisms involved in the conversion processes.
3. The complexity of the interactions between them.
4. The biological activities are regulated by the quantity and quality of the organic manure added.
5. The environmental conditions under which the processing takes place (including climate and soil type).
6. The status of the soil community.

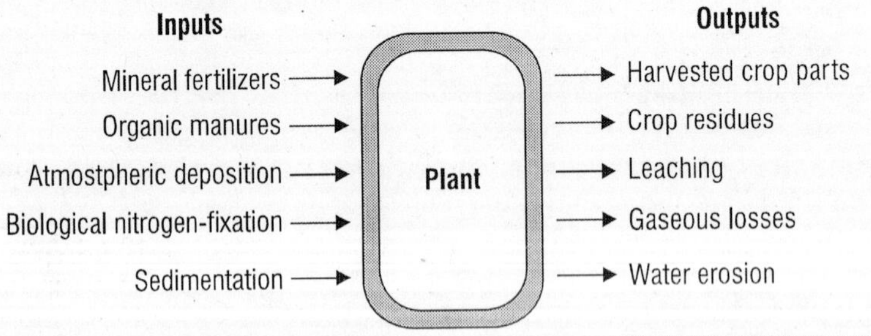

FIG. 1.1: The plantnutrient balance system

Increasing insight into the ways in which these factors regulate soil biological activity constitutes the main progress in achieving a predictive capacity for biological management of soil fertility. Most attention has been given to the use of organic mulches to supply nutrients such as nitrogen and to the possibilities of utilizing *resource quality control* as a tool for managing nutrient availability.

Resource quality refers to the regulatory effect of the chemical composition of an organic resource on its rate and pattern of decomposition and nutrient release.

Early studies of decomposition processes demonstrated the importance of N concentration and C: N ration as determinants of the N-supplying capability of plant residues.

Studies in natural ecosystems subsequently demonstrated that other indices such as the lignin concentration or the lignin: N ratio provided a better predictor of N release patterns.

Berg and McClaugherty (1987) suggested that N is not released from forest litters until decomposition of lignin commences, and others have also stressed the importance of the more recalcitrant components as regulators of decomposition rate and nutrient release.

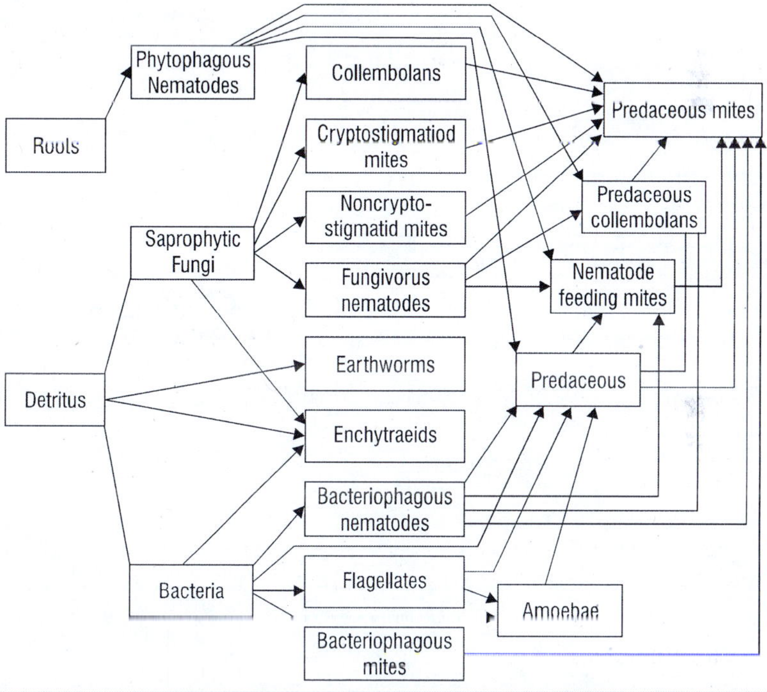

FIG 1.2: A crop residue (detritus)-based soil food web from a calcareous soil under arable cropping in The Netherlands. Under conventional tillage the bacteria-based food web is dominant; under reduced tillage the fungal and earthworm paths are more prominent (Verhoef and Brussaard, 1990).

Polyphenols have also been implicated as regulators of N release in materials that, despite being nutrient-rich and low in lignin, are slow to release N. It has been proposed that the polyphenols interact with N to form stable polymers that are resistant to breakdown and therefore delay N release.

The above theories have been used to evaluate organic residues as sources and suppliers of nutrients within cropping systems.

TABLE 1.2: Chemical Composition, Decomposition Rate, and N-Release Rate for Various Agricultural Residues

Material	Lignin (%)	Polyphenol (%)	C/N	kDM[a]	kN[a]
Gliricidia	12	1.62	13	0.127	0.152
Rice straw	5	0.55	42	0.106	0.052
Maize stover	7	0.56	43	0.085	0.077
Lecuaena	13	5.02	13	0.062	0.055
Acioa	48	4.09	28	0.010	0.001

[a] Decomposition constant (week^{-1}) for dry matter (DM) and nitrogen (N), respectively.
From Tian, G. et al., 1992. *Soil Biol. Biochem.*, 24: 1051-1060.

ENVIRONMENTAL REGULATION OF ORGANIC MATTER MANAGEMENT

Organic matter chemistry – resource quality –is not the only factor needed to make reliable predictions of the effects of organic matter inputs. The physicochemical environment has control over biological processes from a higher level in the hierarchy than resource quality. Much of this regulation such as that deriving from the climate lies outside the control of the farmer. There are practices, however, that can be manipulated so as to *environmentally tune* the influence of organic inputs on soil biological processes. These include the location and timing of the application of organic matter. It is well established, for instance, that incorporation of residues in the soil, as opposed to surface mulching, can accelerate decomposition processes significantly and thus alter the dynamics of nutrient availability. The breaking up of the residues during tillage may be a factor in this effect, as well as the transfer of the organic matter into the more favorable environment below the soil surface.

It is well established that cultivation changes the composition, diversity, abundance, activity of the soil community and a variety of farming practices involve the management of organic inputs. For instance, there is now a considerable body of data relating to differences in soil biological populations and processes in minimum-tillage systems as compared with conventional tillage. These effects have been shown to be dramatic in relation to many groups of soil fauna, including such keystone groups as earthworms, with subsequent benefits gained in low-till systems to the physical structure and water regime of the soil. The potential exists to utilize these practices to manipulate the biological community and thence to determine the outcome of decomposition processes. A major challenge is to determine pragmatically the extent to which these "fine-tuning" factors of microenvironment and community structure need to be taken into account when developing management systems.

ORGANIC-INORGANIC INTERACTIONS

Nitrogen

The use of organic inputs of different qualities to improve the efficiency of nutrient transfer to the crop will not be a practice of any significance if the total amount of nutrient available is insufficient to satisfy the needs for production. Significant input of N to replace that removed in harvest may be achieved in farming systems incorporating N-fixing species. The amount of recyclable N can be enhanced in tree-based systems such as traditional fallow rotations, savanna-based mixed farming systems, or modern agroforestry practices. These systems have the additional advantage of mobilizing other nutrients such as P and K by deep capture by tree roots from lower soil layers.

Phosphorus

In a wide range of tropical environments the most critical element limiting production is phosphorus. While mycorrhizal infection can increase the efficiency of P uptake, there is no biological means of increasing the absolute amount of available P in an ecosystem. P in the form of inorganic fertilizer is thus essential for increased productivity in a large range of tropical farming systems. Other nutrients may also need to be supplied in this form as well. One approach advocated for this is management that incorporates both inorganic and organic inputs of nutrients.

Nutrient Mixture

There is a long history of agronomic research on crop response to the application of mixtures as compared with single sources of nutrients. It has been observed that mixtures of organic and inorganic fertilizers result in increased yields beyond those achieved with inorganic fertilizer alone; in others the yields are diminished. High-quality inputs, releasing nutrients early in the process of decomposition, will serve to supplement inorganic fertilizers rather than competing with them. The most effective response of the crop to fertilization, whether from organic or inorganic sources, may come if the maximum demand for nutrients from the plant coincides in time and space with the availability of nutrients in the soil.

NEED FOR ORGANIC INPUTS

Organic farming is the path way that leads us to live in harmony with nature. Organic agriculture is the key to a sound development and a sustainable environment. It minimize environment environmental pollution and the use of non-renewable natural resources. It conserves soil fertility and soil erosion through implementation of appropriate conservation principles.

Several reasons have been emphasized for the need of organic agriculture, like limited land holdings, poor socio-economic conditions of farmers, rise in input cost etc. The broadest view shows two major reasons viz., population and environment, emphasized the ultimate need for ecofriendly technologies. FAO estimates that by the year 2000 the global population will be 25% higher than in the mid 1980s and that 90% of this population increase will be in the developing countries like India. It will be necessary to increase the supply of food and other agricultural products to meet their needs. At the moment, many of the methods being used to increase production are damaging to natural resources and the environment and farmers are supposed to invest heavily into inputs to improve yields and productivity. Further, the produce from organic farming has become inevitable. The food containing pesticides and other chemicals are increasingly made obvious by many research studies revealing the presence of pesticide residues in eatables. About one million people suffered from pesticides poisoning and 20,000 die every year due to the toxic effects of these chemicals world-wide but the crop loss due to pest is still 15,000 crores, apart from killing a portion of the human population by the same pesticide use. This perspective is also one of the needs for organic approach to farming. In order to avoid the deleterious effects of synthetic chemical fertilizers and pesticides, organic agriculture is needed as an alternative to provide ecologically safe methods of farming.

LONG-TERM EFFECTS

Organic matter have long term effects. The strongest evidence for nutrient transfer from input to plant is obtained in studies using isotopically labeled material. In the few examples available, the N use efficiency of plant residues by a first crop is low, in the order of 15% for legume residues and 5% for cereal straw residues, but there is much variation. This is broadly comparable with, or slightly lower than, the N use efficiency of inorganic fertilizer in the same environment. In the examples given there is, however, a consistently greater partitioning of N from the organic material to SOM than to the plant. This suggests that residual effects are likely to be more important than those in immediate seasons. There is a substantial body of evidence to substantiate this point from long-term experiments in Africa and elsewhere (Swift et al., 1994b).

Optimal development of integrated nutrient use systems will be achieved when it is possible to predict the outcome of a given strategy for use and management of nutrient resources (both organic and inorganic) over the long and short term.

TABLE 1.3: Partitioning of N added in labeled plant residues to the subsequent Crop and to soil organic matter

Authors	Crop (%)	Organic matter (%)
Ladd et al., 1981	11-17	72-78
Janzen et al., 1990	14	21-40
Ng Kee Kwong et al., 1987	11-14	73-84

Source: Myers, R.J.K. et al., 1994. The *Biological Management of Tropical Soil Fertility,* Woomer, P.L. and Swift, M.J., Eds., John wiley & Sons, London, 81-117.

THE FARMING SYSTEMS CONTEXT

The history of agricultural development over the last several decades contains many examples of failures in adoption or sustenance of technologies that were judged scientifically sound (Lal, 1987). The resource management technology may be failure due to water-management schemes in the inland, minimum tillage practices, poor take-up of improved clearing methods, soil conservation methods, and spread of alley cropping. The other reasons for failure to adopt or persist with technology are social, cultural or economic factors as to technical deficiencies. This necessitates an approach to technology development and transfer than integrates the criteria used by the farmer in deciding the actions, as well as those assumed in the scientist's models.

Soil management is only one small component of a farmer's spectrum of activity. The knowledge and perceptions of soil-based constraints should be combined with a holistic appraisal. Farmer's should have balanced approach between various alternatives based on whole range of available environmental, biological, and economic information.

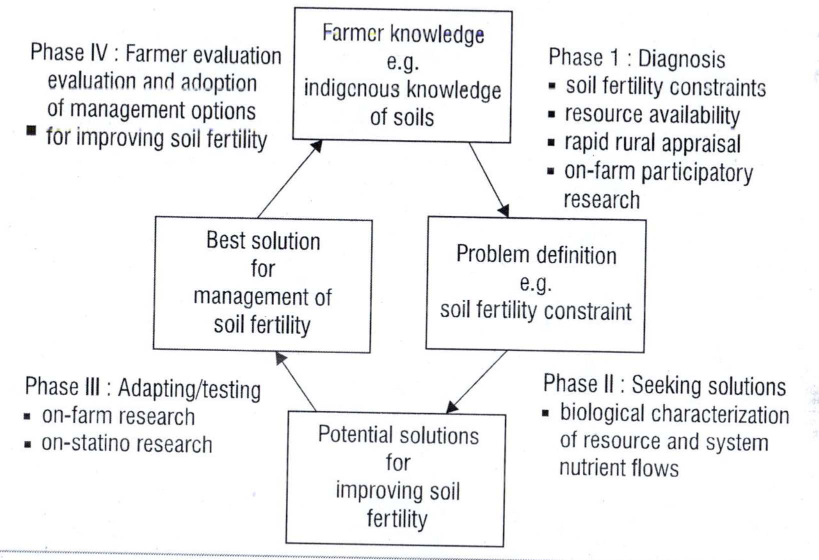

FIG 1.3: The *farmer-back-to farmer* approach to agricultural problem identification and solution (Rhoadoo and Booth, 1902; Moran, 1987) as applied to soil fertility research.

Technological change is thus more likely to succeed in terms of both adoption and of impact on production when it is adapted to farmers' circumstances rather than merely optimized in terms of scientists' criteria. This is most likely to be achieved when scientists and farmers work together throughout the research and development process.

Fig. 1.3 illustrates stages in a research processes that is initiated by joint diagnosis of soil fertility problems. This information, based both on farmers' perceptions and on biophysical and socioeconomic characterization of the farming system. It should be used to guide scientific investigation of the opportunities for improvement in soil management practices.

The process-and-system approach places farmer decision making at the center of the technology development process (Fig. 1.2 and 1.3), highlighting the need for detailed characterization of the environmental context that influences those decisions.

The capacity for increasing and sustaining productivity through improved soil fertility management is subject to a variety of limits of physicochemical, biological economic, social, and cultural dimensions. The determination of these limits, the establishment of interactions – including hierarchical relationships, and the opportunities for modifying them is one of the major purposes of system characterization.

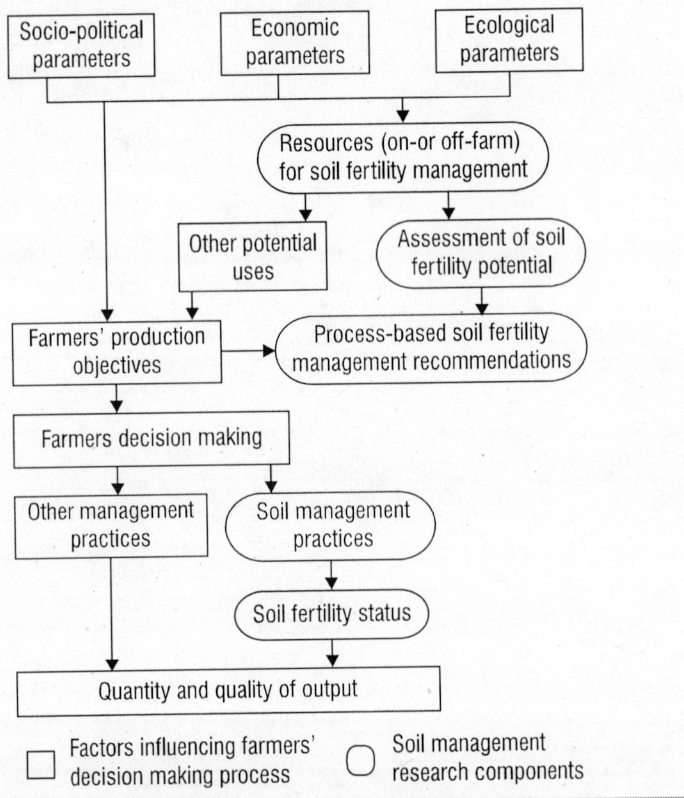

FIG. 1.4: Integration of the research results for biological management of soil fertility with system level information on farmers' decision making.

CHAPTER 2

Organic Sources and Dynamics

Organic matter is the food for most of the flora and fauna responsible for soil biological activities. Proper and ensured management of soil organic matter is the prime requirement for an efficient soil biological system.

Farmers since ancient times have recognized significant benefits of soil organic matter to crop productivity. These benefits have been the subject of controversy for centuries and some are still debated today. The following list includes many of the recognized benefits of soil organic matter.

1. It serves as a slow-release source of N, P, and S for plant nutrition and microbial growth.
2. It possesses considerable water-holding capacity, and thereby helps to maintain the water regime of the soil.
3. It acts as a buffer against changes in pH of the soil.
4. Its dark colour contributes to absorption of energy from the sun and heating of the soil.
5. It acts as *cement* for holding clay and silt particles together, thus contributing to the crumb structure of the soil, and to resistance against soil erosion.
6. It binds micronutrient metal ions in the soil that otherwise might be leached out of surface soils.
7. Organic constituents in the humic substances may act as plant-growth stimulants.

The term **soil organic matter** is generally used to represent the organic constituents in the soil, excluding undecayed plant and animal tissue, their partial decomposition products, and the soil biomass. Thus, this term includes:

1. Identifiable high-molecular-weight organic materials such as polysaccharides and proteins.
2. Simpler substances such as sugars, amino acids, and other small molecules, and (iii) humic substances.

Soil organic matter is frequently said to consist of humic substances and non-humic substances. Non-humic substances are all those materials that can be placed in one of the categories of discrete compounds such as polysaccharides and sugars, proteins and amino acids, fats, simple organic acids, and so on; humic substances are the other, unidentifiable components.

The term **humic substance** refers to an operationally defined, heterogenous mixture of naturally occurring organic materials. These substances are ubiquitous in nature and arise from the decay of plant and animal residues in the environment. Humic substances are generally classified into humic acid, fulvic acid, and humin on the basis of their solubility in water as a function of pH.

ORGANIC MATERIALS USE

Organic manures, including animal manures, crop residues, green manures, and composts were traditionally and preferentially used in developing countries until the 1960's when inorganic chemical fertilisers began to gain popularity.

The shift away from organic recycling practices also served to reemphasize the value of, and need for, organic amendments for short-and long-term improvement of cultivated soils and maintenance of soil productivity. Without regular additions of adequate amounts of organic materials to soils, there is often a concomitant increase in nutrient runoff, erosion, and gradual deterioration of soil physical properties.

Environment pollution has also become an international concern. Proper processing and recycling of organic wastes and residues as resources for agriculture can greatly reduce environmental pollution. Additional benefits include improved public health, conservation of resources, and better appearance of both urban and rural communities.

Composting to Enhance the Usefulness and Acceptability of Organic Materials

Composting is one way in which some of the problems associated with the utilization of various organic wastes (e.g., odors, human pathogens, and storage and handling constraints) can be resolved. Composting is an ancient practice whereby farmers have converted organic wastes into resources that provide nutrients to crops and enhance soil tilth, fertility, and productivity. Through composting, organic wastes are decomposed, nutrients are made available to plants, pathogens are destroyed, and malodours are abated. Composts provide a more stabilized form of organic matter than do raw wastes, and can vastly improve the physical properties of soils. For example, addition of sludge compost to sandy soils will increase their ability to retain water and render them less droughty. In heavy-textured clay soils, the added organic matter will increase permeability and water infiltration and clay soils have been shown to reduce soil compaction, lower the bulk density, and increase the rooting depth.

The value of organic materials as fertilisers and soil conditioners is often misunderstood and has been the source of some controversy. The simplest and most common means of estimating the value of organic amendments is by assessing the current market value of the potentially available plant nutrients they contain. Usually this is done in terms of their macronutrient content, i.e., N, P, and K.

Special Management Practices for Utilization of Organic Materials on Land

Special management practices may be needed if we are to obtain the full value of organic wastes as soil amendments and as sources of plant nutrients. For example, the value of organic materials as fertilisers is increased if nutrient release through decomposition and mineralization coincides with the crop's nutrient demand curve. The nutrient release pattern of organic materials can be controlled to some extent by proper timing of application, pretreatment methods such as composting, and by the method of application. For example, certain organic materials may decompose more slowly if left on the soil surface or concentrated as bands in the soil than if mixed thoroughly in the tillage layer.

BULKY ORGANIC MANURES/ANIMAL MANURES/ CROP RESIDUES/COMPOSTS

Agricultural wastes (animal manures and crop residues or mixtures of the two) are now considered quite an important component for horticultural crop production especially that based on organic systems. Agricultural wastes can now be considered to include crop after harvest and primary processing, tree residues organic/plant residues from social forestry, animal excreta and processing left over from the slaughter houses and agro-industrial wastes. Thus, agricultural wastes comprise all organic wastes produced and disposed of or used in primary agricultural production. Of course, these materials are essentially by-products of production rather than wastes in the strict sense of the word. With the advent of abundant and relatively inexpensive chemical fertilizers, agricultural manures became agricultural wastes, and what was formerly an asset is now often regarded as a disposal problem. Traditionally, cattle, pigs, and poultry were fed on free range or housed in relatively small numbers. Today, in most developed countries, intensive animal production is now being practiced where increasing numbers of livestock are being fed on fewer larger farms.

TYPES OF BULKY ORGANIC MANURES

Farmyard manure (FYM), farm compost, town compost, night soil, sludge and green manures are bulky in nature and supply large quantities of organic matter but small quantities of plant nutrients in comparison to the inorganic fertilisers.

Farmyard Manure

Good-quality farmyard manure is perhaps the most valuable organic matter applied to a soil. It is the most commonly used organic manure in most countries of the world. It consists of a decomposed mixture of cattle dung, the bedding used in the stable and any remnants of straw and plant stalks fed to cattle. It is one of the most important agricultural by-products. Unfortunately more than 50 per cent of the cattle dung produced in India is burnt as fuel and is thus lost to agriculture.

TABLE 2.1: Average nutrient content of bulky manures

Manure	Percentage content		
	Nitrogen (N)	Phosphoric acid (P_2O_5)	Potash (K_2O)
Animal refuse			
Cattle dung, fresh	0.3-0.4	0.1-0.2	0.1-0.3
Horse dung, fresh	0.4-0.5	0.3-0.4	0.3-0.4
Sheep dung, fresh	0.5-0.7	0.4-0.6	0.3-1.0
Nightsoil, fresh	1.0-1.6	0.8-1.2	0.2-0.6
Poultry manure, fresh	1.0-1.8	1.4-1.8	0.8-0.9
Raw sewage, fresh	2.0-3.0	-	-
Sewage sludge, dry	2.0-3.5	1.0-5.0	0.2-0.5
Sewage sludge, activated dry	4.0-7.0	2.1-4.2	0.5-0.7
Cattle urine	0.9-1.2	tr.	0.5-1.0
Horse urine	1.2-1.5	tr.	1.3-1.5
Human urine	0.6-1.0	0.1-0.2	0.2-0.3
Sheep urine	1.5-1.7	tr.	1.8-2.0
Wood ashes			
Ash, coal	0.73	0.45	0.53
Ash, household	0.5-1.9	1.6-4.2	2.3-12.0
Ash, wood	0.1-0.2	0.8-5.9	1.5-36.0
Farm, factory and habitation wastes			
Rural compost, dry	0.5-1.0	0.4-0.8	0.8-1.2
Urban compost, dry	0.7-2.0	0.9-3.0	1.0-2.0
Farmyard manure, dry	0.4-1.5	0.3-0.9	0.3-1.9
Filter-press cake	1.0-1.5	4.0-5.0	2.0-7.0
Plant residues			
Rice hulls	0.3-0.5	0.2-0.5	0.3-0.5
Groundnut husks	1.6-1.8	0.3-0.5	1.1-1.7
Straw and stalks			
Pearlmillet	0.65	0.75	2.50
Banana, dry	0.61	0.12	1.00
Cotton	0.44	0.10	0.66
Sorghum	0.40	0.23	2.17
Maize	0.42	1.57	1.65
Paddy	0.36	0.08	0.71
Tobacco	1.12	0.84	0.80
Pigeonpea	1.10	0.58	1.28
Wheat	0.53	0.10	1.10
Sugarcane trash	0.35	0.10	0.60
Tobacco dust	1.10	0.31	0.93
Tree leaves, dry			
Calotropis gigantea	0.35	0.12	0.36
Careya arborea	1.67	0.40	2.20
Cassia auriculata	0.98	0.12	0.67
Dillenia pentagyna	1.34	0.50	3.20

Manure	Percentage content		
	Nitrogen (N)	Phosphoric acid (P_2O_5)	Potash (K_2O)
Madhuca indica	1.66	0.50	2.00
Pengamia pinnata	3.69	2.41	2.42
Pterocarpus marsupium	1.97	0.40	2.90
Terimalia chebula	1.46	0.35	1.35
Terminalia paniculata	1.70	0.40	1.60
Terminalia tomentosa	1.39	0.40	1.80
Xylia dolabriformis	1.37	0.30	1.61
Green manures, fresh			
Cowpea (*Vigna catjang*)	0.71	0.15	0.58
Sesbania aculeata	0.62	-	-
Cluster-bean (*Cyamopsis tetragonoloba*)	0.34	-	-
Horse-gram (*Dolichos biflorus*)	0.33	-	-
Mothbean (*Vigna aconitifolia*)	0.80	-	-
Greengram (*Vigna radiata*)	0.72	0.18	0.53
Sunnhemp (*Crotalaria juncea*)	0.75	0.12	0.51
Blackgram (*Vigna mungo*)	0.85	0.18	0.53

COMPOSTED MANURE

Another method of augmenting the supplies of organic matter is the preparation of compost from farmhouse, and cattle-shed wastes of all types. Composting has been advocated and adopted extensively during the past 25 years. Composting is the process of reducing vegetable and animal refuse (rural or urban) to a quickly utilizable condition for improving and maintaining soil fertility. Studies conducted in India and elsewhere have shown that good organic manure similar in appearance and fertilizing value to cattle manure can be produced from waste materials of various kinds, such as cereal straws, crop stubbles, cotton stalks, groundnut husk, farm weeds and grasses, leaves, leaf-mould, house refuse, wood ash, litter, urine-soaked earth from cattle-sheds and other similar substances. These materials are rich in cellulose and other readily decomposable carbohydrates and have a carbon-nitrogen ratio of 40 or more to 1. The direct application of such undecomposed, low-nitrogen organic matter as manure brings about a temporary deficiency of mineral nutrients (specially nitrates and ammonium compounds) in the soil by stimulating the growth of microorganisms, which in turn, compete with crop plants for available nitrogen, phosphorus and other elements. Hence, before using them as manure, it is necessary to compost or partially decompose them. This process lowers the carbon-nitrogen ratio to about 10 or 12 to 1.

URBAN COMPOST

In recent years, large-scale composting of town-refuse and night-soil in properly constructed trenches away from human habitations has been taken up successfully by the municipal bodies of many large and small towns. Trenches, 1 to 1.2 m wide, 75 cm deep and of convenient length, are filled with successive layers of night-soil, town refuse and earth, in this order. The compost gets ready in about three months.

BULKY ORGANIC MANURES

Farm Yard Manure (FYM)

The manure is produced in the farm chiefly with animal excreta. It is also called as stable manure, barn manure, dung and cattle manure. The manure consists of a mixture of cow dung, the liquid excreta or urine, the bedding material used in the stable and any remnants of straw and plant stalks fed to cattle. The quantity of dung excreted by an animal depends on several factors such as age, breed and condition of the animal and the nature of food eaten by it. The average excretion of urine per cattle per day is 7-8 litres. Average daily dung excretion pattern is give in Table 2.2. On an average well rotten FYM contains 0.5% N, 0.2%, P_2O_5 and 0.5% K_2O.

Table 2.2. Daily excretion by animals

Animal	Dung (kg/ha)
Cattle	
Young stock	4.90
Adult female	11.30
Adult male	15.60
Buffaloes	
Young stock	5.70
Adult female	14.80
Adult male	14.90

Manure preparation: Trenches of 6 m long, 2 m wide and 1 m deep are dug. The material consisting of dung and urine soaked in it is collected daily and placed in these trenches. Additions are normally applied in sections of one meter length. When each section is filled upto a height of about 0.5 m above the ground level, the top of the heap is rounded off to the shape of a dome and plastered with slurry of cow dung and earth. Before plastering, it is important to apply 4-5 buckets of water in the pit. Plastering conserves moisture and nitrogen and also prevents housefly nuisance. The manure pit should be protected from sun and rain. The manure becomes ready for use in about 4-5 months after plastering. It is possible to obtain 5-6 tonnes of good quality manure per year per head of cattle. Normally 10-25 cart loads of manure is applied per hectare. A cartload of manure measuring 9 cubic meters weighs about 0.5 tonnes.

Compost

Decomposed plant residues/farm wastes are known as farm compost. The waste materials that are available in urban areas can be made into compost called urban compost. The processes of composting facilitates narrowing down the C:N ratio of the material to 10 or 12:1. The nutrient content of compost is 1.01% N, 0.5% P_2O_5 and 0.8-0.9% K_2O. Applications of partially or fully decomposed materials quicken the nutrient release to the crops.

Methods of composting: There are two methods of compost making: aerobic and anaerobic.

Aerobic Method

A pit of one m deep, 2 m width and of convenient length is prepared. Organic wastes consisting of a mixture of crop wastes, weeds, leaves etc., that are collected in the farm are spread to a layer of 0.2 to 0.3 m deep. Cow dung solution is uniformly sprinkled to the top of the refuse. If the refuse is too dry, water is sprinkled over the material equal in weight of the refuse. Urine, earth ashes and bone meal may also be spread in a thin layer. This process is continued till the heap rises to about 0.5 m over the ground. After a month, the heap is turned over, mixed well and refilled adding water if necessary. Two more turnings are given, one at the end of the second month and the other after the third month. Sufficient water should be added in each turning over. Decomposition is ordinarily completed within 4-5 months. In places of heavy rainfall or where the water table is high the compost heaps may be prepared above ground level instead of pits.

Anaerobic Method

A pit of convenient size (4.5 × 1.5 × 1.0 m) is prepared. Every day collections of mixed farm residues are spread in the pit to a thin layer. A mixture of fresh cow dung, ash and water is sprinkled and compacted. The pit is filled till the raw material stands about 0.5 m above the ground. Later it is plastered to a thickness of 2.5 cm with mud and cow dung. Plastering helps to prevent housefly nuisance. Decomposition is completed within 4-5 months.

ENRICHMENT OF MANURE AND COMPOST

The importance of organic manure is very well established. However, the non-availability of these materials in adequate quantities restricts the farmers to apply the manure at its recommended dose. An alternate method has been developed to reduce the quantity, at the same time maintaining its manurial value. The method is known as enrichment of FYM/compost. By this technique, the nutrient availability of compost can be enhanced by addition of inorganic fertilizers while composting.

Enriched Farmyard Manure

Well decomposed and finely sieved FYM is required at 750 kg/ha. Along with this, super phosphate (at the recommended dose of P_2O_5 for each crop) is thoroughly mixed. The heap is plastered with red earth-cow dung paste. Adequate care is required to seal the cracks whenever they appear. After about a month, the manure is ready for application. At the time of application, recommended dose of N and K as urea and muriate of potash respectively are added and mixed thoroughly. The enriched manure should be applied in the seed furrow. This is particularly recommended for the dry lands.

Enrichment of Compost with Microbial Inoculants

Compost prepared by traditional method is usually low in nutrients and there is need to improve its quality. Enrichment of compost using low cost N fixing and phosphate solubilising microbes is one of the possible ways of improving nutrient status of the product. It could be achieved by introducing microbial inoculants which are more efficient than the native strains associated with substrate materials. Both the nitrogen fixing and phosphate solubilising microbes are more exacting in their physiological and ecological requirements and it is difficult to meet these requirements under natural conditions. The only alternative is to enhance their inoculum potential in the composting mass. Studies conducted at IARI, New Delhi, showed that inoculation with *Azotobacter/Azospirillum* and phosphate solubilizing culture in the presence of 1 % rock phosphate is a beneficial input to obtain good quality compost rich in nitrogen (1.8 %). The humus content was also higher in materials treated with microbial inoculants.

Super Digested Compost

Application of super phosphate at 5 % over the raw waste materials during composting improves the manurial value of compost. Application of this compost to acidic, calcareous and heavy soils increases the P availability.

Sheep and Goat Manure

The droppings of sheep and goat provide a valuable manure. Penning of sheep and goat in the fallow field is the common practice in India. It has been estimated that by stocking 1000 sheep for a night will add about 2 tonnes of droppings. The manure contains 3 % N, 1 % P_2O_5 and 2 % K_2O.

Poultry Manure

The excretion of birds contains both solid and liquid parts and hence there is no urine loss. The decomposition rate of the poultry manure is very quick. The manure should not be exposed to sun since there is 50 % N loss within a month. The fresh droppings contain 75 % moisture, 1.5 % N, 1.2 % P_2O_5 and 0.5 % K_2O.

SEWAGE AND SLUDGE

Sewage has two components, the solid portion called sludge and the liquid part called sewage water. Major portion of the suspended solids and effluents (99%) in the sewage is removed by settling. The sludge is latter collected separately, dried and used as manure. Under Indian conditions, the sludge on an average contains 1.5 – 3.5 % N, 0.8-4 % P_2O_5 and 0.3 – 0.6 % K_2O.

Once the solid portion or sludge is removed, sewage water is known as treated effluent. This can be used for irrigation. Since the sewage water is in concentrated form, it is important to dilute it with water in the ratio of 1:1. It has been reported that irrigation with sewage effluent may result in improved soil structure and water holding capacity by decreasing the bulk density and hydraulic conductivity. Though there are many beneficial effects of sewage effluent, it may also contain harmful bacteria. The municipal sewage, particularly, may contain high amount of heavy metals.

NIGHT SOIL POUDRETTES

Night soil contains appreciable amount of N. It can be profitably converted into useful manure known as POUDRETTE. Dehydration of night soil can be done as such, or admixture of absorbing materials like soil, ash, charcoal and saw dust. Among the materials, saw dust is the most effective one which can be very well used. Saw dust has high dehydrating capacity as well as the property of absorbing foul smell. About 40-50 % of saw dust could be added with the night soil. By this way, it is possible to get a dry, acidic poudrette which may contain 2-3 % N. Odourless manure can also be obtained by admixing night soil with equal volume of ash and 10 % of powdered charcoal. The poudrette produced by this way contains 1.3 % N, 2.8 % P_2O_5, 4.1 % K_2O and 24.2 % lime.

CONCENTRATED ORGANIC MANURE

Oil Cakes

After extraction of oil from the oilseeds, the residues are called as oil cakes. The edible oil cakes from groundnut and sesame are mainly used as cattle feed. The non-edible oil cakes viz., *neem, karanj, mahua,* etc. are used as organic manures. The powdered oil cakes should be incorporated into the soil before planting for quick decomposition. Effect of oil cake is improved when applied afresh compared to dry form. It has been shown that neem cake application reduces the pest incidence. Oil cakes contain high amount of N and small quantities of P and K they have low C:N ratio (3-15) which helps in releasing the nutrients quickly. Unlike bulky organic manures, they have no residual effect. Oil cakes are sometimes adulterated with earth, sand, hulls or husks and the manurial value is lowered.

Meal Group of Manures

Bone meal, horn and hoof meal and blood meals are examples for meal group of manures. These are quick acting and can be used in a wide range of soils and crops.

Bone meal is obtained by crushing the bones of animals. When treated with steam, it is known as "steamed bone meal". This is a good source of phosphorus. It contains about 1-2 % N and 25-30 % P_2O_5. Bone meal is best suited for acidic soils.

Hoof and horn-meals are obtained by grinding hoof and horns of dead animals. They contain 10-15 % N, 1.0 % P_2O_5 and 2.5 % lime.

Blood meal has high N content (10-12 %) and is obtained from the slaughter houses. Since the C-N ratio is very narrow (1: 3-4), it decomposes and supplies nutrient rapidly for crops.

TABLE 2.3: Nutrient content of important oil cakes

Oil cakes	Nutrient content (%)		
	N	P	K
Karanj(Pongamia glabra)	4.0	1.0	1.3
Mahua(Bassia latifolia)	2.5	0.8	1.8
Neem(Melia azadirachta)	5.2	1.0	1.4

Crop Residues

Residues left out after the harvest of the economic portions are called crop residues/straw. In the developing countries like India, they are mostly used as cattle feed. In the developed countries, harvesting is done using combine harvester and hence the straw cannot be used as cattle feed. They are generally burnt in the field itself. Straw has good manurial value since it contains appreciable amount of plant nutrients. On an average, cereal straw and residues contain about 0.5 % N, 0.6 % P_2O_5 and 1.5 % K_2O. The crop residues can be recycled by way of incorporation, compost making or mulch material.

Agro-Industrial Wastes

Agro-industrial wastes are available in substantial quantities at processing sites and can be effectively utilized as manure.

Rice Husk

It is the major by-product of the rice milling industry. Unhulled paddy grain constitutes 20-25 % of husk. It is a poor source of manure and the nutrient content is very low (0.3–0.4% N, 0.2-O.3% P_2O_5 and 0.3–0.5% K_2O). Rice husk should be incorporated into the wet soil and can be used in saline and alkaline soils to improve the physical conditions. It can also be used as a bedding material for animals.

Bagasse

The most important by-product of sugar industry is bagasse. It is mainly used as fuel in boilers of sugar factories. It can be used as manure raw or after composting. It contains 0.25 % N and 0.12 % P_2O_5.

Pressmud

It is a by-product obtained during the process of sugar manufacturing. It contains about 1.25 % N, 2.0 % P_2O_5 and 20-25 % organic matter. Addition of pressmud is highly useful to acidic soils since it contains high amount of lime (upto 45 %)

Tea Wastes

In the tea industry, tea wastes are available during the course of tea production, processing and storage. Tea wastes are used for extraction of caffeine. The decaffeinated tea wastes can be used as a manure. Nutrient content of the spent tea waste is 0.3-0.35 % N, 0.4 % P_2O_5 and 1.5 % K_2O.

Coir Waste

It is a waste product from the coir industry and mostly dumped near the road sides. Coirpith contains high amount of lignin (30 %), cellulose (26%) and wide C:N ratio (112:1). To reduce the bulk and C:N ratio, composting is recommended. The composted coirpith contains 1.26 % N, 0.06 % P, 1.20 % K with C/N ratio 24:1. The lignin content is reduced to 4.8 % due to composting.

GREEN MANURING

Green manuring can be defined as a practice of ploughing or turning into the soil un-decomposed green plant tissues for the purpose of improving physical structure as well as fertility of the soil. From the time immemorial the turning in a green crop for improvement of the conditions of the soil has been a popular farming practice. Green-manuring, wherever feasible, is the principal supplementary means of adding organic matter to the soil. It consists in the growing of a quick-growing crop and ploughing it under to incorporate it into the soil. The green-manure crop supplies organic matter as well as additional nitrogen, particularly if it is a legume crop, which has the ability to fix nitrogen from the air with the help of its root-nodule bacteria. The green-manure crops also exercise a protective action against erosion and leaching.

The adoption of green manuring depends upon the agroclimatic conditions. Broadly the following two types of green manuring can be thought of.

Green Manuring In Situ

In this system, green manure crops are grown and buried in the same field which is to be green-manured, either as a pure crop or as an intercrop with the main crop. In India, various methods of growing green manure crops in situ are followed to suit local conditions. For the proper decomposition of the green manure, it is necessary that the green material should be succulent and there should be adequate moisture in the soil. Plants at the flowering stage, contain the greatest bulk of succulent organic matter with a low carbon/nitrogen ratio. The incorporation of the green-manure crop into the soil at this stage allows a quick liberation of nitrogen in the available form.

Green Leaf Manuring

Green-leaf manuring refers to turning into the soil green leaves and tender green twigs collected from shrubs and trees grown on bunds, waste lands and nearby forest areas. The common shrubs and trees used are Glyricidia (*Glyricidia maculata*), *Sesbania speciosa*, *Karanj* (*Pongamia pinnata*), etc.

Green manuring *in situ* is followed in northern India while green leaf manuring is common in eastern and central India.

POULTRY MANURE: AN EFFICIENT ORGANIC MANURE FOR RICE

Rice crop accounts for nearly 45 per cent of our country's cereal production.

It is extensively grown in several States encompassing the zones with varying levels of productivity. The declining growth trends noticed with respect to rice productivity especially in the nineties has made the scientists and farmers to explore new strategies to avert the situation.

Rice is considered to be one of the least fertilizer use efficient crops especially when cultivated in irrigated eco-system. Usually farmers go for application of higher doses of fertilizers in order to meet nutrient requirement of the crop by taking into consideration the inherent losses of nutrients during cultivation. However, in recent years, the ever increasing cost of inorganic fertilizers has become a major prohibitive factor for extensive use of inorganic fertilizers like the contamination of natural resources and deterioration of soil health in rice growing ecosystems have been realized by farming community.

Considering the above factors, integrated approach to mix the right amount of organic manures with inorganic fertilizers has been advocated to rice growers. Among organic manures, farm yard manure (FYM) is the most commonly used. There are also several other sources of organic manures, which could be used in the integrated nutrient management practices. In this context, several organic sources (poultry manure) were evaluated at Agricultural Research Station, Siruguppa, Karnataka. The results of the study suggest that PM was the best among all the other organic manures. In another study, different levels of PM (1 to 4 t/ha) in comparison with FYM and without organic manures was evaluated. The results revealed that application of 1 t/ha of PM with 75% NPK (112:75:75 kg/ha) was on par with 100% NPK (150:75:75 kg/ha). The PM has direct as well as residual effect. Permanent plot experiment indicated that 1 t/ha of PM is almost equal to 7 t/ha of FYM in improving the rice yields.

Poultry manure contains 3.03% of nitrogen, 2.60% of phosphorus and 1.4% of potassium, whereas FYM contains only 0.94% nitrogen, 0.56% of phosphorus and 0.72% of potassium. Apart from this, its better nutrient content, PM is also less voluminous than other organic sources thus, helping in transportation from distant places. The cost of nutrients in PM works but to be cheaper when compared to other sources of organic manures, thus helping the farmers to get more monetary returns per unit area.

FORMATION MECHANISMS OF COMPLEX ORGANIC STRUCTURES IN SOIL HABITATS

Soil contains natural reservoir of biotic debris. It consists of plant remains, dead animals, and microorganisms that are subject to continuous turnover. With time, dead biomass is either mineralized or transformed to yield diverse complex organic substances. Transformation proceeds in two stages (Hayes, 1991):

1. The **first stage** involves degradative processes that lead to the formation of substances. The more recalcitrant lignin components are selectively preserved (Kogel-Knabner, 1993).
2. In the **second stage**, the substrates and the preserved products are

further transformed by synthetic processes to result of humus formation.

Formation of humus is a biochemical process. The degradation of biotic debris is catalyzed mainly by microbial enzymes. The subsequent synthesis of the humic material can be mediated by both enzymes and abiotic catalysts that are present in the soil. Soil enzymes originate mostly from microorganisms and to a lesser extent from plants, and clay minerals or metal oxides constitute abiotic catalysts. Polycondensation reactions between reducing sugars and amino acids play a role in humus formation. One of the most important catalytic reactions in soil is oxidative coupling that links phenolic products of the degradative processes into humic polymers through enzymatic reaction and mineral catalysis. Oxidative coupling can also result in the binding of xenobiotics to humus. This reaction is usually accompanied by a detoxification effect and may be catalyzed by either enzymes or minerals. Organic chemicals can be bound to soil through various mechanisms involving either physical sorption or covalent binding. By definition, the adsorption process is reversible; in practice, however, xenobiotics are increasingly difficult to desorb as they become gradually sequestered in inaccessible microsites within the soil matrix. The compounds that are bound through stable covalent linkages are believed to an integral part of humus, thus representing little or no threat to the environment.

ORGANISMS INVOLVED IN ORGANIC MATTER FORMATION

Soil contains various living and non-living components, without this the organic matter would not exhibit such richness. Soil has following components:

Living organisms	less than 1% of the total volume of the soil
Minerals	about 50%
Organic matter	3-6%
Air the water	up to 50%

Trophic groups of living organisms in soil

1. Photoautotrophic higher plants, algae and photosynthetic bacteria
2. Photoheterotrophic bacteria
3. Chemoautotrohpic bacteria
4. Chemoheterotrophic organisms

They include all soil animals, both vertebrates and invertebrates, most bacteria, and most fungi. These organisms are responsible for the biological and biochemical processes of humus formation. Chemoheterotrophs are the only organisms that can perform the two critical functions necessary to the humification process:

1. An active role of transforming raw materials into humus.
2. A passive one of serving as a source of biomass after death.

Chemoheterotrophs (microorganisms) play the most important active role in humification. Plants, however, are most important in passive processes because they serve as the major source of biomass. The tree other groups of organisms play only a passive role in the humification process.

Chemohetertrophic Amimals

These include beetle larvae, centipedes, millipedes, and termites in tropical soils. They are primary decomposers and break up plant litter into smaller pieces.

Saprophytic Bacteria, Fungi and Protozoa

They are also important in the primary decomposition process because these organisms produce extracellular enzymes that catalyze the dissolution of outer protective tissues. Secondary decomposition is performed exclusively by microorganisms, which include most of the soil saprophytic bacteria and fungi. These microorganisms specialize in selectively degrading various large organic molecules such as lignin and cellulose.

The products of microbial degradation can also be linked into humic polymers by the action of the same microorganisms. The microorganisms involved in humification are those that synthesize the oxidoreductive enzymes required to mediate the processes of synthesis.

Oxidoreductive Enzymes

They are designated as either peroxidases or mono-phenol monooxygenases. Peroxidases are found in most higher plants and numberous bacteria and fungi include two distinct classes:

Classes of Mono-phenol monooxygenases	
Laccases	Laccase activity was first detected in the Japanese lacquer tree *Rhus vernicifera.* Microbial laccases were subsequently found in many fungi (e.g., Agaricales, *Aspergillus funigatus, Rhizoctonia pratiocola, Trametes versicolor,* and others) and in actinomycetes like *Actinomyces albocrustosurs, A galbus, and Streptomyces* spp.
Tyrosinases	Tyrosinase activity has been detected in fungi such as *Agaricales (Aspergillus nidulans)*, brown-rot fungi, *Psalliota arvense*, and *Russula nigricans* and in various actinomycetes.

Both peroxidase and laccase activity have also been detected in soil extracts, apparently as a result of the production of the extracellular oxidoreductases.

Synthetic Reactions in Soil

The microorganisms responsible for the humification process are ubiquitous, therefore, the synthetic reactions are common in soil. The extent of these reactions is limited only by proliferation of the oxidoreductase producers, and the ability of these microorganisms to proliferate appears to be practically limitless. Various studies have shown that essentially all microbial species can be found in most soils, differing only in numbers and proportions. Even in deep layers of soils (up to 500 m) where only anaerobic conditions are expected to prevail, the composition of the microflora is essentially the same as that observed on the surface. Also, the transformation of organic compounds in both types of environments differs little except in the extent of the reactions.

DEGRADATIVE PROCESSES

Humus formation is a slow decomposition process. It undergoes a constant slow at a rate of about 2-5% a year. Though this process is slow, even than the content of humus in soil remains fairly constant over considerable periods of time due to its systematic replenishment. Plant materials decompose fairly and rapidly, with half-lives varying from days to months. Only a small portion of the total plant input, however, is converted to humic substances. During the first year, about 70% of the plant material is degraded to CO_2, and about 5-10% forms new biomass. Thus, 75-80% of plant substances undergo an initial rapid phase of plant litter decomposition. The material remaining is decomposed more slowly, and only the products of this second phase are incorporated into soil humus.

Types of Plant Material for Decomposition

Alexander (1977) divided these substances into six broad categories:

The decomposition in the soil may be the *preferential decomposition* (decomposition of easily mineralizable materials) or the *selective preservation* (decomposition of refractory plant components).

Plant components (cellulose, hemicelluloses, and proteins) are easily mineralized while lignins and lipids are refractory to decomposition. During the period of rapid decomposition, the former are quickly mineralized to CO_2 by microorganisms as a source of carbon and energy. Nitrogen contained in proteins is mineralized to ammonium and nitrates.

TABLE 2.4: Plant materials which decompose in soil. The type of material with their availability

Plant material	Availability in %
Cellulose	15-60% of the dry weight
Hemicelluloses	10-30%
Lignins	5-30%
Water-soluble sugars, amino acids, and aliphatic acids	5-30%
Ether and methanol-soluble fats, oils, waxes, and pigments	
Nitrogen- and sulfur-containing proteins	

The Pathway of Lignin Decomposition

Lignin degradation, however, is critical for humification to occur because the process generates phenols and phenolic acids, which are fundamental constituents of the core humus polymers. The pathway of lignin decomposition to soil is not yet fully understood. Lignin has a complicated structure of phenyl propanoid units connected irregularly by C- O-C and C-C side chain and direct ring-to-ring linkages. The CO_2 released during lignin degradation is the product is the product of cometabolism rather than the result of direct microbial mineralization. Compared to cllulose, very little (less than 1%) lignin carbon is incorporated into new biomass.

Microorganisms for Lignin Biodegradation

White-rot fungi are well known of the lignin biodegraders, because they contain peroxidases that remove electrons from lignin aromatic rings with completely etherified hydroxyl groups. This reaction yields cationic radicals that are then stabilized by the cleavage of C-C bonds in the lignin side chains. The radicals then react with O_2 or water and become stabilized by the formation of hydroxy or keto derivatives.

Because lignin is highly resistant to biodegradation, it may stabilize reactive components like cellulose, hemicellulose, or proteins by forming chemical or physical linkages to these molecules. A similar protective effect is produced by soil minerals and humus polymers that anchor unstable plant constituents by various adsorption forces or chemical binding.

Highly degradable proteins, for example, may be protected against rapid biodegradation by their nucleophilic addition to aromatic polymers through free NH_2 or SH groups and/or by adsorption on humus or clay.

Acquired Stability

The plant components may not be completely mineralized during the initial rapid decomposition phase. Therefore, many relatively unstable plant constituents, such as poly saccharides, proteins, soluble sugars, ad amino acids, can survive in the soil for a sufficient length of time to participate directly in humus formation.

SYNTHETIC PROCESSES FOR HUMUS FORMATION

The Role of Oxidative Coupling in Humus Formation

Humus (humic acid, fulvic acid, and humin) is a complicated structure. All main fractions of humus have fundamentally similar structures. They have a large aromatic polymer that in nature occurs as micelles. Incorporated into such polymers are various phenols, quinines, nitrogen heterocycles, and benzoic acids having reactive carboxyl, hydroxyl, carbonyl, and thiol groups that can act as binding sites for carbohydrates and amino acids. Humus also contains aliphatic moieties, often hydrophobic in nature, and abounds in stable free radicals that give oxidative properties to humus and enable the attraction of ionized or protonated compounds.

Theories of Humus Formation

One of the oldest theories proposes that humic polymers are lignins or other resistant plant materials that have undergone minor modifications under soil conditions (Waksman and lyer, 1932).

Plant materials serve exclusively as a source of energy suggests that the role of microorganisms is limited to the production of various amino compounds and phenols that, once discharged into soil, polymerize form humus. It seems possible that each of these processes can, to a certain extent, contribute to humus formation (Bollag and Loll, 1983).

Modern theories of Humification

This theory is based on the fact that a central role is played by the process of polymerization. The phenolic compounds derived from lignins (ferulic acid, pyrogallol, orcinol, syringic acid, and catechol) are considered to be the most important substrates for these polymerization reactions. This hypothesis was confirmed by many researchers. They have successfully synthesized humic-like substances by polymerizing one or more phenol-derived compounds. The synthetic polymers produced in this way are quite similar to natural humus, particularly with respect to elemental analysis, exchange capacity, total acidity, resistance to microbial degradation.

Phenolic compounds may be polymerized during the process of humus formation through oxidative coupling reactions that, as with most biological processes, require catalysts. This function is fulfilled by a series of biotic catalysts (the oxidoreductive enzymes) produced by soil microorganisms and abiotic catalysts (clay minerals and metal oxides) present in soils. Humus itself also exhibits catalytic properties due to the presence of free radical components.

The mechanism of coupling requires the oxidation of phenolic compounds involving the loss of a proton and an electron from the hydroxyl group. This loss results in the production of free radicals or reactive quinines that covalently couple to form chains of humic polymers. Such coupling has been shown by Liu et al., (1981) who used a simple model involving one of the components of humus, namely syringic acid, to demonstrate the formation of various oligomers, ranging from dimers to hexamers, by a laccase isolated from the fungus *Rhizoctonia praticola*. A similar approach was taken by Simmons et al. (1988). Mass spectroscopy and nuclear magnetic resonance (NMR) techniques were used to identify the products of an enzymatic reaction in which horseradish peroxidase was used to catalyze the oligomerization of guaiacol. Such studies have contributed to an understanding of mechanisms involved in humus formation.

The syringic acid and guaiacol (polymeric forms) are similar to the structure particularly with regard to the oxygen bridges and the direct carbon-carbon linkages between aromatic rings. Liu et al. (1985) demonstrated that it is possible to incorporate a per tide unit similar to that depicted in Stevenson's structure into a simplified synthetic model of humic acid. This incorporation could also be mediated by oxidoredus tive enzymes, such as the laccase isolated from *Rhizoctonia proticola*.

CONCLUSIONS

Regular application of organic manures and recycling of crop residues, need to be essential part of management to maintain and sustain productivity. The organic matter is not only the source of nutrient itself but also provide food and congenial environment to the useful microbs There is a need of policies and strategies for preparation and distribution of bulks and their uses in agriculture and modern technology. This assumes greater significance in the coming years as with rapid urbanization the bulky organic wastes are increasing. On the other side, increasing interest in organic farming is creating immense possibility of their eco-friendly utilisation

CHAPTER 3

Vermiculture

Soil fauna play a prominent role in regulating soil processes and among these the termites and the earthworms play a vital role in maintaining soil quality and managing efficient nutrient cycling. In organic farming practices the soil is considered to be a living component with is physical, chemical and biological characteristics.

Various aspects of earthworm activities in nature, and benefits that can be derived with mass culture, *viz.*, for cast production, vermicomposting and commercializing live material, indicate needs of developing mass earthworm culture. This is popularly referred as vermiculture. Vermiculture, as a subject, involve background of various aspects on biology etc. of earthworm. All these are scattered in literature.

Varmicompost is a method of making compost with the use of earthworms, which generally live in soil. They eat biomass and excrete it in digested form. This compost is generally called vermicompost or wormicompost. It is estimated that 1800 worms, which is an ideal population for one sq meter can feed on 80 tonnes of humus per year.

Earthworms form a major component of the soil system and these organisms have been efficiently ploughing the land for millions of years and assisting in the recycling of organic nutrients for the efficient growth of plants. Advent of chemical fertilizers and biocides and their large scale application have changed the structure of soils and have, in most cases, eliminated soil organisms. Thus, human interference in natural ecosystems through modern practices of agriculture, silviculture and horticulture involving large scale application of chemicals in the form of fertilizers and pestiides has been responsible for the destruction of endemic soil fauna. For millions of years before the so called green revolution, earthworms have been ploughing the soil and fertilizing them.

Earthworms are one of the numerous range of burrowing organisms which improve soils. Fertile soils harbour bacteria, fungi, actinomycetes, protozoans, proturans symphylids, pauropods, spring-tails, pseudoscorpions, insects and their larvae, millipedes, centipedes, slugs, snails, snakes, rodents and the earthworms. These fauna along with the flora contribute to soil health. The presence of earthworms in soils is generally indicative of the presence of other faunal representatives contributing to soil fertility. It would therefore certainly be appropriate to call earthworms as indicators of soil health.

The greatest asset to our Country is that we have about 509 species of earthworms and that we need not depend on the exotic spices of earthworms. During the early nineties a New Zealand planarian flatworm *Artioposthia triangulata* had been reported to be eating up earthworms and spreading fast across England, Ireland and Scotland. This parasite had the potential to cause a plague of eliminating earthworms from soils. The chief cause attributed to the arrival of these flatworms into Scotland was by the inadvertent import of potted plants from New Zealand, the soils of which probably might have contained flatworm eggs. Thus transfer of organisms of any kind including the earthworms can directly or indirectly cause irreparable damage to local biodiversity as could be envisaged from this example.

EARTHWORMS AS INDICATORS OF SOIL FERTILITY

Earthworms can generally be called as biological indicators of soil fertility, for soil with earthworms most definitely support healthy populations of bacteria, fungi, actinomycetes, protozoans, insects, spiders, millipedes and a host of others that are essential for sustaining a healthy soil.

Earthworms act on litter and the absence of earthworms is characterized by the presence of litter mats on surface sol. Soils inhabited by earth worms have surface casts to a range of about 98 tons dry weight ha-1 yr-1 in woodlands to about 133 tons dry weight ha-1 yr-1 in grasslands, maximum production however occurring during the wet periods. Soil fertility results from the maintenance of soil conditions such the decomposition processes operate at a level adequate to release plant nutrients from the litter at a rate that will sustain optimum growth. Availability of earthworms in soils have always promoted plant growth. Scientists have reported the production of grass, whet and clover to increase many fold by the presence of earthworms, while increase in growth of maize, paddy, sugarcane, vegetables and ornamental plants have also been reported.

Soil Enrichment

Earthworms that burrow deeply into the mineral strata and return, periodically, to cast faecal material at the soil surface facilitate the transport of certain elements to the surface from deep in the profile. There is abundant evidence that concentrations of exchangeable calcium, sodium, magnesium, potassium and available phosphorus and molybdenum are higher in earthworm casts than in the surrounding soil. The finding is confirmed by similar reports from U.S.A., New Zealand, Europe, Africa, India and elsewhere.

IN addition to the physical mixing of the soil by burrowing activities, soil enrichment is achieved by speeding up mineralisation of organic matter 2-5 times by the earthworms.

- Earthworm ingestion causes in increase in surface area of the organic wastes
- Ingestion removes senescent bacterial colonies and stimulates new bacterial growth
- Nitrogenous excretions from the worms enrich the soil formed from

organic wastes

- Earthworm burrowing enhances the oxygen penetration
- Mineral nutrients are released through enhanced microbial mineralisation, and
- Earthworm feeding increases the interaction among microflora, thereby improving the flow and exchange of nutrients.

Earthworms also eliminate pathogens in the waste using two mechanisms:

- their gut microflora outcompete the pathogens; and
- they produce bacteriostatic substances.

Several valuable compounds are also produced through the earthworm microflora interaction. These include vitamins (such as B_{12}) and plant growth hormones (such as gibberellins).

EARTHWORM: THE NATURAL BIOREACTOR

Bhawalkar (1989) designated earthworms as the *Natural Bioreactor.* Earthworms play a key role in soil biology as versatile natural bioreactors. They effectively harness the beneficial soil microflora, destroy soil pathogens, and convert organic wastes into valuable products such as biofertilisers, biopesticides, vitamins, enzymes, antibiotics, growth hormones and proteinous worm biomass:

1. Earthworm's gut is an effective tubular bioreactor, with raw materials (feed) entering from one end (mouth) and the product (castings) coming out through the anus.
2. They maintain a stable temperature through novel temperature regulation mechanisms; thus accelerating the rates of bio-processes and preventing enzyme inactivation caused by high temperatures.
3. Enzymes are active at a very narrow pH range, and earthworm bioreactors efficiently maintain the highly nonlinear pH parameter. The calciferous glands in them contain carbonic anhydrase, which catalyse the fixation of carbon dioxide as calcium carbonate and prevent any fall in the pH of the body fluid.
4. Earthworm's gizzard is a novel colloidal mill in which the feed is ground in to particles smaller than 2 microns, giving thereby anenhanced surface area for the microbial processing.
5. Earthworm bioreactors have an in-house supply of enzymes such as protease, lipase, amylase, cellulase, lichenase and chitinase. These enzymes biodegrade complex biomolecules into simple compounds utilizable by the symbiotic gut microflora.
6. Earthworms have a built-in oxygen plant, which can separate aerial oxygen by chemical absorption into their blood haemoglobin which has a strong affinity for oxygen. The haemoglobin is fully oxygenated at an oxygen partial pressure as low as 19 mm Hg, which saturates only 50 per cent of human haemoglobin.
7. Earthworms promote growth of microorganisms in their gut by

providing ideal conditions therein. Bacteria and actinomycetes, which are important in waste-degradation increase exponentially along the entire length of the tubular bioreactor, reaching densities, 1,000 times greater than in the surrounding soil. However, *E. coli*, *Salmonella* and otherpathogens aredestroyed by competition from the active microflora and the action of intestinal secretions (Satchell, 1983). Most human pathogens are anaerobic and cannot survive in the highly aerobic microenvironment created by the earthworms.

8. Earthworms produce castings containing balanced plant nutrients and immobilized microflora which continue to function in the soil. Vermicastings are also rich in vitamins, enzymes, antibiotics and growth hormones.

Earthworms digest some of their gut microflora (SCP) and build up their biomass. Worm proteins (70 per cent of dry biomass) are balanced with all essential amino acids and have a potential for use in animal nutrition.

INTERACTION OF VERMICOMPOST - EARTHWORM-MULCH - PLANTROOT (VEMP)

VEMP interaction is essential for an effective utilization of the soil ecosystem in agro-practices. Use of green manures in organic farming is also essential. The possibility of interactions of the root hairs with nitrogen fixing organisms may be increased by the VEMP interaction. It is very well known that earthworms prefer lower levels of lignin and tannin, and in case of green manuring younger plants are recommended for ploughing-in for an effective release and uptake of nutrients.

VAM is effective in cases of wheat and other crops and the population of VAM is very well enhanced in the presence of earthworms, as earthworms are reported to be vectors of viable propagules of VAM. The uptake of phosphate, though facilitated by VAM, dispersal of these spores and conservation of such microbes is facilitated by earthworms and other soil dwellers.

RECYCLING OF WASTES THROUGH VERMITECH

Vermitech is useful in recycling the agro-wastes through and the effective utilization of VEMP reduces. Nutrients presenting vermicompost are readily available and the increase in earthworm populations on application of vermicompost and mulching lead to the easy transfer of nutrients to the plants providing synchrony in interfered ecosystems. The availability of microbes in the compost and the nutrients in soisl tilled by earthworms are definitely more due to the formation of drilospheres. Vermicompost, therefore, leaves no doubt on its potential for good yields, the yield however depends on the management practices followed by the farmer.

EARTHWORMS AND PLANT GROWTH

The effect by earthworms on plant growth may be due to several reasons apart from the presence of macronutrients and micronutrients in their secretions and in vermicasts in considerable quantities. Certain metabolites produced by earthworms may also be responsible to stimulate plant growth. It is considered that earthworms release into the soil certain vitamins and similar substances. Though the NPK value of the vermicasts is always lower than any standard chemical fertilizer, several experiments have proved that wormcasts can promote lush growth of plants. This probably may be due to plant growth promoters like cytokinins and auxins present in casts.

ADVANTAGES OF ORGANIC MANURE AND THE ROLE OF EARTHWORMS

The green revolution is the result of intensive agriculture. Presently, an extensive use of organic manures may boos agricultural production and can feed to the growing population of under developing countries.

With the intensive use of chemical fertilizers, farmers can harvest three crops in a year with good irrigation facilities, but there was no thought about its adverse effect in the long run on the soil condition in general and on the environment in particular for nearly four decades. To restore the soil composition, farmers are now showing inclination to revert to organic farming. With this changing trend, a comparative picture of the two practices gives better scope for adoption of organic farming to suit to the existing conditions. Along with that it is necessary to produce the organic manure in large quantities from the available sources of organic matter to maintain the high level of produce in the available agricultural land.

Organic manure can be prepared from natural organic matter. These materials are the reservoirs of the nutrients that are required for the plant growth. These nutrients are in the form of complex, bound organic chemicals with strong bonds in most of these plant and animal debris. The entry of these chemicals in the form of simple organic or inorganic chemical into the plant system is the main basis of biogeochemical cycle. If these chemicals are judiciously made to enter the cycle of events in a regular systematic form in nature, then there will be no dearth for nutrient supply to the plants in a natural way and with this, there will be cleaning of the environment rather than polluting it.

Microbes are the **primary decomposers** while earthworms are the major **secondary decomposers**. Earthworm manure is formed from the dead tissues of plant and animals and thus is naturally the source of macro and micro nutrients in limited quantities. The weather conditions and the soil type in the tropical countries do not favour the restoration of carbon resource in the soils. The application of organic manure replenishes organic carbon to the impoverished soils. The presence of high level of oxidisable organic carbon helps in the slow release of nutrients from the manure and curbs the leaching of nutrients. The beneficial activity because of the nutrient availability.

What is Vermicompost

Vermicompost is a stable fine granular organic matter, when added to clay soil loosens the soil and provides the passage for the entry of air. The mucus associated with the cast being hygroscopic absorbs water and prevents water logging and improves water holding capacity. Thus, in the sandy soils where there is the problem of water retention, the strong mucus coated aggregates of vermicompost hold water for longer time.

In the vermicompost, some of the secretions of worms and the associated microbes act as growth promoters along with other nutrients. It improves physical, chemical and biological properties of soil in the long run on repeated application. The organic carbon in vermicompost releases the nutrients slowly and steadily into the system and enables the plant to absorb these nutrients. The soil enriched with vermicompost provides additional substances that are not found in the chemicals. The multifarious effects of vermicompost influence the growth and yield of crops.

Vermiculture and Pollution Hazard

Earthworm can minimize the pollution hazards caused by organic waste by enhancing waste degradation. The experiences of the farming community have shown that the exclusive use of chemicals to keep away the famine has turned out be a mirage. It is like giving alms for a day, instead of providing an opportunity to a person to earn for his life-time to escape from poverty.

The ideal earthworm species of earthworms help to convert waste into wealth. If this operation is made successful by the own efforts of the farming community, one can certainly by hopeful of converting the wasteland into a fertile land. It is this hope of the possibility of producing the needed topsoil or humus by earthworms that has encouraged this effort. This may help in reaching the goal of green revolution from a different angle.

ADVANTAGES OF VERMI-COMPOST

1. It enable easy management.
2. Vermicomposting wastes at source or place where one wishes or has facilities, compost can be more profitably utilized and commercialization for economic gains.
3. Management of live population in field too first necessitates development of suitable conditions, which are not easy due to modern agriculture practices. Even if these are made available, a large culture of suitable earthworm species, *viz.*, approximately on tonnage level = about 1 million live individuals per hectare land are required. All such requirements necessitate development of earthworm culture (vermiculture).
4. Huge quantities of domestic, agricultural and rural industrial organic wastes can be recycled for various usages. This also reduces pollution.

The other applications of vermicompost in paddy fields is at the time of tillering, panicle initiation, flowering and drain fillingis essential. Unlike the chemical fertilizers, the loss of nutrients by leaching and other phenomena related to nutrient loss are minimal in case of VEMP interactions. In Azolla applied systems, the decomposition and release of nutrients maybe hastened by the application of vermicompost. The effectiveness of blue-green algae with vermicompost is not much understood compared to other biofertilizer.

In sugarcane, paddy and other crops, vermicompost can be effectively utilized as a carrier medium for *Azospirillum*, *Rhizobium* and phosphate solubilisers, which are recommended with FYM (Farm Yard Manure). These also provide very good association for enhancing their interactions with plant roots. Experiments conducted on the effect of earthworms and/or vermicompost and vermiwash through vermitech on the cultivation of vegetables like tomato, brinjal and Okra have yielded very valuable results.

ADVERSE EFFECTS ON CROPS

Earthworm activity can also to a less extent damage crops by either seizing fallen leaves of live plants and pulling them down or by uprooting delicate seedlings. Casts may show up as ugly structures in ornamental lawns or may form an obstruction on golf courses or football pitches. Earthworms have also been occasionally reported to transmit many parasites and diseases of plants and animals.

ECONOMIC VIABILITY

Vermin-technology is popular because it is a simple methodology with low investment. It technology does not need sophisticated infrastructure. All that is required is some pits or containers to initially decompose organic waste and tanks made out of stone slabs or wooden planks, trenches or for any convenient material just to protect earthworms from pests and predators. Therefore, the investment on this is directly related to the choice of the material used to build tanks. Stone slab is subjected to availability while plastic bins would be less durable and expensive. Thus, tanks built using cement and bricks could suit all situations. However, trenches without any cement are popular in low rainfall areas. The cost of bricks and amount spent on construction vary from place

VERMICULTURE PROCESS

It is very simple and easy to method. It needs careful stepwise programme. However, like all live animal or culture of any biological material, first step is to start purely as a test experiment on a small scale. This enables acquire personal experience(s) on various aspects. For beginners, hobbyist and farmers, simple procedure should be first to start on experimental scale and record every detail of steps followed and responses on earthworms. The care should be observed that how these fair in particular medium with increase or decrease in their numbers, waste matter conversion, cast production and composting etc. Following are the important levels of planning:

1. Site selection
2. Availability of decomposable organic waste, its daily quantity and quality, alternate organic material, transportation and stocking etc.,
3. Marketable outlet and requirement with future scope.
4. Collection and study of know how on earthworms to be cultured.
5. Collection, procurement of suitable species.
6. Testing of suitability of species on performance in available organic

waste.

7. Maintenance of seed culture for eventual large scale culture.

Other important considerations should also be kept in mind, like funding, availability of labour, quality of waster and packaging etc. etc. are to be kept in mind. These aspects comprise essential steps in any project formulation. In terms of wider utilization of vermiculture, first step is selection of suitable species for culture and eventual vermi-composting.

SELECTION OF SUITABLE SPECIES

The selection of suitable species for vermiculture is based on the requirement, *viz.*, for composting, poultry and animal feed, marketing for fish or other forms of culture and sale for fish baits. As yet demands in India for fish bait is meagre and lack for wanting development of fresh water sport fisheries, including for selected fish species (like Eel Fish) culture and also large scale culture. For all culture feeds, care has to be taken that earthworm feed does not harbour parasites and pathogens.

The selection for suitable species is based on biological and ecological parameters, like habitat characteristics, distribution in soil or feed media and trophic functions. Usually, following three types of earthworms which can be used in vermiculture:

1. Epiges
2. Endoges
3. Aneciques

Epiges

These are small in size. They have uniform body colouration and live on surface litter or dung. These have short life cycle with high reproduction and regeneration rate. They can tolerate disturbance and have active gizzard; but have limited period of activity. These are phytophagous. These species are good bio-degradators and can release good nutrients in the soil. They do not re-distribute nutrients. Thus have insignificant role in humus formation and are not good for use in field conditions for soil reclamation, unless careful manipulations are done

Endoges

These are small or large in size. They have weak pigmentation and found in topsoil layer of organic and mineralised matter. Burrow branchings are horizontal and worms moderately tolerate disturbance. They may show some preference in selection of feed substrate (buffered and predictable) conditions. These are **geophagous**. Life cycle is of intermediate duration. They have potential for soil improvements because of high efficiency in energy utilization from poor soils, therefore, can be used in field with some manipulation.

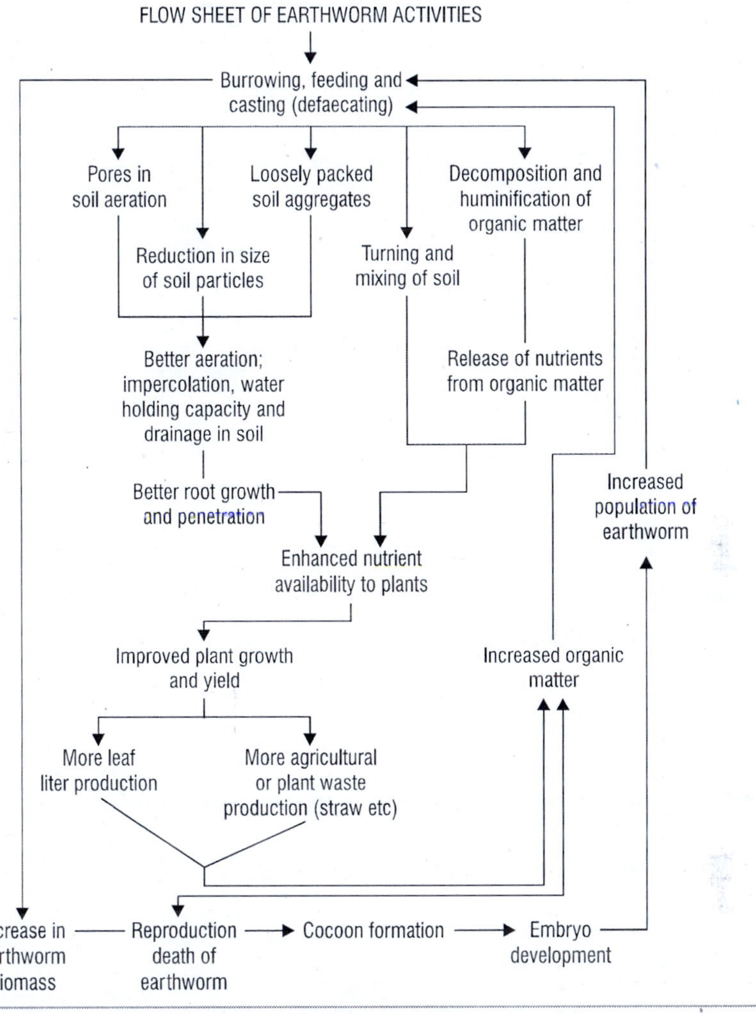

FIG. 3.1: Action of earthworm for production of vermicompost.

Aneciques

They are large in size. They have pigmentation only at anterior and posterior ends. They are largely nocturnal, deep burrowing and pull leaves or litter matter into soil (burrows) and are **phytogeophagous** and tolerance to disturbance is poor. Their reproduction rate is low. They may have useful role in mixing nutrients as from deep burrow come up to surface for casting.

BASIC CHARACTERISTICS OF SUITABLE SPECIES

Farmers or gardener needs only those species, which can quickly convert organic waste to vermi-compost. Therefore, a worm should have following characteristics:

1. Worm species should be tolerant to disease.

2. The culturing techniques should be simple enough to adopt.
3. Worm should be efficient convertor of plant or animal bio-mass to body proteins, so that its growth rates are high.
4. It should have high consumption, digestion and assimilation rate (composting qualities).
5. The worm should have wide adaptability (tolerance) to environmental factors (capability to live in varying temperature conditions).
6. The worms should have feeding preference and adaptability for wide range or organic material (high and rich organic matter).
7. The worm should produce large numbers of cocoons that should not have long hatching time, so that multiplication and organic matter conversion is fast.
8. Growth rate, maturity from young one to adult stage should be fast.
9. Worm should have compatibility or tolerance with other worms (as with possibility of mixture of species by ameteurs) as would add to productivity of biomass (worms) and conversion rate at different strata (layers) of organic matter, *i.e.*, faster composting. This feature, *i.e.*, composting with different layer feeders have so far received very little attention.
10. Worms should be disease resistant.
11. Worms on introduction in substrate, should have least inactivity period (= vermistabilization period).
12. The worms should feed near the surface of organic matter.

FIXING EARTHWORMS FOR IDENTIFICATION

1. Paralyse the worms by placing them in water and gradually adding alcohol.
2. When they are narcotized transfer them to a petri plate having 10% formalin solution by holding the worms straight for a few seconds. Worms stiffen.
3. After 24h, transfer them into 70% alcohol.

TRANSPORT OF FIXED WORMS TO LABORATORY

1. Dip cotton pads in alcohol (70%) and place fixed worms on these pads.
2. If there are more than one species mark the numbers on small pieces of paper with pencil and stich them posterior end of the worms.
3. Roll the cotton pads and tie with thread.
4. Place the rolled pads in a polythene cover and seal it to prevent drying of cotton pad.
5. Place the polythene cover in paper covers and transport to laboratory.

DESCRIPTION OF SUITABLE SPECIES

In whole world, only 3-4 species are in wide use. In India, only two peregrine species are being extensively used for vermiculture namely *Eisenia foetida* and *Eudrilus eugeniae.* Several species are also available but they have to be tested on various grounds like, their inbuilt adaptability to climatic conditions, per existing distributional range, local availability and in view of needs for better waste conversion rates etc. The important species for Indian sub-continent are as follows:

Family: Lumbricidae

Bimastos parvus (= Allolobophora (Bimastos) parvus Eisen)

They are distributed in Kashmir (Gorai, Srinagar); Himachal Pradesh (Kasauli, Barog, Simla, Chail, Narkanda, Kothai); Punjab (Ferozepur and other areas?); Rajasthan (Pratapgarh); and Uttar Pradesh (Nainital and other areas?). They are brownish red body with saddle shaped clitellum, dorsally covering 6 or more segments, from XXIV or XXV to XXX; length 25-40 mm; diameter 1-2 mm; total segments 85-111 usually 90. Other characters morphological and are differentiable by experts only for confirmation. In many temperate countries, allied species, namely, (1) *Allolobophora subrubricunda* and (2) *Allolobophora chloritica*, (3) *Dendrobaena rubida*, (4) *Eisenia hortensis* and *Lumbricus rubellus* are used in vermitechnology.

These species, barring perhaps *E. hortensis* are more or less established in temperate regions of India. In such regions, vermitechnology with these species deserves attention, as live individuals would be more readily available in nature and would be better suited to local conditions. Only one has to search and know local distribution or occurrence in different areas.

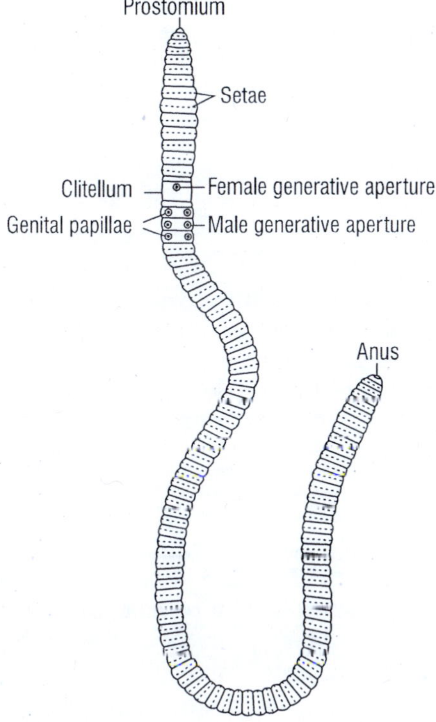

FIG. 3.2: Ventral view *of Metaphire posthuma*.

Eisenia foetida (Sav.)

Eisenia foetida is perhaps world's most widely used earthworm in vermitechnology for vermicomposting. They are distributed throughout India in nature. They found in sewage dumps (sewage filter mesh grills), sludge, tanks and filters. They are marketed under different names like Red worm, Pink worm or Purple worm, Tiger worm, Brandling worm or even vermiculture or vermicomposting worm. Living worms appears coloured (red, brown or purple or even darker). Careful observation indicates that they have dorsally coloured bands and are often two per segment. Ventrally body is pale. On maturity, clitellum spreads over 7-9 segments in length over XXIV, XXV or XXVI to XXXII body segments. On these, ridges wall develop on maturity. They are quite hardy and can tolerate wide fluctuations of temperature and humidity. The culturing process of species is quite easy due to feeding adaptability or adjustment to organic matter. They can live on wide variety of degradable/decomposable organic wastes. On these, the average conversion efficiency ranges upto 7 mg/worm/day. They have fast growth rate. Mature individuals can attain up to 1500 mg body weight. Mature adults attain body weight upto 1.5 gm and attain reproduction capability within 50-55 days of hatching from cocoon. A mature worm on average can produce one cocoon every third day and from each cocoon on hatching (within 23 days) emerge from 1 to 3 individuals. However, generally, survival of all emerged individuals is not seen. Still, on average, numbers can multiply or increase as high as 30 times. Duration of life is 70 days. They do not show movement outside feed substrate (decomposing organic waste).

Family: Eudrilidae

Eudrilus eugeniae (Kinb.)

This species is perhaps fastest growing and second most widely used earthworm in vermitechnology for composting and for use as vermi-protein. These species are originally distributed in Equatorial West Africa. They are presently distributed (may now be naturally occurring also) in most parts of the world and have been introduced for various usages in vermitechnology. It is common in parts of even America and Europe. In parts of America, species is common and is popularly known as *Night Crawler*. In India, common in many vermiculture establishments, particularly in Southern India this species has ben recommended for vermiculture. It is distributed in Travancore, Pune and North Konkan. The living worms appear brown and red to dark violet. The general colouration is like animal flesh. Their length ranges from 32-140 mm, diameter is 5-8 mm and total body segments range 145-196. In mature adults, clitellum spreads over 5-6 segments numbering XIII or XIV to XVIII. Spermathecal and female pores are fused commencing after 13/14 body segments. Male pores commence before 17/18 body segments. It grows fast (assimilation and conversion to biomass) and is better than other species and ranges upto 12 mg/day. It maximum body weight ranges up to 4.3 mg/individual. They attain maturity within 40 days, and a week later individuals commence cocoon laying, almost 1 cocoon/day or upto 4 cocoons/3 days for 46 days, depending upon appropriate conditions. Total life span in laboratory is estimated to range 1 to 3 years. Cocoon incubation period that at times, number of individuals per cocoon may range 1-5. They are low temperature tolerant, however, they can tolerate high temperature as well (over 48°F in room conditions, *i.e.*, under indirect sun exposure). Harvesting of the species from culture heaps is little difficult. However, simple solution is to up turn the heap, leave it for 1-3 hours, all worms move to lower layer. From top layers, compost can be collected and worms can be replaced in culture media or harvested for appropriate utilization. Their reproductive activity gradually decreases and so composting activity decreases. In such state is advisable to utilise worm for animal feed or just for burying in fields (a nitrogenous waste).

Family: Megascolecidae

They are available throughout world. So far only six species of this family are in use for vermitechnology work. These are *Amynthas diffrigens, Lampito mauritii, Metaphire anomala, Metaphire birmanica, Perionyx excavatus* and *Perionyx sansbaricus,* which is endemic to India. It is essential to know about their distribution in of India. For all taxonomic characteristics consultations with specialists is necessary.

Lampito mauritii (Kinb.)

Species is also referred in old documents as *Megascolex mauritii* (Kinb.). They are distributed throughout India (excepting Uttar Pradesh, wherein present distribution is not known). They are reported occurring in Kapurthala (Punjab), many areas of Bengal, Gujarat, Maharashtra, Madhya Pradesh, South Rajasthan, Andhra Pradesh, Tamil Nadu, Goa, Kerala, Andaman Islands, Lakshdweep Island, etc. They are also common in many parts of South and South East Asia. The body colouration is dark yellow except at anterior end, which has purplish tinge. The total body length 80-210 mm with a diameter of 3½-5 mm. They have 166-190 total body segments. Clitellum spread over to four body segments, on XIV to XVII and is ring shaped. Male pores on papillae, female pores near each other in pairs on XIV segment and spermathecal pores, in three pairs.

Metaphire anomala Mich. (= Pheretima anomala)

They found in Sibpur (Calcutta). We still need their distributional status. They are expected to be distributed in Orissa, Bihar, South-East Madhya Pradesh, Andhra Pradesh, Tamil Nadu, Karnataka and Kerala, also in parts of Maharashtra. The body colourations have not yet been studied. The total body length 80-90 mm, diameter 5 to 5.5 mm and total body segments 130 (variation range?). Clitellum, ring shaped covering three body segments, i.e. XIII-XVI with XIV ventrally having setae. Spermathecae pores absent but copulatory papillae in pairs present. The details on performance of species in vermiculture work are needed.

Metaphire posthuma (= Pheretima posthuma)

This is common species in India. Unfortunately other details on vermitechnological usages, if any, to-date yet to be studied.

Perionyx excavatus E. Perr.

In India, this genus is spread from eastern or western Himalayas downwards in regions of heavy rainfall. Now perhaps, species are widely distributed and two exotic species, *Perionyx excavatus* and *Perionyx sansbaricus*, believed to introduced from Australia or New Zealand. Only *P. excavator* is in use for vermitechnology in many parts of the world. In India, the species is common in Eastern Himalayas (Dibrugarh, Sadiya and Darjeeling), Western Himalayas (Kumaon Dist., Sahastradhara, Dehradun, Simla) Pilibhit, Bengal (Calcutta, Rajasthan, Sibpur), Maharashtra (Bombay) and Little Andaman Islands. The body colouration on dorsal surface (upper part) deep purple to reddish brown and lower side pale. The total body length ranges from 23 to 120 mm with a diameter of 2.5 mm. They are chitellum ring shaped, covering five or less segments on XIII or part, XIII to XVII. Male pores in a depression (line or area). Spermathecal pores in 7/8 and 8/9, nearly at same distance as male pores. The life cycle duration is of ± 46 days, growth rate is 3.5 mg/day, body weight (max), 600 mg. Reproduction rate is fast. They attain maturity within 21-22 days and reproduction (cocoon laying) commences by 24th day. The average cocoon production is 1.1/worm/day. This is almost double the reproduction rate of most commonly used worms in vermiculture. The incubation period reportedly ranges to ± 18.7 days and mean number of hatchings is 1 to 3 per cocoon.

Perionyx sansbaricus Michaelson

They are widely distributed in many parts of India. The species is found in natural environments of Baroda, Igatpuri, Manmad, Wathur near Mahabaleshwar and Londa near Castle Rock (Maharashtra) Khandwa and Kala Khund near Indore, Coonoor and Kotagiri in Nilgiris, Kodaikanal and New Delhi and vicinity area. The dorsally body colour is purple which is darker at anterior portion. On lower side (ventrally) body is pale. The body length is 32-63 mm with a diameter of 2.5 to 3.5 mm. The total body segments range 84-108. Clitellum is ring shaped, extends on 5 segments, over XIII-XVII body segments. Area of male pores variable (XVIII). Spermathecal pores near the middle line-6/7, 6/8 and 8/9.

The other species which are used for vermicultire are, *Octochaetus (Octochaetoides) surnensis* Mich., *Ramiella bishambari* (Steph.), *Dichogaster bolaui* (Mich.), *Dichogaster affinis* (Mich.), *Dichogaster curgensis* (Micha.), *Dichogaster saliens* (Bedd.), *Ramiella bishambari* (Steph.), *Erythrodraeodrilus suctorius* (Steph.), *Ocnerodrilus (Ocnerodrilus) occidentalis* (Eisen.), *Moniligaster perrieri* (Mich.), *Drawida willsi* (Mich.), etc.

MAINTENANCE OF BASE CULTURE

Base or mother or seed culture is used for large-scale culture. These should be carefully maintained and should be checked for health. The selection of feed of an individual is necessary. It should have any chemical that is lethal to worms. The procurement of organic waste (feed) should also be checked for seepage and contamination. For initial multiplication best feed substrate is cow dung manure. It should be at advance stage of decomposition and should be devoid of parasites.

Vermicastings as Inoculant

Earthworms coccons are invariably associated with vermicastings, which include also beneficial microflora, humus and undigested organic matter.

It is desirable to use vermicastings of burrowing-type earthworms, to inoculate new fields. Live, mature worms may not withstand the transport shock and invariably die when introduced into an altered environment.

However, vermicastings containing the beneficial microflora and the worm coccons trigger biological activity in the news soil when covered with organic mulch and properly watered. Earthworms hatched in the newly triggered environment adapt themselves effectively and very soon establish an equilibrium of earthworm population, suitable to thrive there.

VERMICOMPOSTING MATERIALS

The biologically degradable and decomposable organic wastes are used in vermiculture and vermicomposting. Commonly used composting materials are as follows:

Animal Dung

Cattle dung, sheep dung, horse dung, goat dung and poultry dropping etc may be used for this purpose. In use of animal dungs other than cattle dung, various preliminary testing and precautions for pathogens and responses to earthworms are necessary. The uses of horse dung should be done carefully because tetanus virus is common in horse dung and is lethal to human beings.

Agricultural Waste

Agricultural waste obtained after harvesting and threshing may be used. They include stem, leaves, husk (excepting paddy husk), peels, vegetable waste, orchard leaf litter, processed food wastes, sugarcane trash and baggase; and processing wastes.

Forestry Wastes

These are plant products such as wood shavings, peels, saw dust and pulp. All these besides various types of forest leaf litter can be used. The unutilized forest waste such as leaf litter amy also be used for vermicomposting.

City Leaf Litter

The burnt leaf litter from avenue or residential areas may be used, however, reports are not available. If it is used, this would keep cities clean and would provide useful product. The leaf litterof mango, guava, grasses and certain weeds may be used, but we need more information on this aspect.

Waste Paper and Cotton Cloth etc.

These are decomposable organic waste. These if are not being recycled for other useful products, can be recycled with vermicomposting.

City Refuge

City refuge or garbage on daily production basis comprise important items of city factors and a considerable portion of city refuge can be sorted and recycled or composted. Most of household as kitchen waste with little manipulation can be used for vermicompose.

Biogas Slurry

After recovery of biogas, if not required for agricultural use, viz., in conventional composting can be used for vermicomposting.

Industrial Wastes

The industrial wastes like waste from food processing, distillery etc. can also be used in vermiculture with some manipulations.

PRELIMINARY TREATMENT OF COMPOSTING MATERIAL

Initially, proper collection, sorting or separation of compostable, non-compostable and non-biodegradables like plastics, stone, glass, ceramics and metals should be done. Heavily contaminated wastes (even in kitchen wastes, heavily spicy wastes) with chemicals should be separated. The cleane matter selected for composting should be heaped and large lumps should be broken. The separated matter should be be spreaded in a layer upto 1 foot and to be exposed to sun for a day. This helps in killing several unwanted organisms and removes foul smell. The mixing of daily organic waste produce may be done with somewhat pretreated leaf litter in approximate ratio ranging from 10 to 40% of the waste to be vermicomposted.

Pre-Treatment of Leaf Litter and Agricultural Waste

They are heaped on ground and exposed to sun. Heaps are then be beaten with some stick or bamboo to break into smaller pieces.

Materials like crop plant stems and bagasse etc. are required to be cut into smaller pieces for enhancing decomposition and vermicomposting processes. The non-marketable vegetable produce which are sometimes infected with insect pests should be chemically treated. The common example to these is insect pest popularly known as brinjal borers which effect several vegetables.

These insects can be controlled by chopping waste into smaller pieces up to 4" size and spread over ground. This layer be then sprayed with 4% aqueous solution of some Neem insecticidal formulation. This would gradually kill insect pests stages without making waste unfit for vermicomposting. Neem formulation can also be prepared by collecting Neem fruits, drying and taking out seed kernel as we take out almonds by removing hard outer covering. 20-25 gm Neem seed kernel be wrapped in a cloth over night and squeezed in half liter of water with repeated dipping and squeezing with turning movements. Solution so obtained is insufficient concentration to kill most insect pests stages.

SMALL SCALE OR INDOOR VERMICOMPOSTING

It is done under covered areas (with a shade). It is should be protected from climatic adversaries like high rains, prolonged spells of high or low temperatures (from less than 10° F to more than 45° F) and predators like ants, rodents and large insectivorous birds are abundant. Small scale indoor vermiculture and composting is easy and can be adopted.

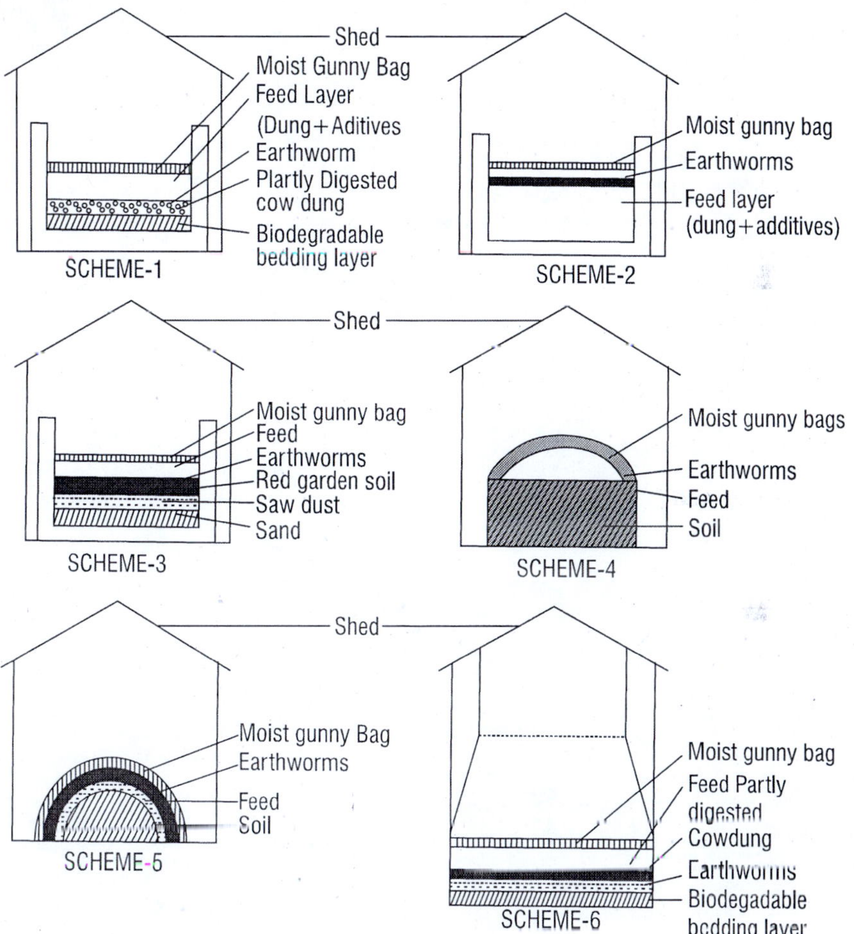

LARGE SCALE OR OUTDOOR VERMICOMPOSTING

Simple promotion of vermic activity in fallow fields after harvesting crop(s). Large or small quantities of plant wastes to be vermicomposted, after pre treatment be heaped and mixed with small quantities of mature cow dung manure (FYM).

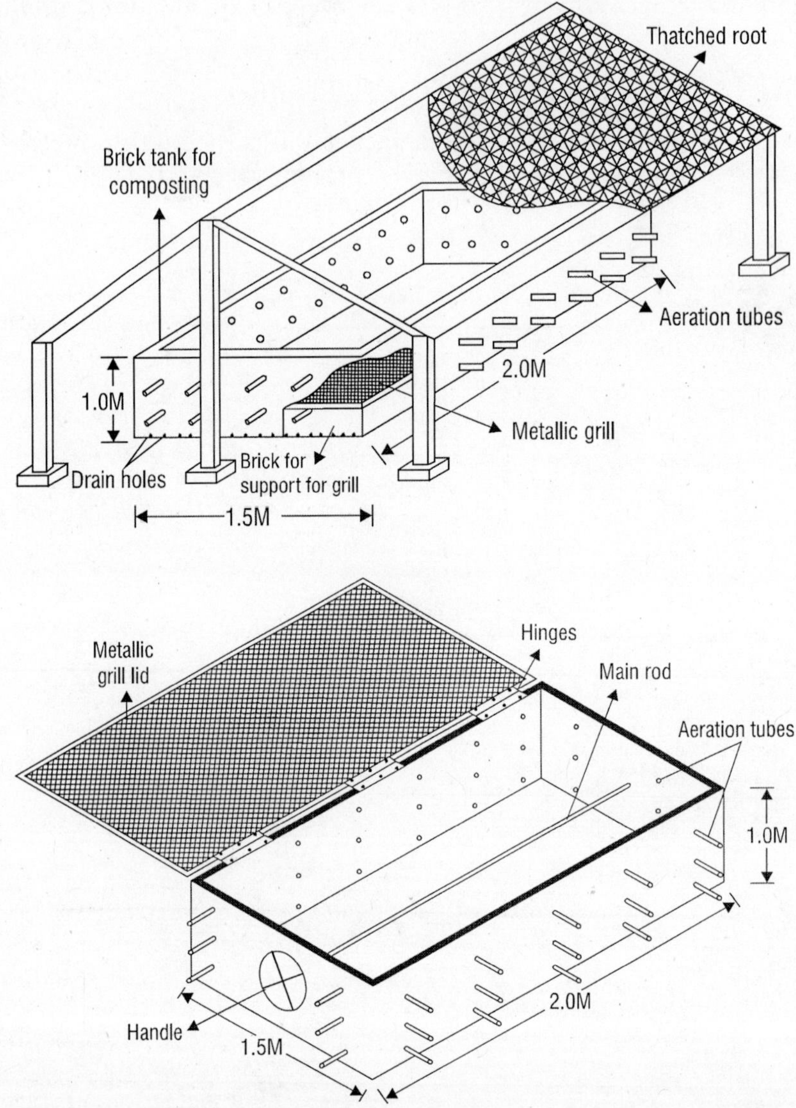

Then add bacterial inoculum for enhancing decomposition process and eventual faster vermicomposting. These heaps be sprinkled with little water (roughly 5 ltrs per 20 kg wastes) and covered with a hesian cloth or left on ground. Whole waste heap be kept moist for 3-7 days and periodically it is upturned and mixed. After 2-7 weeks, heap comes in a stage that vermicomposting is fast. This waste is now ready for vermicomposting.

REQUIREMENTS FOR VERMICOMPOSTING

Container

The shape and size of the vermicomposting container depend on the requirement, that is quantity of waste to be composted and number of live earthworms we want to culture. On an average, 2000 adult earthworms can be maintained in containers of 1 m^2 dimension. These with appropriate conditioning of composting material would convert approximately 200 kgs wastes every month. Roughly in a container of 2.23 × 2.23 mt. About 10 kgs of earthworms can convert approximately 1 ton per month. However, to have optimal conversion normally only upper 9-12" layer is composted. This should be softly scrapped off.

Bedding Material

This is the lower most layer of earthworm feed substrate that is required to be vermicomposted. For this any biodegradable matter is used like banana stem peels, coir pith, coconut leaves, sugarcane trash, stems of crops, grasses or husk. Waste or discarded cattle feed can also be used for bedding.

Moisture Content

Moisture content during vermicomposting should be maintained between 30-40 per cent. If moisture is high, dry cow dung manure or leaf litter should be mixed with substrate.

Temperature

Requirement for optimal results is 20-30° C. However, survival of earthworms is even at lower temperatures and upto 48° C air temperature. Obviously with little provision of shade, temperature within worm feed substrate (material to be vermicomposted) can be reduced. For this it is desirable that substrate should not be tightly packed in containers.

Successful Adoption of Vermiculture

The potential of earthworms to improve poor and reclaimed soils has prompted attempts to introduce worm populations in farm land. In New Zealand, large scale inoculations of earthworms into pasture land has been practiced for many years. These trials led to significant improvement in soil conditions and grass yields as reported by Stockdill (1982). The reported development of a turf-cutting machine to distribute earthworms from areas with large populations to pastures short of worms is encouraging.

Dutch polders reclaimed from the sea suffer from a lack of earthworms. Earthworm inoculations have been done in them to accelerate the process of soil maturation. It was found that without worms, a layer of organic matter is built up on the soil surface, and the tramping of livestock compacted the soil below. Grass roots penetrated only the shallow layer of organic thatch and the ripping action of grazing animals separated the turf from the soil. Earthworms broke down the organic matter on the surface. Infrared photography revealed that the plots with worms were warmer at night and cooler during the day than those without worms. On the polders, earthworms increased grass yield by 10 per cent, enough to pursuade farmers to abandon their traditional practice of regular ploughing and reseeding to maintain production.

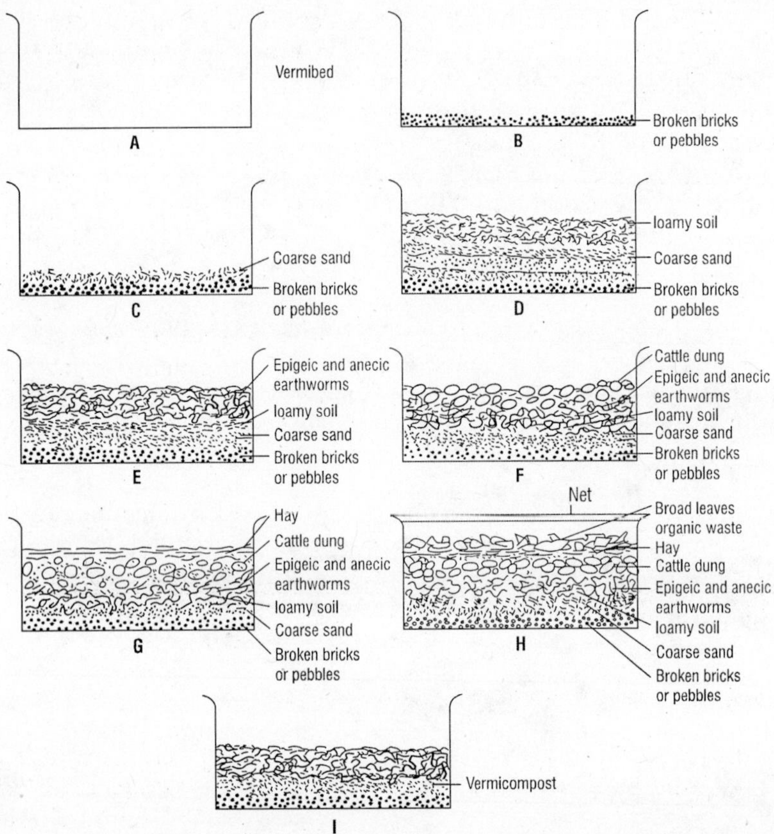

The colonization of a calcareousstrip-minein the U.S.A. by introduced earthworms increased incorporation of organic matter and nutrient availability. Curry and Cotton (1983) found that well-managed, reclaimed peat could be much accelerated through the deliberate introduction of earthworms, especially the deep burrowing species.

The former Soviet Union has also realized the potential of earthworms. In the Republics of Turkmenia and Uzbekistan the desert soils require irrigation to produce high yields of arable crops. Irrigation changes the soil that is inhospitable to earthworms into one where they thrive. Scientists in the Soviet Union have assisted the slow natural spread of earthworms by a programmed inoculation, by transporting sacks of soil from a river-bank richly endowed with worms in convoys of lorries.

Bruce Marshall, a farmer in the uplands of Scotland, improved the yields on his farm by inoculating it with worms. The larger earthworm population has attracted moles, whose criss-cross runs through the soil eliminated the need to install artificial drainage. Marshall's land now supports almost 50 per cent more sheep. He thinks that earthworms are valuable in conservation, providing food for nesting waders, game birds, foxes and badgers.

Researchers at the Field Laboratory of the University College of Wales studied the rehabilitation of reclaimed opencast mines. Owing to an improper soil structure, the reclaimed fields were prone to waterlogging and damage by trampling. Inoculation of earthworms and supply of organic matter proved vital to the regeneration of these fields.

Although a variety of earthworm inoculation techniques have been developed, all rely on collecting earthworms from the field. This is very laborious, often requires specialist equipment or expertise and favours the collection of surface dwelling earthworms rather than deep burrowing species. Dependence on field collection has meant that the technique of earthworm inoculation could not be exploited as widely as its potential would warrant. Its future development would depend on advances in the intensive production of appropriate species through proper culture techniques. Much of the research on intensive culturing of earthworms (vermiculture) has focused on *Eisenia foetida*. Little is known about the growth and reproduction of soildwelling earthworms under intensive conditions.

Scientists at the Open University, U.K. are tackling the problem of producing large quantities of burrowing earthworms under artificial conditions. By defining the optimum needs of worms under culture as well as field conditions these workers expect to keep the worms in high population density.

INITIATION OF VERMICULTURE IN INDIA

The use of *Perionyx excavatus* was the beginning of pioneer attempt in India to employ earthworms for the degradation of solid waste. Later *Eudrilus eugeniae* was brought to the laboratory in University of Agricultural Sciences, Bangalore, India to test the possibility of their establishment in this subcontinent, which has similar climatic conditions as that of its place of origin (Nigeria). Since 1982, *Eudrilus eugeniae* has been promoted for waste degradation. When compared to *Perionyx excavatus*, this species has surpassed both in feeding and reproductive rate. In the beginning of the experiments, it was thought that the response may be short lived, as many organisms show a tendency to perform extremely well in a new environment till they get established. Such a change in their efficiency was not observed in these last 15 years. A series of experiments has shown that it is an ideal species to work on agricultural and agrobased industry waste. Their sensitivity to density pressure and meandering nature demand larger area of space for their establishment. *Eisenia fetida* also entered the laboratory from Germany and is preferred species for composting activity in the urban areas (urban waste).

The University of Agricultural Sciences, Bangalore, India initiated vermiculture in India and they propogated the knowledge of vermiculture to farming community. The university released Press Note by the Technology Release Committee in the year 1984. Till 1990, the technology did not induse much interest in many people. At initial stage, several controversies were developed about the composting and selection of species of earthworms.

Most of the researchers were of opinion that indigenous species of earthworms should be used for composting and not the exotic species like *Eisenia fetida* and *Eudrilus eugeniae*. The group that opposed the use of exotic species in this country lacked the knowledge of their feeding habits, habitat preference, their reproductive potentials and ability to survive in captivity to accept them for vermicomposting. But a detailed and in-depth study of the biology of 15 out of the 60 species of earthworms collected during the survey of Karnataka State helped me to put forward my views that a very few species of earthworms can be used for composting purpose and a few more as co-inhabitants to hasten up the process of decompositing along with other soil organisms.

In India, which was under colonialisation for several centuries, exotic species of plants and animals have been transported either intentionally or unintentionally. The northern part of the country has more exotic species of earthworms than the southern peninsula.

Successful Applications in India

Composting of Waste

Vermiculture technology is being harnessed to set up units for cost-effective treatment of various nontoxic organic solid and liquid wastes from the cities, dairies, sugar and distillery units, pulp and paper mills, tanneries, fermentation industries and food-processing units. Apart from environmental protection through the pollution abatement, these units serve as production centers of vermicastings which are being used as earthworm culture because of the beneficial microflora and worm cocoons associated with them.

Use in Agriculture

Vermicastings produced through the burrowing earthworms are being used by thousands of farmers in diverse agroclimatic conditions of India to change from chemical to sustainable agriculture without a loss of yield. The produce is of better nutritional qualities, without pesticide residues and with better keeping qualities-enhancing exportability. The farmers are witnessing soil improvement, as indicated by measurements of soil pH, nutrients availability, electric conductivity and water holding capacity. Eroded as well as saline soils are being developed through vermiculture biotechnology, with noticeable results within a year.

CONCLUSION

Earthworms are useful not only for making compost and waste management but they also have a utility in the context of developing countries as a ready recknor for knowing the soil health at farmer level.They are one of the easily available biological indicators of soil activities and this natural facility need to be used for monitoring and maintaining the biological health of soil.

CHAPTER 4

Biological Fertilizers

Over the past century, many of the basic scientific principles of plant nutrition and soil fertility have been explained. In developed countries, inorganic chemical fertilizers have been widely accepted as a major source of improving and maintaining soil fertility. Plants need sufficient nutrients in proper balance for normal growth and development. Sixteen chemical elements are known to essential for plants. Except carbon, hydrogen and oxygen, rest 13 elements are taken up by plants from the soil. There is a continuous removal of nutrient from soil for crop production and yield. Depletion of soil nutrients poses a major threat to sustainability of crop production and underlines the need for maintaining it by tapping other plant nutrient sources.

The term biofertilizers includes selective microorganisms like bacteria, fungi, and algae which are capable of fixing atmospheric N or convert insoluble phosphate in the soil into forms available to plants. Biofertilizers are cost effective, eco-friendly and renewable sources of plant nutrients to supplement chemical fertilizers. Biofertilizers also play a vital role in maintaining long term soil fertility and sustainability.

SOIL BIOTA IN SUSTAINABLE AGRICULTURE

Agricultural ecosystems, unlike the natural ecosystems, are human manipulated ecosystems and regular and sometimes intense disturbances are a major part of economically viable agriculture management systems. Sustainable agriculture is a complex issue associated with producing food while maintaining our biophysical resources including soil, water and bitoa with no adverse impacts on the wider environment. It should:

1. maintain or improve the production of 'clean' food
2. maintain or improve the quality of landscapes which includes soils, water, biota and aesthetics.
3. have minimal impact on the wide environment.
4. be economically viable
5. be acceptable to society.

There are different concepts of sustainable agriculture, but none is generally accepted. It embraces several forms of non-conventional agriculture that are often called organic, alternative, ecological or low input. However, from economic and ecological perspectives, two basic criteria must be met if agriculture is to be sustainable in teh long term. These are:

1. A reduction in the use of non-renewable resources
2. Protection of the environment

Agricultural practices in developed countries have become highly dependent upon fossil energy inputs in the form of fertilizers, cultivation, herbicides and pesticides. Inputs of man made chemicals could be reduced if agricultural practices were targeted towards capturing the benefits that could be derived from improved management of the biota. Such benefits include.

1. the enhanced supply of nutrients via management practices that encourage nutrient recycling from organic residues.
2. improvements in soil structure via practices that encourage the activity of earthworms and other burrowing soil animals and
3. Improvements in plant health via practices that promote biological control of pests and diseases.

Additionally, these management practices that promote biological control of pests and diseases.

Additionally these management practices could help to reduce the off farm effects resulting from loss of man made nutrients (e.g., losses of N. P. and S through leading), that damage the environment through pollution of ground water supplies.

BIODIVERSITY OF SOIL BIOTA

The soil biota contains representatives of all groups of microorganisms and fungi, green and blue green algae, and all but a few entirely marine phyla of animals. In a fertile temperate soil, the biomass may exceed 20 t ha^{-1} and its diversity has been compared with that of coral reefs, although it is probably more complex.

Possible Management Strategies

1. Increase soil organic status by organic matter retention / incorporation, continual crop cover and decreased cultivation to minimize mineralization.
2. Improve nutrient status of the organic resource by introducing legume crops. Cut and carry of legume residues to ensure mineralization occurs at Peak Plant demand.
3. Avoid applying soluble or readily mineralized nutrient sources at times when they may be lost from the soil profile.
4. Choose rotations that maintain VAM inoculation.
5. Strategic use of fertilizers may be necessary to maintain crop production where organic and soil microbial processes are unable to meet the crop demand.

Root activity and Rhizosphere transformations are intensively investigated to develop plant growth promoting products, induction of defense, release of metabolities, bioremediating principles etc.,

IMPORTANCE OF BIOLOGICAL SOURCES OF NUTRIENTS

When wee say *farming,* we really talk about the growing of plants for our food. To know about plants the way they are, live and grow are thus the things we need to understand.

For that we should first look at the figures on the nutrition on the nutrition of plants in the table below:

I. **Elements from the Atmosphere:** They represent 95 to 98% of a plant's dry weight. Vital elements all constitutive: *carbon, oxygen, hydrogen, nitrogen.*

II. **Elements from the Soil**
   a) They represent 2-5% of a plants' dry weight.
   b) 12 vital elements:
   c) 2 non-constitutive
   d) Potassium and chlorine; and 10 constitutive:
   e) Phosphorus, boron, calcium, magnesium, sulfur, iron, manganese, molybdenum, copper, zinc.
   f) 18 trace-elements
   g) 4 non-constitutive
   h) Lithium, sodium, rubidium, cesium; and
   i) 14 constitutive
   j) Fluorine, silicon, selenium, cobalt, iodine, strontium, barium, aluminum, vanadium, tin, nickel, chromium, beryllium, bromine.

The above table shows as that plants feed themselves in two distinct ways, from the atmosphere and from the soil. Elements taken from the atmosphere are; carbon 44%, hydrogen 6% and nitrogen (via the microbial world) 2 or 3%. It thus appears that the bulk of a plants food is taken from the air. It is easily available for plants and there is plenty of it. It also appears that we have a lot to do to control pollution.

The balance, a mere 5% of a plant's diet, comes from the soil. This consists of small quantities of various minerals. It can be concluded that plants feed quantitatively from the atmosphere and qualitatively from the soil. To feed plants quantitatively from the soil, as is being done in conventional agriculture, is thus going against their physiology.

The elements taken from the soil are of two kinds, constitutive and non-constitutive of plant matter. The constitutive elements are absorbed by plants either in an oxidized or chelated way. These transformations are brought about by the microorganisms of the soil.

The non-constitutive elements (e.g., potassium, lithium, etc) are borrowed from the soil, so to say, and simply provide the plants with electric charges needed for their various growth processes. These elements are returned to the soil at the maturity stage. Only green plants and other living parts such as the seeds, contain minute quantities of these elements.

It thus appears that all constitutive elements pass through the biological cycle and are thus entering the plants in an organic way. The only elements to enter plants in an atomic way are not constitutive of plant matter. Plants have thus an organic diet.

For example, a crop of rice at an average of 2 tons per acre consists of 1900 kg. of air (elements from the atmosphere) and of 100 kg. of soil (elements taken from the soil). One of the elements taken from the soil by the rice plant is zinc, at about 8 ppm. However, in conventional farming the prescription is to apply 12 kg of zinc sulphate to the acre (Plumbers or farmers that is the question). The conclusion is that conventional farming promotes odd practices that go against the nature of the plants or are aimed not at enriching the soil but at enriching others.

We can thus see very clearly that plants feed organically from the soil. These facts and this knowledge allow us to set aside, once and for all, the notion that *plants have mainly or purely a mineral diet.*

Overcoming Soil Deficiencies-Nature's Way

The soil contains most of the nutrients needed by plants. There are exceptional cases wherein soils may be naturally deficient in one mineral or the other but this is a rare phenomenon. Nature usually corrects herself, by growing accumulator plants for instance. Wild thistles seem to thrive in soils deficient in phosphorus, potassium and selenium, through on analysis we find that thistles have a high content of these minerals! Mushrooms accumulate zinc! Thus for zinc in rice fields think of wild mushrooms. By growing accumulator plants as a green manure or associated with the crops, we can assist nature in recreating a balance. Of course a farmer cannot always wait until his soil has regained balance. Hence it might be necessary to add rock powders containing the element in short supply. Another way to overcome deficiencies is to collect leaves of various plants, young leaves, manure leaves and old leaves, compost them and apply in small quantities of plants. Or else simply buy a little of the mineral from a drug store and add it to the compost pile. The fact however is that the deficiencies are usually brought about either by wrong fertilization or wrong cultivating methods like mono-cropping or by using hard chemicals that kill the micro-organisms. In no way does it mean that this or that element is, as such, lacking in the soil but that there are no soluble forms of it available.

In fact the core of the question of farming is to find out *howto stimulate and enlarge the microbial population in our soils.*

The come back to the elements from the soil, how they are taken up by plants and in what quantities, let us take a look at table 2 alongside.

Therefore, most of the soils contain enough elements to provide us with bountiful and nourishing food.

Is it not strange that farmers are advised to apply up to 12 kg. of zinc per acre for rice? Agricultural science could study the factor that make for a high quality of agricultural soils. High quality and high productive soils can still be found in the natural state. We need to protect at all cost whatever remains of these natural margins. To study their agricultural methodology and to and to learn how all problems of agriculture have already found a solution there, is what is truly needed.

Table 4.1: Essential elements and their organic sources

Element	% of dry matter	Manner of uptake	Origin
			I. ATMOSPHERE
Carbon	44	Photosynthesis	Carbonic gas from air
Oxygen	44	Photosynthesis	Carbonic gas and water
Hydrogen	6	Photodecomposition	Rain water
Nitrogen	1 – 4	Biological fixation by microbes of the soil	Nitrogen gas from the air
			II. SOIL
Potassium	0.1 – 2.5	Element "on loan" from the soil solution and returned to the soil at the end of the plant's life cycle.	From the mother rock reserves 20-80 t/ha. Potassium from micr-oflora reserves up to 60 kg/ha.
Calcium	0.1 – 3	Acid Soils: Decomposition of the mother rock by climate, roots and microbes Alkaline soils: Decarbonation of mother rock by microbes	Calcium from the mother rock reserves 15-1200 t/ha. Calcium from humus reserves 300 kg -3/ha.
Magnesium	0.1 – 0.9	Chelatino or oxidation of magnesium by microbes	Magnesium of the mother rock reserves 300 kg-3/ha
Phosphorus	0.1 – 0.9	From the mother rock: oxidation by soil microbes. Liberated from humus by soil microbes. From clay particles by microrhiza filaments which recover retrograde phosphates.	Phosphorus from the mother rock reserves 600 kg-15 t/ha. Phosphorus from humus reserves 150 kg-1.5 t/ha
Sulfur	0.1 – 0.5	Mother rock: oxidation by soil microbes. Mineralisation of the humus by soil microbes.	From mother rock reserves 300-900 kg/ha. From humus reserves 120 kg. 1.2 kg/ha. From atmospheric pollution possible 10-100 kg/ha coming through the leaves as SO_2.
Chlorine	0.01 – 0.9	Element "on loan" from the soil solution and returned to the soil at the end of the plant's life cycle	From mother rock reserves 300-600 kg/ha. Cl from the world of microorganisms reserves 30-60 kg/ha
Iron	500 ppm	Mother rock chelation by soil microbes Liberation of organic iron from the humus	From the mother rock reserves 20-300 kg/ha. From humus reserves 3-15 kg/ha
Boron	500 ppm	Mother rock: chelation by soil microbes. Liberation of organic boron from humus	From the mother rock reserves 12-300 kg/ha. From humus reserves 3-15 kg/ha.
Manganese	500 ppm	Mother rock: chelation by soil microbes. Liberation of organic Mn from the humus.	From the mother rock reserves 300 kg-15 t/ha. From humus reserves 3-15 kg/ha
Copper	20 ppm	Mother rock: chelation by soil microbes. Liberation of organic copper by soil microbes from the humus.	From the mother rock reserves 3-300 kg/ha. From humus reserves 1-5 kg/ha.
Zinc	20 ppm	Mother roci: chelation by soil microbes. Liberation of organic zinc by soil microbes from the humus	From the mother rock reserves 30-1500 kg/ha. From humus reserves 1-2 kg/ha.

Element	% of dry matter	Manner of uptake	Origin
Molybdenum	1 ppm	Mother rock: chelation by microbes. Liberation by soil microbes from the humus.	From mother rock reserves 3-30 kg/ha. From humus reserves ± 1 kg/ha

BIOFERTILIZERS FOR SUSTAINABLE AGRICULTURE

Biofertilizers have for long witnessed shifting fortunes in agriculture. While the technology of using them has been successfully exploited in many developed countries, competent exploitation in developing countries has been hampered by several factors, including lack of trained personnel, lack of appreciation of the benefits of inoculation and absence of suitable industrial support. A major problem, especially relevant to our biofertilizer programme has to do with relations between research institutions and industry. A transfer of technology between the research institutions and industry is essential. It must be promoted as effectively as possible through cooperation between institutional research, which is usually public and industrial research and production, which are either private or public. At the same time, we should not overlook the significance of ensuring and maintaining a high quality standard of the product. The limited shelf-life, particularly of bacterial bio-fertilisers dictates that product streams must be handled with a quick delivery system at low temperatures. A strong extension and training programme, actively supported by research and industry is the need of the hour. We should recognize that in adopting a rational approach to the use and management of natural resources in sustainable agriculture, the microbial fertilisers hold vast potential for the future.

In developing countries, the most important challenge is to produce sufficient food for the growing population from inelastic land area. Products of biological origin can be advantageously blended to replace a part of the energy-intensive inputs. It is in this context, biofertilizers can provide to the small and marginal farmers an economically viable lever for realizing the ultimate goal of increasing productivity. These microbial systems siphon out appreciable amounts of nitrogen from the atmospheric reservoir and enrich the soil with this important but scarce nutrient. The crop-microbial-soil ecosystem can, therefore, be energized in sustainable agriculture with considerable ecological stability and environmental quality. Although the potential of biofertilizers in crop production systems has been well documented and substantiated, the major reason for the shifting fortunes of these biological inputs lies in the lack of an organised industrial back-up, an effective quality control system, a powerful extension machinery and a broad research base.

Biofertilizers are inputs containing microorganisms, which are capable of mobilizing nutritive elements from non-usable form to usable form through biological processes. They are less expensive, eco-friendly and sustainable. The beneficial microbes in the soil, which are the greater significance to horticultural crops are biological nitrogen fixers, phosphate solubilizers and mycorrhizal fungi which are the phosphate scavengers.

Nitrogen is one of the chief and important constituents of protein and nucleic acid molecules that play the basic role in cell metabolism, growth, reproduction and transmission of heritable characters. Therefore, without a constant supply of this unavoidable element, life can not go on.

Biological nitrogen fixation (BNF) is the reduction of atmospheric nitrogen to ammonia by microorganisms in the soil. It involves a highly specialized and intricately evolved interactions between soil microorganisms and higher plants for harnessing the atmospheric nitrogen. It is a fascinating biological phenomenon studied extensively to provide low-cost nitrogen and improve crop productivity. The nitrogen fixing organisms associated with horticultural crops are the Rhizobium species, living in symbiotic relationship with the leguminous plants and free living fixers belonging to the *Azotobacter* family and the *Azospirilla* living in association with the root system of crop plants.

Some soil microorganisms play an important role in improving soil fertility and crop productivity due to their capability to fix atmospheric nitrogen, solubilize insoluble phosphate and decompose farm wastes resulting in the release of plant nutrient. The extent of benefit from these microorganisms depends upon their number and efficiency, which however, is governed by a large number of soil and environmental factors. When the number and activity of specific microorganism called microbial inoculant or biofertilizer is used to hasten biological activity to improve availability of plant nutrient. One of the essential plant nutrients on which successful agriculture depends to a great extent is nitrogen. High crop requirement, susceptibility to gaseous and leaching losses have made nitrogen the most in demand fertilizer material followed by phosphorus. Although air has about 80% of nitrogen, plants cannot make use of atmospheric N. The discovery of bacteria which had the ability to reduce non-usable (atmospheric nitrogen) into usable form (ammoniacal) had been a major breakthrough in the field of agricultural research. Of late, increasing attention is being paid to harmness the potential benefits from these renewable sources of plant nutrients because of the following reasons:

1. Depleting soil fertility due to widening gap between nutrient removal and supplies.
2. Depleting feedstock/fossil fuels and increasing cost of fertilizers.
3. Growing concern about environmental hazards.
4. Increasing threat to sustainable agriculture.

A number of products are now available that are generally referred to as soil and plant additives, of non-traditional nature. These products include:

1. Microbial fertilizers and soil inoculants contain unique and beneficial strains of soil microorganisms.
2. Microbial activators that supposedly contain special chemical formulations for increasing the numbers and activity of beneficial microorganisms in soil.
3. Soil conditions that claim to create favourable soil physical and chemical

conditions which result in increased growth and yield of crops.

4. Vermincompost helps in improving soil health and fertility.

Nitrogen fixing organisms can be provided to the farmers in the name of microbial inoculants otherwise termed as biofertilizers. The biofertilizers containing biological nitrogen fixing organisms are of utmost important in agriculture in view of the following advantage:

1. They help in the establishment and growth of crop plants and trees.
2. They enhance biomass production and grain yields by 10-20%.
3. They are useful in sustainable agriculture.
4. They are suitable in organic farming.
5. They play an important role in agroforestry / silvipastoral systems.

After the advent of the acetylene test for assessing nitrogen fixing potential, rapid progress has been made in identifying nitrogen fixing organisms. Out of a large number of microorganisms possessing the property of nitrogen fixation, only a few such as *Rhizobium*, *Azotobacter*, *Azospirillum*, BGA, Azolla, etc. have been commercially exploited. The reaction in biological nitrogen fixation is essential the same as in production of chemical fertilizers (Haber-Bosch process) i.e. the catalytic reduction of dinitrogen (N_2) to ammonia (NH_3).

HISTORY OF BIOFERTILIZER RESEARCH

Although the beneficial effect of legumes in improving soil fertility was known since ancient times, the field of biological nitrogen fixation opened up with the discoveries by Boussingault and Hellreigel in 1886. Nobbe and Hiltner produced for the first time a laboratory culture of Rhizobia under the name "Nitragin" in 1885. Starting with Rhizobia, a vigorous search for other N fixing microorganisms began and soon it was found that there were other non-symbiotic bacteria such as *Azotobacter*, which could fix atmospheric nitrogen. It was followed by a third group called blue green algae. Subsequently, microorganisms capable of solublizing phosphate were discovered. In 1970, a new group of bacteria called *Azospirillum* was identified. Vesicular-Arbuscular Micorrhizae (VAM) are the latest introduction in the list of biofertilizers which mobilize phosphorus.

In India, systematic research on biofertilizers started with the first study of N.V. Joshi in 1920. This was followed by extensive research by Gangulee, Sarkaria and Madhok on the physiology of the nodule bacteria and its inoculation for increasing crop yields. Important milestones in production, development and promotion of biofertilizers in India are given in Table 4.2.

Table 4.2: Some milestones in research, production and promotion of Biofertilizers in India

Year	Events
1920	First study on Legume-*Rhizobium* symbiosis by N.V. Joshi
1934	Earliest documented production of *Rhizobium* inoculant by M.R. Madhok
1956	First commercial production of Biofertilizer
1957	Study on solubilization of phosphorus by micro-organisms by Sen and Pal
1964	Spurt in demand of Biofertilizer for soybean particularly in Madhya Pradesh

Year	Events
1970	Scope for use of charcoal, lignite and FYM as alternate carriers to peat reported by V. Iswaran
1977	Use of ISI mark for *Rhizobium*
1983	Setting up of National Project on Development and Use of Biofertilizers by Ministry of Agriculture, Government of India
1986	Setting up of National Facility for BGA Collection at IARI, New Delhi by Department of Biotechnology, Government of India
1990	Setting up National Facility for *Rhizobium* Germplasm Collection at the Division of Microbiology, IARI, New Delhi by Department of Biotechnology, Government of India

Microorganisms induce many biochemical transformations in the soil. These include mineralization of organically bound forms of nutrients, exchange reactions, fixation of atmospheric nitrogen and various other changes leading to better availability of nutrients already present in the soil. The group of microorganisms responsible for nitrogen fixation, phosphorus solubilization and compost decomposition are being put to beneficial use in the form of biofertilizer.

Among the major part nutrient elements, nitrogen is an example of deficiency in sufficiency considering the presence of 80,000 tonnes of atmospheric nitrogen of hardly 100 kg ha^{-1}. All living beings with the exception of those members endowed with *nif* genes depend on fixed nitrogen to meet their requirement. Chemical fertilizers like urea, ammonium sulphate etc. are examples of industrial nitrogen fixation. There also exist atmospheric fixation (lightning etc.), combustion fixation (industry and automobiles), ozonization and biological fixation. Biological nitrogen fixation contributes maximum (67.5%) towards enriching earth surface with nitrogen.

Microorganisms capable of assimilating nitrogen are marked with *nif* genes. They synthesize nitrogenase enzymes responsible for converting otherwise inert N_2 to plant usable NH_3. Such common useful organisms are *Rhizobium*, *Azotobacter*, *Azospirillum* and Blue Green Algae (BGA). There are some microorganisms capable of solubilizing insoluble soil phosphorus while some collect available phosphorus from remote places out of reach of plant root hairs by sending elongated filaments (e.g., vesicular arbuscular mycorrhiza). Few of the heterotrophic organisms decompose cellulose rapidly. Such beneficial organisms are domesticated in suitable carriers, which on application to soil augment crop growth and yield. These carriers are slangishly called biofertilizers or appropriately called bioinoculants or microbial inoculants or microbial fertilizers.

Biofertilizers do not come under the purview of the definition of the term fertilizers, as they do not contain substantial quantity of plant nutrients as other fertilizers like urea, diammonium phosphate, muriate of potash etc. contain. The term biofertilizer is a misnomer but is in wider use. They should be termed as inoculants, after the name of the microorganisms they contain, viz *Rhizobium* inoculant, *Azospirillum* inoculant or blue green algae inoculant.

Bioinoculants or biofertilizers are different from chemical fertilizers. Biofertilizers on application remain in soils, multiply and keep benefiting the growing crops. They do not get depleted as in the case of fertilizers and, therefore, if the optimum soil conditions prevail, population of added microorganisms builds up and frequent application of biofertilizers can be avoided.

Table 4.3: Various sources of Nitrogen fixation on the earth surface

Nitrogen fixation	10^6 tonnes year^{-1}	Per cent contribution
Industrial fixation (Haber-Bosch)	40	15.3
Atmospheric fixation (Lightning etc.)	10	3.8
Combustion (Industry, automobiles)	20	7.6
Ozonization	15	5.7
Biological fixation	175	67.3

Source: Technical Handbook on Symbiotic Nitrogen Fixation, FAO Publication, (1983).

Table 4.4: Comparative overview of chemical fertilizer and biofertilizer*

Factors	Chemical Fertilizers	Biofertilizers
Production	Industrial, centralized	Biological, small scale or decentralized
Process	Chemical	Biological
Raw materials and energy budget	Fossil fuel and others imported, based on non-renewable energy sources, energy intensive	Atmospheric nitrogen for nitrogen fixers, unavailable P for PSM and VAM, organic residues for compost decomposing organisms. Indigenous and local, based on renewable energy sources very low energy bill
Pressure on foreign exchange	Exist	Nil
Global Production	About 40 x 10^6 tonnes	175 x 10^6 tonnes biologically fixed nitrogen annually added to earth surface
Efficiency	40-45% for nitrogenous fertilizer for upland crops and less than 33% for rice, 25-33% for phosphatic fertilizers. Heavy N losses due to leaching, volatilization and denitrificaton. P availability decreased due to fixation.	About 90% efficiency. Losses due to leaching, fixation are negligible.
Effect on subsequent crop	For nitrogen – nil or low	Residual effect for nitrogen and phosphorus
Pollution effect	Exists due to indiscriminate use	Pollution free
Cost	High cost input Rs. 6.0 for one kg N, Rs. 14.0 for one kg F_2O_5	Low cost input – 20 paise per kg N through *Rhizobium*, 50 paise through BGA
Targeted	All crops	*Rhizobium*-legumes; BGA-*Azolla*-Rice; *Azotobacter, Azospirillum*-Most of cereals, cotton, sugarcane.
Shelf life	Long	Short for bacteria, long for BGA
Accessability	Affordable section	Small and marginal farmers
Irrigation	More useful to irrigated field	Useful for both irrigated and dryland farming

Factors	Chemical Fertilizers	Biofertilizers
Soil health	Indiscriminate use deteriorates the soil health	Improves the soil health

*Partly adopted from Venkataraman, G.S. and Shanmugasundaram, S. (1992).

In India, *Rhizobium* biofertilizers specific for different legumes and blue green algae are the most popular among farmers. *Azospirillum* and *Azotobacter* are at an intermediate stage of acceptance while phosphate solubilizing microorganisms (PSM), VAM, *Azolla,* cellulose decomposing organisms etc. are at preliminary stage. The common beneficial organisms which may be used for preparing biofertilizers and the crops benefited are presented in Table 4.5.

Table 4.5: Positive response in crops yield by the use of biofertilizer

Contributing plant nutrients	Microorganisms	Crops benefited
A. Nitrogen	Symbiotic organisms *Rhizobium* (Symbiosis with legumes)	*Pulse legumes:* chickpea pea, lentil, moong, urad bean, cowpea, pigeonpea
		Oil legumes: soybean groundnut
		Fodder legumes: berseem lucern
	Azolla (*Azolla* sp *Anabaena azollae* symbiosis)	Rice
	Associative symbiotic organisms *Azospirillum*	Sorghum, pearl millet, rice, wheat, finger millet, maize, *kodo-kutaki* tomato, chilly
	Non-symbiotic organisms Heterotrophs – *Azotobacter*	*Vegetables:* onion, brinjal, tomato, cabbage
		Flower: chrysanthemum marygold
		Cereals: wheat, rice, maize, sorghum, sugarcane
	Photoautotrophs – BGA	Rice
Phosphorus	P. Solubilisers and mineralizers Fungi- *Aspergillus penicillum* Bacteria-*Bacillus pseudomonas*	For all crops
	P absorbers (Root fungus symbiosis)	
	Vesicular Arbuscular Mycorrhiza (VAM) Ectomycorrhizae Endomycorrhizae-Glomus	Linseed, maize, wheat, soybean
Zinc	Root-fungus symbiosis VAM	Linseed, wheat

Source: Verma, L.N. and Bhattacharyya, P. (1990).

ECONOMICLA AND ENVIRONMENTAL BENEFITS

The production of fertilizers depends on nonrenewable energy sources and is energy intensive. The energy requirements for the production of one kg fertilizer nitrogen 11.2, phosphorus 1.1 and potash 1.0 KWH. The requirement of petroleum based energy source in the production of biofertilizer is almost negligible. Organisms like *Rhizobium*, *Azospirillum* etc. assimilate 20-40 mg nitrogen g^{-1} of carbon. The biofertilizer production cost is very low so is the selling price. It has been estimated that one kg of fertilizer nitrogen costs more than rupees six while biofertilizer nitrogen costs only twenty paise through *Rhizobium* and fifty paise through Blue green algae (BGA). On nutrient basis, one tonne of *Rhizobium* biofertilizer is equivalent to 100 tonnes of fertilizer nitrogen (considering 50 kg N fixed ha-1 by application of 500 g *Rhizobium* biofertilizer) and one tonne of soil base BGA is equivalent to two tonnes of fertilizer nitrogen (considering 20 kg of N fixed ha^{-1}) by application of 10 kg BGA (Verma and Bhattacharyya 1991).

Table 4.6: Average requirement of energy per metric tonne of product in 10^3 BTU's

Energy Sources	Nitrogen Urea 46% N (a)	Phosphate TPS 46% P_2O_5 (b)	Potash KCT 60% K_2O (c)
Natural gas	30,552	920	1,356
Electricity	2,334	2,600	1,063
Fuel oil	26	730	1
Imported steam	6,152	360	-
Total	39,064	4,610	2,420

(a) based on ammonia plants using centrifugal compressors, (b) including rock production and energy recovery from sulphuric acid manufacture, (c) based on shaft mining
Source: FAO Publications, (1985). Fertilizer 85/4.

Biofertilizers are cost effective as well as environment friendly. They have a favourable influence on soil health. Biofertilizers are compatible with chemical fertilizers and agrochemicals like pesticides and herbicides without demanding extra or special care in their handling. They are safe to crops and users both.

The use of biofertilizer in various cropping systems may save 20-40 kg ha^{-1} of fertilizer nitrogen and 10-20 kg ha^{-1} phosphoric acid per cropping season. The success in the transfer of *nif* operon from nitrogen fixing bacterium *Klebsiellapneumoniae* into non-nitrogen fixing bacterium *Escherichia coli* has shown promising possibilities like achieving nodule formation in non-nodulating plants, industrial preparation of nitrogenase enzymes and their use in nitrogen fixation by-passing the use of non-renewable energy sources.

Biofertilizers are available for almost all crops and for three nutrients – nitrogen, phosphorus and zinc – and, therefore, the biofertilizer consumption is expected to increase many folds in coming years.

DEVELOPMENT OF BIOFERTILIZER INDUSTRY

The prominent role of legumes in soil fertility was first noted by J.B. Boussingault (1834), a French farmer, substantiated by Hellreigel (1886) in Germany by identifying root nodule as advantageous growth and Biejerinck (1888), a Dutch Scientist, who discovered that nodule is formed by bacteria, now called Rhizobia. Beginning of legume inoculation was made by Nobbe and Hiltner (1895) in Germany by using extracts of crushed root nodules under trade name *Nitrogin*™.

The early method of a application of inoculum involved top dressing of new fields with large quantity of soils transported from places where the legume grew well. This cumbersome soil transfer method was probably the reason for the birth of the present day inoculant production industry. Instead of the bulk transfer of soil which involves spreading of soil born pathogens, weed seeds, ineffective rhizobia etc., the present method supplies *Rhizobium* biofertilizers in small packets commercially.

The first inoculants were pure cultures of *Rhizobium* growing on solid medium and these were marketed in 8-10 ounces glass bottles. Since then, the methods of producing inoculants commercially have undergone a significant evolution to meet the demands of present day utilization of biological nitrogen fixation technology in agriculture. *Rhizobium* biofertilizers are commercially produced in both developed where fertilizer N is affordable and developing countries

The earliest documented evidence of the production of biofertilizers from isolated *Rhizobium* strains in India is that Madhok (1934). India in the largest producer and consumer of biofertilizers in the world, Madhya Pradesh, Tamil Nadu and Gujarat being the major producing states (see appendix).

GROUPS OF BIOFERTILIZERS

Organisms used for biological fertility management, are either free living or having symbiotic association with plants. They directly or indirectly contribute nutrition to crop plants.

Free living	This group includes numerous species of blue algae and certain bacteria such as Azotobacter.
Symbiotic association	The bacteria, Rhizobium, live symbiotically on the roots of legumes. *Azolla*, a water fern, fixes atmospheric N in symbiotic association with blue-green algae (BGA), *Anabaenaazollae* in rice ecosystem.
Associative symbiotic	One of the species of Azospirillum, *A. lipoferum* has associative symbiosis with higher plant systems.

Based on the type of microorganisms, the biofertilizers can also be classified as follows:

1 Bacterial biofertilizers e.g., Rhizobium, Azospirillum, Azotobacter, Phosphobacteria.
2. Fungal bio fertilizers: e.g., Mycorrhiza.
3. Algal bio fertilizers: e.g., blue green algae (BGA) and Azolla.
4. Actinomycetes biofertilizers: e.g., Frankia.

Biofertilizers are mostly cultured and multiplied in the laboratory. However, bluegreen algae and azolla can be mass-multiplied in the field.

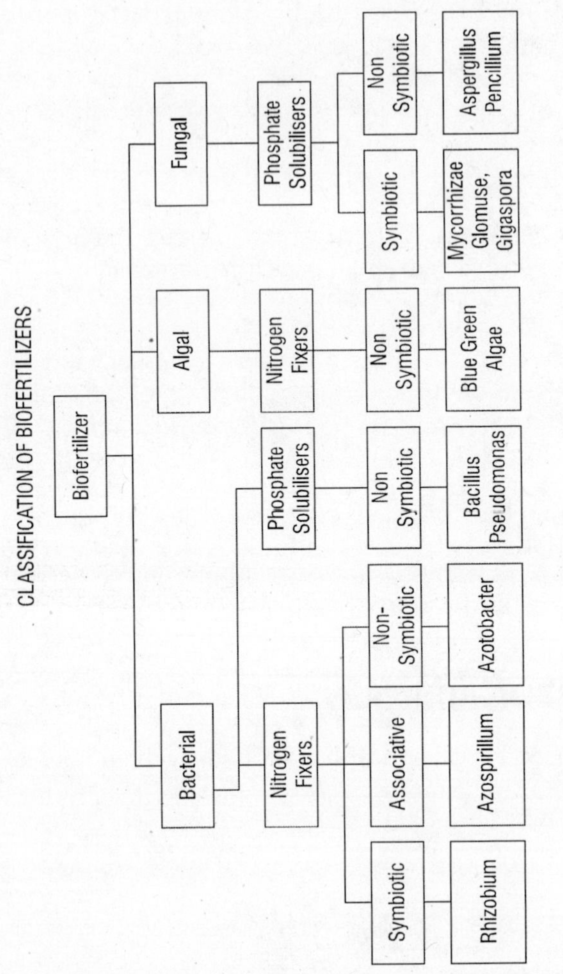

Fig. 4.1: Groups of biofertilizers.

Table 4.7: Some nutrient equivalent of biofertilizers and organic / green manures

Component	Input Level	Nutrient Equivalent
I. BIOFERTILIZERS		
Rhizobium	Inoculant	19-22 kg. N
Azotobacter and Azospirillum	Inoculant	20 kg. N
Blue Green Algae	10 kg/ha	20-30 kg. N
Azolla	6-12 t/ha.	3-4 kg./t
II. ORGANIC/GREEN MANURES		
Organic Manure (FYM)	Per tonne	3.6 kg. N + P205 + K2O (2:1:1)
Green Manure (Sesbania)	Per tonne	4.4 kg. N*

Component	Input Level	Nutrient Equivalent
Green Manure (Sesbania)	45 days crop	50-60 kg. N for HYV Rice
Cowpea Intercropped with Castor	Legume Buried after 6 weeks	30 kg. fertilizer N on castor
Leucaenia Loppings	88 kg N in Leucaenia	25 kg. fertilizer N on Sorghum
Sugarcano Trash	5 t/ha.	12 kg. N/t
Rice straw + Water Hyacinth	5 t/ha.	20 kg. N/t

* = For tall rice varieties, Effect expected to be higher on HYVs.
Source: H.L.S Tandon (1992).

CONCLUSION

The management of soil fertility through biological fertilizers is a vital component of sustainable agriculture. The hypothesis of sustainable agriculture is of recent origin and denotes the maintenance of crop productivity at levels necessary to meet the requirement of increasing population without deteriorating the environment and depleting the natural resources. Thus, it envisages an integration of traditional techniques with appropriate modern advances. The need for integrated nutrient supply system and its assumed importance towards sustainable agriculture need to be investigated. The great potential of biofertilizer can be exploited for this purpose.

CHAPTER 5

Microbal Inoculants for Nitrogen Fixation

About 80 per cent of the atmosphere is nitrogen, plants are unable to use it as nutrient since it exists in inert form. Biological nitrogen fixation is the conversion of atmospheric nitrogen by living organisms into forms that plants can use. The process is carried out by a group of bacteria and algae, which fix or convert the elemental nitrogen into assimilable forms either in association with plants or in free living state. The conversion of inert nitrogen is facilitated by the enzyme, nitrogenase, present in the nitrogen fixing microorganisms. The enzyme nitrogenase is coded by a complex of genes known as *'nif'* and to fuel the process bacteria need energy which is obtained either from the organic materials in soil or from the products of photosynthesis of the associated plants.

Of the different systems of biological nitrogen fixation, the *Rhizobium* – leguminous plant association has been reviewed in detail. As such, this chapter covers only the other major systems.

NITROGEN FIXING ORGANISMS AND PLANT ASSOCIATIONS

Nitrogen is the most important plant nutrient required in large quantities for crop production. The industrial production of nitrogenous fertilisers is mainly by Haber-Bosch process where nitrogen and hydrogen are combined at high temperature (400-500° C) and pressure (200-1000 ATM). This process consumes large amounts of energy obtained mainly from fossil fuels. The increasing cost and diminishing availability of this costly input is a major concern in many countries. The continuous use of inorganic nitrogen also endangers soil health and causes problems of nitrate pollution. Search for alternative ways of increasing crop yields has obviously brought the subject of biological nitrogen fixation (BNF) to the forefront. The energy economics is also in favour in BNF because it consumes only half the energy consumed by industrial nitrogen fixation. BNF is highly attractive for developing countries, which do not have enough reserves of fossil fuels and are unable to pay high prices for their purchase

The nitrogen fixing microorganisms form either nonsymbiotic, associative symbiotic or symbiotic relationship with plants. Non-symbiotic or free-living organisms fix nitrogen in soil and their numbers are generally more in rhizosphere of plants. They can be categorised into obligate aerobic, facultative or anaerobic organisms. Although a number of bacteria are reported to be nitrogen fixers in this group, the benefits of inoculation have been proved only with *Azotobacter* and the blue green algae. Associative symbiosis refers to an intracellular relationship between plant roots and a dinitrogen fixing bacterium, *Azospirillum.* They are found in cortical cells and protoxylem vessel sin crops such as *Digitaria, Panicum,* maize, sorghum, wheat, barley, pearl millet and a number of others. In case of the symbiotic nitrogen fixers, the bacteria called rhizobia fix nitrogen in nodules on the roots of leguminous plants. In the symbiotic relationship, the bacteria receive the products of photosynthesis as energy source and, in return, they fix nitrogen from the air for their host. The symbiotic association of cyanobacterium, *Anabaena azollae* and the water fern *Azolla* is a potential source of nitrogen for high yielding rice crop. *Frankia,* an actinomycete form nodules called 'actinorhiza' in woody species of 18 genera, including *Casuarina.*

Table 5.1: List of nitrogen-fixing micro-organisms in arable soils

Type of association	Microorganisms involved
Asymbiotic/Associative symbiotic	
Bacteria	
Aerobic	*Azotobacter, Beijerinckia, Derxia, Azomonas, Azotococcus, Methanosinus,*
Acetobacter	
Microacrophili	*Azospirillum, Herbaspirillum, Coryne-bacterium, Thiobacillus*
Faculative	*Klebsiella, bacillus, enterobacter, Escherichia, citrobacter, Pseudomonas*
Anaerobic	*Clostridium, Propionibacterium, Desulfovibrio, Rhodospirillum, Rhodospirillum, Rhodopseudomonas*
Blue green algae	*Nostoc, Anabaena, Aulosira, Calothrix, Tolypothrix, Chlorogloea, Aphanothece, Schizothrix, Scytonema, Stigonema, Anabaenopsis, Campylonema*
Symbiotic	
Bacteria	*Rhizobium* with legumes and *Parasponia Azotobacter* with *Paspalum notatum*
Actinomycetes	*Frankis* with woody plants
Algae	*Anabaena azollae* with *Azolla, Anabaena/Nostoc* with *Cycads, Nostoc* with *Gunnera Lichens* (alga + fungus)

AZOTOBACTER

Among the heterotrophic free-living nitrogen fixing bacteria, *Azotobacter* is the most intensively investigated genera. *Azotobactor* are present in neutral or alkaline soils and *A. chroococcum* is the most commonly occurring species in arable soils. *A. vinelandii, a. beijerinckii, A. agilis, A. insignis* and *A macrocytogenes* are other reported species. The number of *Azotobacter* rarely exceeds 10^4 to 10^5 g^{-1} of soil. The isolated cultures of *Azotobacter* fix about 10 mg nitrogen g-1 of carbon source under *in vitro* conditions. Apart from its ability to fix atmospheric nitrogen, *Azotobacter* is also known to synthesis biologically active growth promoting substances such as indole-acetic acid (IAA), gibberellins and B-vitamins in culture media. Many strains of *Azotobacter* also exhibited fungistatic properties against plant pathogens such as *Fusarium, Alternaria* and *Helminthosporium.*

The population of *Azotobacter* is generally low in the rhizosphere of crop plants and in uncultivated soils. The occurrence of this organism has been reported from the rhizosphere of a number of crops such as rice, wheat, maize, cotton, sugarcane, bajra, vegetables and plantation crops. A species of *Azotobacter, A. paspali* form specific association with the grass, *Paspalum notatum* which is abundant in many acidic sandy soils of Brazil (Dobereiner and Campelo, 1971).

BEIJERINCKIA

The distribution of *Beijerinckia* is restricted to tropical and subtropical regions. They are capable of growing on a wide range of pH from 2.9 to 10.0 and higher population of this organism is reported from soils having a pH 4.9 to 7.4.

Beijerinckia indica is the dominant species in acidic soils. *B. fluminensis, B. mobilis* and *B. derxii* are the other species reported from tropical soils. A survey on the distribution of *Beijerinckia* in several Brazilian states demonstrated the presence of the organisms in 97 per cent of the soil samples under sugarcane and 60 per cent of the soil samples from other vegetation. Rhizosphere occurrence of *Beijerinckia* has been reported in rice, sugarçane, forage grasses such as *Digiteria decumbens, Panicum purpurescens, Cynodon dactylon* and *Setaria sphacelata,* plantation crops such as coconut, arecanut, cashew, cocoa and pepper and pearl millet.

AZOSPIRILLUM

Though this organism was first described as *Spirillum lipoferum* by Beijerinck in 1925, its potential to fix nitrogen was realized only in 1975 by Doberiener and Day. They isolated the organism from the roots of *Digitaria* and described the plant *Spirillum lipoferum* association (Dobereiner and Day, 1976). Tarrand *et al.,* (1978) renamed the organism as *Azospirillum* and described two species, *A. lipoferum* and *A. brasilense.* Though three more species, *A. amazonense, A. halopraeferens, A. irakense* were added to the genus, the world wide distribution and benefits of inoculation have been proved mainly with the first two species. Of the two physiological types within the genus *Azospirillum,* one group has an oxidative metabolism and the other has the ability to ferment certain sugars, producing acid. In addition to their nitrogen fixing ability, certain strains denitrify under anaerobic conditions and could also assimilate NH_4, No_3 or No_2.

Azospirillum is a common soil inhabitant of tropics. The organisms form associative symbiosis with many plants particularly with those having the C_4-dicarboxyliac pathway of photosynthesis (Hatch and Slack pathway). They grow and fix nitrogen on salts of organic acids such as malate, succinate, pyruvative or lactate. The development of white, dense and undulating fine pellicle on semi-solid malate medium is the characteristic of *Azospirillum.*

OTHER BACTERIA

A number of bacteria capable of fixing nitrogen have been isolated from the rhizosphere and roots of tropical grasses and cereals. They include *Klebisella pneumoniae, Klebsiella oxytoca, Enterobacter cloacae, Enterobacter agglomerans, Citrobacter freundi, Alcaligenes faecalis, Pseudomonas* sp. *Escherichia intermedia, Bacillus polymyxa, Bacillus macerans and Bacillsu circulans.* Wani (1985) isolated *Derxia gummosa, Erwinia herbicola, Enterobacter* sp. and *Bacillus polymyxa* from a survey in 200 sites in the traditional millet growing areas in Northwestern India. Kundu and Tauro (1986) found the association of *Bacillus, Pseudomonas* and *Klebsiella* with the roots of wheat, barley and pearl millet. A new bacterium, *Acetobacter diazotrophicus* was recently isolated from certain sugarcane varieties which were capable of obtaining large quantities of nitrogen through biological fixation. The bacterium behaved as an endophyte and possessed unique physiological properties for a diazotroph such as tolerance to low pH, high salt and sugar concentrations, lack of nitrate reductase activity and nitrogenase activity which tolerates short term exposure to ammonia.

Mishustin and Yemtsev (1982) reported variation in the distribution of different species of anaerobic *Clostridium* such as *C. pasteurianum, C. butyricum* and *C. acetobutylicum* in different agroclimatic zones in USSR. Nitrogen fixing ability of different species varied from 2.6 to 9.5 mg nitrogen g^{-1} of glucose. Application of carbon containing compounds, particularly plant residues, stimulated anaerobic nitrogen fixation in flooded soils sown to rice.

ASYMBIOTIC BLUE GREEN ALGAE

The rice fields in India and South East Asia provide a congenial environment for the growth of blue green algae (BGA) comprising of unicellular, colonial and filamentous forms. The nitrogen fixing BGA mainly belong to the orders Nostocales and Stigonematales. Most of the nitrogen fixing BGA possess heterocysts which are large, thick walled empty cells. Evidences suggest that heterocysts are the sites of nitrogen fixation. There are also reports of nitrogen fixation by non-heterocystous forms. The results of an All India Survey which covered 2213 soil samples from rice fields revealed the occurrence of nitrogen fixing BGA in 33 per cent samples (Kaushik, 1985). The distribution pattern of algae in different states were: Uttar Pradesh (87%), West Bengal (60%), Orissa (43%), Tamil Nadu (10%), Kerala (9%) and Kashmir (7%).

SYMBIOTIC BLUE GREEN ALGAE

Blue green algae form symbiotic association capable of fixing atmospheric nitrogen with fungi, liverworts, ferns and flowering plants. The association between heterocyst forming nitrogen fixing blue green alga – *Anabaena azollae* and the aquatic fern *Azolla* can be a potential source of organic manure and nitrogen in rice production. The fern forms a green mat over water with a branched stem, deeply bilobed leaves and roots. The dorsall fleshy lobe of the leaf contains the algal symbiont within a central cavity. *Azolla* can be applied as green manure by incorporating in fields prior to rice planting. It can also be grown simultaneously with rice seedling after transplanting and incorporated into soil by hand. *Azolla pinnata* is indigenous in Asia, but recently other species such as *A. caroliniana, A. microphylla, A. filiculoides* and *A. mexicana* have been introduced for agricultural use because of higher biomass production.

The alga-fungus association, known as lichen occurs on soils, rocks and plant tops. *Nostoc, Calothrix* and some other genera of BGA are involved in the symbiosis. Nitrogen fixation has been proved by ^{15}N technique in genera of lichens such as *Colema, Sterecaulon, Leptogium, Lichina* and *Peltigera.* Some species of mosses and liverworts have the lower surface of the thallus inhabited by species of *Nostoc.* Yet another example of symbiosis is the occurrence of *Anabaena* and *Nostoc* in distinct zones in the cortex of coralloid nodules in roots of higher plants belonging to *Cycadaceae.*

FRANKIA

Angiosperms belonging to the several general such as *Casuarina, Alnus, Myrica, Dryas, Coriaria, Ceanothus, Discaria, Elaeagnus, Hippophae, Shepherdia,* etc. form symbiotic associationwith actinomycete, *Frankia.* Among the nodulated non-legumes, those species which belong to the genus *Casuarina* have the potential to stabilize eroding land surface and to improve the nitrogen status of improverished soils in addition to providing timber, firewood or charcoal. The nodules first appear as small lateral swellings on the roots but later develop into new lobes at their apices, so that a clustered coralloid structure is formed. In the host cells, the endophyte is seen in the form of hyphae, vesicles and bacteria like cells. Convincing evidence of nitrogen fixation has been obtained in the species of *Alnus, Hippophae, Casuarina, Shepherdia* and *Ceanothus* when studies were conducted with nodulated plants using ^{15}N labeled gas.

RHIZOBIUM

Agrobacterium (*A. radiobacter, A. tumefacions*) has also been included in the species: *Rhizobium radiobacter,* but this splitting of *Rhizobium* at the generic level into *Agrobacterium* is not accepted by most of the microbiologists.

Rhizobium is relatively more effective and widely used. *Rhizobium,* in association with legumes, fixes atmospheric N. The legumes and their symbiotic association with the rhizobium bacterium result in the formation of root nodules that fix atmospheric N. Successful nodulation of leguminous crop by rhizobium largely depends on the availability of a compatible strain for a particular legume. *Rhizobium* population in the soil depends on the presence of legume crops in the field.

The presence of *Rhizobium* bacteria in the root nodule can be seen under high power microscope by crushing a washed nodule between two glass slider. This kind of bacteria, preferentially, infect legumes as compared to non-legumes. Structurally, *Rhizobium* are rod shaped but great variation can be observed during their life cycle. These are coccoid, very small, highly motile and ellipsoidal forms. The bacteroids are usually irregular with X, Y, star and club-shaped forms. They are 0.5 to 0.9 micron by 1.2 to 3.0 micron in size. All the species are non-sporing and non-acid fast. These symbiotic bacteria (*Rhizobium*) are difficult to cultivate in ordinary culture media but they grow on manitol agar. *Rhizobia* are susceptible to antibiotics (produced by other microorganisms of soil) and also to the action of bacteriophages. Fungicides, herbicides and other plant protectants are also toxic to *Rhizobia*.

The most widely used biofertilizer is *Rhizobium*, which colonizes the roots of specific legumes to form tumour like growths called root nodules. It is these nodules that act as factories of ammonia production. Rhizobia have the ability to fix atmospheric nitrogen in symbiotic association with legumes and certain non-legumes like *Parasponia*. A mature nodule consists of a central 'bacterial zone' surrounded by several layers of cortical cells. The process of N-fixation is wholly dependent on the activity of the enzyme nitrogenase which is located within the bacteriods. The volume and number of bacteroids have a direct positive relationship with the amount of nitrogen fixed.

In the absence of legumes, the population of rhizobium in the soil diminishes. Most strains of rhizobia native to the soil are effective in N fixation. Some of them may be ineffective and produce ineffective nodules. These nodules are smaller in size, soft and whitish in color. Their N fixing capability is much low. The effective strains produce large, pink and firm nodules. Their N fixing capacity is far greater. Artificial seed inoculation is often needed to restore population of effective strains of rhizobium near the root zone to ensure high yields. Each legume requires a specific species of rhizobium to form effective nodules (Table 5.2).

Table 5.2: Cross inoculation group of Rhizobium

***Rhizobium* species**	**Cross inoculation group**	**Legume types**
R. leguminosarum	Peas	Pisum, Vicia
R. phaseoli	Beans	Phaseolus
R. trifolii	Clover	Trifolium
R. meliloti	Alfalfa	Melilotus, Medicago, Trigonella
R. lupini	Lupini	Lupinus, Orinthopus
R. japonicum	Soybean	Glycine
R. sp.	Cowpea	Vigna, Arachis

There are specific *Rhizobium* for every legume and inoculation with efficient strains of *Rhizobia* is essential for the nitrogen gains and better crop yields. Inoculation of pulse crop with appropriate cultures can give yield increases of the order of 15-30 per cent. Thus *Rhizobium*-legume symbiosis can generally meet 80 per cent of the N requirement of the legume crop. A starter does of N as chemical fertilizer is required (equivalent to about 20 per cent requirement of legume crop) for good crop yields.

The *Rhizobium*-legume association can fix upto 100-300 kg N/ha in one crop season and in certain situations can leave behind substantial nitrogen for the following crop. The range of nitrogen fixed per hectare per year by different legumes is 100-150 kg for clover, 80-85 kg for cowpea, 100-300 kg for alfalfa, 240-320 kg for faba bean, 90-100 kg for lentil, 150-200 kg for lupins, 50-60 kg for groundnut, 60-80 kg for soybean, 50-55 kg for mungbean and 100-400 kg for pasture legumes.

Many legumes may be nodulated by diverse strains of rhizobia, but growth is enhanced only when nodules are produced by effective strains of rhizobia. It is thus extremely important to match microsymbionts prudently for maximum nitrogen fixation. A strain of rhizobia that nodulates and fixes a large amount of nitrogen in association with one legume species may also do the same in association with certain other legume species. This must be verified by testing. Leguminous plants that demonstrate this tendency to respond similarly to particular strains of rhizobia are considered "effectiveness" group. The effectiveness grouping of leguminous plants are given in Table 5.3.

Table 5.3: Effectiveness Groupings of Legumes (Legumes that tend to respond similarly when inoculated with the same strain of rhizobia)

Rhizobium species	Effectiveness groupings	Leguminous species
Rhizobium meliloti	1.	*Mcdicago sativa, M. falcate, M. minima, M. tribuloides, Melilotus denticulate, M. alba, M. officinalis, M. indica.*
	2.	*Medicago arabca, M. hispida, M. lupulina, M. orbicularis, M. praecox, M. truncatula, M. scutellata, M. polymorphi, M. rotata, M. rigidula, Trigonella foenum-graecum.*
	3.	*Medicago lacmata*
	4.	*Medicago rugesa*
Rhizobium trifolii	5.	*Trifolium incarnatum, T. subterraneum, T. alexandrinum, T. hirtum, T. arvense, T. angustifolium.*
	6.	*Trifolium pratense, T. repens, T. zybridum, T. fragiferum, T. procumbens, T. nigtescens, T. glomeratum*
	7.	*Trifolium vesiculosum, T. herytheum, T. bocconei, T. biossieri, T. compactum, T. leacanthum, T. mutabile, T. vernum, T. physodes, T. lasyurum*
	8.	*Trifolium ruepellianum, T. tembense, T. usambarense, T. steudneri, T. burchellianum var. burchellianum, T. burchellianum var. johnstonii, T. africanum, T. pseudostriatum.*
	9.	*Trifolium semipilosum var. kilimanjaricum, T. masaiense, T. cheranganiense, T. ruepellianum var. lanceolatum.*
	10.	*T. medium, T. sarosience, T. alrestre*
	11.	*T. ambiguum*
	12.	*T. heldreichianum*
	13.	*T. masaiense*
	14.	*T. reflexum*
	15.	*T. rubens*
	16.	*T. semipilosum*
Rhizobium leguminosarum	17.	*Pisum sativum, Vicia vilosa, V. hirsute, V. faba Ve. Tenuifolia, V. tetrasperma, Lens esculenta, Lathyrus aphaca, L. cicera, L. hirstus, L. odoratus, L. sylvestris*
	18.	*Lathyrus ochrus, L. tuberosus, L. szenitzii*
	19.	*Lathyrus sativus, L. clymenum, L. tingitanus*
	20.	*Vicia faba, V. narbonensis*

Rhizobium species	Effectiveness groupings	Leguminous species
	21.	*Vicia sativa, V. amphicarpa*
Rhizobium phaseoli	22.	*Phaseolus vulgaris, P. coccineus, P. angustifolius*
Rhizobium lupini	23.	*Lupinus albicaulis, L. albifrons, L. albus, L. angustifolium, L. arboreus, L. argenteus, L. benthamii, L. formosus, L. luteus, L. micranthus, L. perennis, L. sericeus, Lotus uliginosus, L. americanus, L. pedunculatus, L. strictus, L. strigosus.*
	24.	*Lupinus densiflorus, L. Vallicola*
	25.	*L.nanus*
	26.	*L. polyphyllus*
	27.	*L. subarnosus*
	28.	*L. succulentus*
Rhizobium japonicum	29.	*Glycine max*
Rhizobium spp. (cowpea type)	30.	*Vignaunguiculata, V. sesquipedalis, V. luteola, V. cylindrica, V. angularis, V. radiata, V. mungo, Desmodium sp., Alysicarpus vaginalis, Crotalaria sp., Macroptilium lathyroides, M. atropurpureum, Psophocarpus sp., Lespedeza straita, L. stipulacea, Indigofera sp., Cajanus cajan, Cicer arietinum*
	31	*Phaseolus limensis, P. lunatus, P. aconitifolius, Canavalia ensiformis, C. lineata*
	32.	*Arachis hypogaea, A. glabrata, Cyamopsis tetragonoloba, Lespedezsa sericea, L. japonica, L. bicolor*
	33.	*Centrosema pubescens, Galactia sp.*
	34.	*Lotononis bainesii*
	35.	*Lotononis angolensis*
Rhizobium spp. (lotus)	36.	*Lotus corniculatus, L. tenuis, L. angustissimus, L. tetragenolobus, L. caucasicus, L. crassifolius, L. creticus, L. edulis, L. frondosus, L. subpinnatus, L. weilleri, Dorycnium hirsutum, D. rectum, D. suffrutitosum, Anthyllis vulneraria, A. lotoides*
	37.	*Lotus uliginosus, L. amercanus, L. scoparius, L. anguttissimus, L. pedunculatus, L. strictus, L. strigosus, Ornithopus sativus, Lupinus angustifolius, L. alous, L. luteus.*
Rhizobium spp. (corozilla, *Petalostemon-Onobryohis*)	38.	*Corozilla varia, Onobrychis viciifolia, Petalostemon purpureum, P. candidum, P. microphyllum, P. miuiflorus, P. villosum, Leucaena leucocephala, L. reusa*
Rhizobium spp. (various)	39.	*Daleaulopecuroides*
	40.	*Stroptostyles helvola*
	41.	*Robina pseudoacacia, R. hispida*
	42.	*Amolpha canescens*
	43.	*Caragina arborescens, c. frutescens*
	44.	*Oxytropis sericea*
Rhizobium spp. *Astragilus* sp.	45	*A. cicer, A. falcatus, A. Canadensis, A. mexicanus, A. oriziculatus*

Source: Legumes inoculants and their use. FAO (1985)

The amount of nitrogen fixed by legumes through root nodule bacteria exhibits considerable variation. Fixation is influenced by soil factors particularly nutrient status. For efficient performance under field conditions, legumes require adequate availability of soil nutrients suchas phosphoric acid, potash, calcium etc. In arable soils, which receive liberal dressings of nitrogenous fertiliser, legumes do not turn to atmospheric nitrogen. Estimates of nitrogen fixation by some legumes are presented in Table 5.4.

Table 5.4: Estimates of nitrogen fixed by some legumes

Crop	Nitrogen fixed (kg ha^{-1})
Alfalfa	100-300
Clover	100-150
Chickpea	26-63
Cluster bean	37-196
Cowpea	53-85
Groundnut	112-152
Lentil	35-100
Greengram	50-55
Pigeonpea	68-200
Soybean	49-130
Peas	46
Fenugreek	44

Source: Wani and Lee (1991).

General Taxonomic Characteristics of Legumes

The following family and subfamily descriptions for legumes are in the form of botanical descriptions. The glossary contains a brief description of the subfamilies and information on the approximate number of species and their general distribution.

Characteristics of the family Plants, shrubs, woody vines, and annual and perennial herbs

Leaves usually alternative and compound-bipinnate, simply pinnate, or palmate, rarely simple. Inflorescence variously racemose, in simple racemes, panicles, spikes or heads. Flower structure varies to the extent that 3 subfamilies are recognized; corolla typically 5-parted; stamens 3 to many, mostly 10, free, or united by their filaments in various ways; pistil single, free, Fruit characteristically a pod (legume), dehiscent or indehiscent.

Characteristics of the subfamilies Mimosoideae: Plants, shrubs, woody vines, a few perennial herbs

Leaves pinnate or bipinnate, often with glands on the rachis; leaflet pairs many, some with sleep movements, or touch-sensitive; in certain groups – for example *Acacia* – characteristic broad, modified petioles (phyllodes) simulate leaflets lost after the seedling stage; stipulates in some genera reduced to spines. Flowers regular (actinomorphic), crowded into globose heads, or cylindric spikes, rarely recemose, some inflorescences monoecious; calyx lobes 5, usually valvate; petals 5, equal, valvate in bud, usually united above the base; stamens mostly 10 or multiples thereof, free, jointed at eh base, or united forming a tube. Fruit a pod, straight, curved, or spirally twisted, usually 2-valved, dehiscent, some breaking transversely into 1-seeded segments.

Caesalpinloideae: Plants, shrubs, rarely scandent, rarely herbs

Leaves pinnate or bipinnate, rarely simple; some with translucent dots. Flowers mostly irregular (zygomorphic), usually in showy racemes or panicles; calyx lobes 5, overlapping or separate; petals usually 5, sometimes rudimentary or absent, slightly unequal, the upper petal distinctive and innermost in bud; dehiscing by joined, 10 or fewer, some may be staminodes; anthers indehiscent, some with winged sutures (samara), or fleshy and drupelike.

Papilionoideae: Plants, shrubs, annual or perennial herbs

Leaves mostly palmately 3- or more-foliolate, or odd or evenly pinnate, not bipinnate, or trifoliate, rarely simple; some with tendrils or spines. Flowers very irregular (papilionaceous), in 1 – to many-flowered, terminal or axillary inflorescences; calyx tubular, regularly 5-toothed or lobed; petals 5, unequal, overlapping; standard uppermost and outermost; the 2 wings lateral, intermediate; the 2-keel petals lowermost, inside, usually joined along the lower margin, hiding the ovary and stamens; stamens; stamens 10, rarely fewer, usually united into 2 bundles (dialdelphous) of 9 + 1 or 5 + 5, or in 1 tubular bundle (monadelphous), rarely free (Sophoreae). Fruit a variously shaped pod, straight, curved, winged, or moniliform, usually 2-valved and dehiscent, some transversely jointed (loments), ripe joints separating.

TECHNIQUES TO ENUMERATE RHIZOBIA

Several direct and indirect methods have been found useful in the enumeration of rhizobia both from soil and in broth culture. Many workers have enumerated viable rhizobia in the aliquots directly by the dilution plate count method (Vincent, 1970, Karren and Wollum, 1980 and Hoben and Somasekaran, 1982). Drop plate count technique of Miles and Misra was found to be more convenient in estimating viable rhizobia (Hoben and Somaskaran 1982). Development of fluorescence technique further helped in the precise enumeration of rhizobia directly from the soil (Schmidt *et al.*, 1968; Schmidt, 1974; Bohlool an dSchmidt, 1973; Reyes and Schmidt, 1979). Enzyme linked immunosorbant assay (ELISA) was also used to enumerate soil rhizobia (Nambiar and Anjaiah, 1985).

Plate Count Technique

Vincent (1970) demonstrated the plate count technique to enumerate viable rhizobia by spreading an aliquot on the surface of congo red agar medium. The above method gave good results, when the population of other heterotrophic micro-organisms was not more than ten times the population of rhizobia. He also suggested a rough grading for natural rhizobia based on their appropriate number in one gram of soil as absent or sparse (<25), few (25 to 1×10^3), abundant (1×10^3 to 1×10^5) and very abundant (>1×10^5).

Further, the difficulty in this method was to distinguish a particular *Rhizobium* sp. in the presence of other rhizobia and similar soil bacteria. Attempts were made to develop a specific medium to isolate *Rhizobium* directly from soil using appropriate antibiotics in the growth medium. A selective medium for *Rhizobium* consisting of six antibiotics was developed by Grahan in 1969. This medium was, however, found not useful when the rhizobial counts were low.

Three methods of dilution plating viz., pour plate, spread plate and Miles and Misra drop plate were compared by Hoben and Somaskaran (1982). They reported that all the three plating methods gave close estimates, but Miles and Misra drop plate method has advantages over the other methods because of the economy in materials and labour. Usual plating methods are not applicable to the determination of the rhizobia in soil and un-sterile peat cultures because; 1) no medium is sufficiently selective to exclude the possibility of other forms being confused with *Rhizobium,* 2) where the *Rhizobium* is markedly in the minority in the dilutions at which it would need to be counted, it will be outnumbered or antagonized by other micro-organisms (Data and Vincent, 1962).

Flourescent Antibody Technique (FAT)

Schmid *et al.* (1968) tested the fluorescent antibody technique (FAT) to study soil rhizobia. The effectiveness of FA technique for detection of a particular strain of *R. japonicum* in soil was studied by the examination of contact slides. *Rhizobium* can be seen in soil by contact slides by means of immunoflorescence compound, but the study of their interactions with soil sites is limited by nonspecific adsorption of the FA. Similarly, Bohlool and Schmidt (1973) studied the persistence and competition of *R. japonicum* in soils using FAT. For the study of persistence of rhizobia in soils, microscopic slides were first buried in the appropriate soil overnight to allow the accumulation of soil film on the surface of slides. Slides were rinsed in distilled water and kept submerged in water suspension of 10 days old shaken culture of *Rhizobium* strain USDA 110 to 30 minutes to allow the bacteria to attach to the soil films. Slides were air dried and kept in air tight plastic containers. Slides were removed at different intervals and stained with FA specific for strain USDA 110. Counting efficiency of FA was estimated to be about 30 per cent (Reyes and Schmidt, 1979).

Enzyme linked Immunosorbant Assay (ELISA)

ELISA was also found to be useful in the enumeration of rhizobia. Nambiar and Anjaiah (1985) reported that the loss of viability in the peat carrier does not interfere with the enumeration of live rhizobia by ELISA. A single ELISA microlitre plate can be used for testing eight samples and results can be known in 2 days. One person can process 20 microlitre plates in a day. However, ELISA requires mainly gamma-globulin conjugation step which limits the use of this technique in ecological studies. Non-sustainability of ELISA-test when the population is less than 10^3 cells g^{-1} of soil is also reported. Under such conditions plant infection technique can be employed

(Nambiar and Anjaiah, 1985).

Plant Infection Technique

Most probable number (MPN) by plant infection technique is a frequently used method to enumerate rhizobia in soil and in inoculants when other microorganisms are present. Accuracy of this technique for counting *R. japnonicum, R. meliloti, R. trifoli* and *R. leguminosarum* has been established (Date and Vincent, 1962; Brockwell, 1963; Weaver and Frederick, 1972). But the accuracy of the MPN plant infection method for enumerating cowpea rhizobia has not been well established.

The method consits of inoculating the test plants grown aseptically with aliquots from a dilution series of the sample under examination. The number of rhizobia in the sample can be calculated from the proportion of test plants forming nodules at each dilution. Plant infection test though time consuming was found reliable and accurate to enumerate rhizobia and also to exclude other diverse microorganisms (Brockwell, 1963). Tuzimura and Watanabe (1961) compared the pour plate method and plant infection method to estimate pure culture population. They reported that the plant infection technique does underestimate the number of organisms in pure culture. Similar observations were made by Date and Vincent (1962).

Weaver and Frederick (1972) developed a new effective culturing system for Rhizobia towards determining a most probable number count by using legumes grown in plastic growth pouches. The counts obtained by this method were equal to pour plate counts. However, MPN counts were significantly less than the plate counts in two instances out of 8 samples. But even in these instances, the results were comparable considering the 95% confidence interval on the MPN counts. The differences may be due to slight errors in the dilution sequence or to a soil effect (Weaver and Frederick, 1972). Grassia and Brockwell (1978) stated that the estimation of bacterial numbers using plant infection assay was dependent on the probability of bacteria reaching and infecting the roots of test plant which did not follow a probability distribution.

There are truly no selective media for counting rhizobia directly from the soil except when antibiotic resistant strains are used (Bushby, 1981). But these techniques appear too specialized to offer any prospect of distinguishing reliably between rhizobia from various nodulation groups. Therefore, bacteriologically controlled plant infection test is usually employed (Brockwell, 1982). The progress in the study of *Rhizobium* ecology has been hampered by the lack of suitable methods for quantifying population changes in soil and plant rhizosphere (Bushby, 1981). The plant infection test is suitable not only for counting nodule bacteria in soil but also on inoculated seed, in legume rhizosphere or from any other situation where the rhizobia are outnumbered by other microorganisms (Brockwell,1982).

Growth system for MPN methods

Rhizobium of a particular host legume, is commonly counted by plant infection technique. This technique involved growing the small seeded legume plant aseptically on a seedling agar in enclosed tubes (Vincent, 1970; Brockwell *et al.*, 1975) or partly enclosed tubes (Gibson, 1963) or in plastic growth pouches (Weaver and Frederick, 1972). Large seeded legumes were grown in many systems such as jar assembly (Leonard, 1944; Vincent, 1970) glazed jar (Vincent, 1970; Weaver *et al.*, 1972), plastic growth pouches (Weaver and Frederick, 1972), vermiculite (Grassia and Brockwell, 1978), perlite medium in Leonard's bottle or plastic pots and partly in sterile condition in sand vermiculite mix (Sand: Vermiculite: Grit, 1:2:2 v/v) (Rao and Dart, 1980). Brockwell *et al.* (1975) reported that the dilution nodulation frequency technique (plant infection test) is suitable where a small seeded species like *Glycine ussuriensis* is grown in plant tubes as a test plant for identification and enumeration of *R. japonicum.* Similarly, growing large seeded legumes like chickpea in seedling agar tubes (25 × 200 mm) was also successful when cotyledons were excised under axenic conditions (Rupela and Dart, 1980; Toomson et al., 1984). Growing redgram cultivar ICP-1 (size 10 g/100 seed) in 25 × 200 mm tubes for MPN technique was also successful (Rao and Dart, 1980).

Suitable test host for MPN technique

High symbiotic promiscuity was observed within greengram, groundnut, redgram and blackgram (Dadarwal *et al.*, 1977). Stowers and Elkan (1980) suggested a test host; siratro (*Macroptilium atropurpureum*) for counting rhizobia nodulating cowpea group of legumes such as redgram, groundnut, cowpea, velvet, bean, punabean, lentil and horsegram. However, there was an indication of specificity in redgram *Rhizobium* strains within the cowpea miscellany. Thus there was a possibility of getting higher estimates of cowpea rhizobia with siratro as the trap host than those infective on redgram (Rao and Dart, 1980).

Boonkerd and Weaver (1982) reported on the accuracy of MPN technique as compared to plate counts using five strains of cowpea rhizobia. The number of rhizobia determined by the MPN method using cowpea, siratro and peanut under-estimated the population from 10 to 100 fold. Experiments using soybean as a test host indicated that the MPN method was accurate for *R. Japonicum.* The experimenter using the MPN method should be aware of its accuracy for particular *Rhizobium*-legume symbiosis.

Woomer *et al.* (1988) studied the reliability of MPN technique for enumerating rhizobia in tropical soils. They showed that there is no significant interaction of range of transition (ROT) of the *Rhizobium* legume plant infection count data between species and sites. The acceptability of MPN codes (P> 0.99) was the highest (97 to 99%) with *Vicia sativa, Trifolium repens* and *Glycine max* and lowest (72%) with *Leucaena leucocephala.* The two legumes *Medicagosativa* and *Macroptilium atropurpureum* yielded 87% and 75% acceptable MPN codes respectively. Gino and Brunal (1989) working on the use of MPN to detect soil microorganisms stated that MPN technique has the advantage of requiring small amount of space during incubation period and permitted to be used for any laboratory analysis. One can also analyse greater number of samples with more replications for each sample, making it possible to choose a single step dilution factor most appropriate for the less degree of contamination of the samples.

DISTRIBUTION OF RHIZOBIAL POPULATION

Most of the cultivated soils in India are found to contain native or indigenous rhizobia capable of nodulating cowpea group of legumes in varying numbers (IARI, 1971). Soils contain different rhizobial populations based on legume coverage. This population consists of a number of species, sub species and strains showing wide range of symbiotic effectiveness with a legume. Number of indigenous rhizobia is influenced by soil types, conditions and soil depths.

Wilson (1931) observed species variation in abundance from time to time in New York soils. *Rhizobium leguminosarum* counts have been observed to range from <10 to$>10^5$ per gram of soil, whereas the number of *R. trifolii* ranged from 2500 to 10^6 per gram of soil. Population densities were commonly low in acid soils and submerged soils (Wilson, 1926) and seasonal fluctuations have been observed (Wilson, 1930; Gowda and Rai, 1991). The population of rhizobia even within restricted area was found to be diverse and a few samples could give only a poor idea of its real nature (Purchase and Vincent, 1949).

Native nodulation of tropical legumes was observed in India and many other countries like Australia (Mclead, 1962; Diatloff and Langford, 1975), Africa (Obaton, 1977), U.S.A. (Kremer and Peterson, 1982) and others. Population of *Rhizobium* was greater in soils that supported a rotation with legume for which the *Rhizobium* was a symbiont than in a rotation system in which the legume was not present. Thus more *R. meliloti* cells were observed in a soil recently grown with alfalfa than in a soil in which alfalfa was absent in the previous year (Walker and Brown, 1935). On the other hand, Weaver *et al.* (1972) found no correlation between the abundance of *R. japonicum* and the number of years since soybeans were present in a particular field. Similarly, they observed no correlation between the numbers of *R. japonicum* and soil texture, pH or organic matter levels. Conversely Tuzimura and Watanabe (1961) observed that *R. japonicum* was abundant in the fields which had soybeans the year before, the population declined the following year although it was still significantly higher than in adjacent sites not cropped to soybeans for twenty years. The legume rhizosphere is known to specifically stimulate native rhizobial numbers in many cases (Robinson, 1967; Rao and Dart, 1980).

Tuzimura and Watanabe (1962) have demonstrated that rhizobia of different cross inoculation groups are frequently abundant in the rhizosphere of plants, which they are unable to infect. Not only legumes but also non-legumes will occasionally exert a pronounced stimulatory effect on the root nodule bacteria (Rovira, 1961). Reyes and Schmidt (1979) found that rhizobia are stimulated even in non-legume rhizosphere like wheat equally in comparison to legume rhizosphere.

In the soil regularly cultivated with legumes, the native rhizobia formed effective nodules without inoculation indicating a survival of naturally occurring rhizobia over the seasons in appreciable numbers (Diatloff and Langfor, 1975). Ireland and Vincent (1968) estimated clover rhizobial population for studying quantitative competition for nodule formation by plant dilution count.

Brockwell and Dudman (1968) surveyed the population and effectiveness of *R. meliloti* associated with pastures of *Trifolium subterraneum* in various localities in Australia. Very high populations of 9.8×10^5 per gram of soil was recorded but the mean level of nitrogen fixing ability (effectiveness) was 52% of that of control strain used in the commercial clover inoculants. Escuoler (1972) estimated the numbers of *Rhizobium meliloti, R. trifolii, R. leguminosarum* and *R. lupini* in different locations by plant infection count technique. It was reported that *R. trifolii* was most abundant followed by *R. leguminosarum* while *R. meliloti* and *R. lupini* were less abundant and very wide range of effectiveness was recorded in locations where heavy application of fertiliser nitrogen was practised for cultivating the legumes. Chatel and Parker (1973a) estimated rhizobial populations from the samples of rhizosphere and non-rhizosphere soils at different time intervals and reported higher population of *R. lupini* than that of *R. trifolii.* Similarly, Hamdi *et al.* (1978) enumerated the populations of *R. meliloti* and *R. trifolii* in sixty five soil samples by semiquantitative method. There was no quantitative relationship between size of rhizobial population and soil properties such as pH, EC, lime content etc. and also soil types.

Rao and Dart (1979) enumerated cowpea rhizobial population in alfisols and vertisols by using siratro and redgram as trap hosts. They found that alfisols contain more than 10^4 rhizobia per gram of soil than vertisols. Soil types and depth also affected population. Similarly, population of groundnut rhizobia in cultivated soils of India ranged from 10^2 to 10^6 per gram of soil (Sheth, 1979). Due to inadequate method for identifying indigenous strains in the soil, the dynamics and diversity of *Rhizobium* ppulation in a legume field are poorly understood (Noel and Brill, 1980).

There is a large variability in the *Rhizobium* populations between sites and depths within a field. In one alfisol field the *Rhizobium* population was less affected by soil depth, remaining high (10^4 g^{-1} dry soil) through out the profile with a slight reduction at a depth of 150 to 160 cm (10^3 g^{-1}). In another field, the population declined with depth, the reduction being more pronounced below 100 cm in soil profile. Similar patterns have been observed in vertisol fields (Rao *et al.*, 1981). Generally alfisols contain more rhizobia than vertisols. Some soils of ICRISAT sites had fewer than 100 cowpea rhizobia per gram soil (Rao *et al.*, 1982).

The estimation of cowpea rhizobial populations using *Macroptilium atropurpureum* as host plant gave great variations in numbers between sampling sites in the same field. More consistent numbers were obtained in alfisols (10^4 to 30 x 10^5 g^{-1} soil) than in vertisols (0 to 10^6 g^{-1} soil) according to Rao *et al.* (1982).

Hegde (1983) enumerated the redgram rhizobial population in 14 soil samples from different places of Karnataka using siratro and redgram as trap host. The population ranged from 1.8 x 10^2 to 1.2 x 10^5 per gram of dry soil. Population changed with soil depth and seasonal conditions. Similarly, population varied in different soil types with different cropping pattern. In general, loamy soils were found to have good population of mung bean rhizobia as compared to sandy loam soils. Panther *et al.* (1984) reported high population of mung bean rhizobia in rice fields in Punjab.

Singleton and Tavares (1986) determined the rhizobial population in four different Hawaiian soils with six legume hosts viz., *Glycin emax., Vigna unguiculata, Phaseolus lunatus, Leucaena leucocephala, Arachis hypogea* and *Phaseolus vulgaris. Rhizobium* population ranged from less than 6 x 10^0 to 10 x 10^4 per gram of soil and all the four sols were devoid of *R. japonicum.*

Rhizobium culture

Rhizobium culture has been used for leguminous vegetable crops is in paractice among farmers. *Rhizobium* forms symbiotic association with roots of leguminous crops forming nitrogen fixing nodules on the root system. Rhizobia enter the roots of legumes either through root hairs or directly at the emergence of lateral roots. They form infection threads through which the bacteria enter the root system. The entry of *Rhizobium* into the roots of plans is a perfect symbiosis. Initial stages is symbiosis is the reciprocal molecular conversation, transmission of a gene inducer from legume host to bacterium and the bacteria in response synthesizing a morphogen that is transmitted to the plant, switching the developmental fate of the legume root, the bacteria and formation of bacterioids in the nodules. It is enzyme nitrogenase that reduces atmospheric nitrogen to ammonia. This enzyme has been studied in a very great detail. (Das, 1991)

There are 2 types of rhizobia:

1. Slow growing *Bradyrhizobium* sp.
2. Fast-growing *Rhizobium* sp.

Soil contains native species of *Rhizobium* but all are not capable of forming nodules. Some of the strains are highly specific to certain species. These are called cross inoculation groups. For example, *Rhizobium leguminosarum* can nodulate peas only, while *R. leguminosarum bv. phasioli* can nodulate French bean and other beans and *Rhizobium* sps. Nodulate cowpea. Artificial inoculations help increase the native population of a particular species in the soil when that crop is grown.

The efficacy of *Rhizobium* inoculation has been established in India beyond doubt by the results of the coordinated trials conducted by the Indian Council of Agricultural Research, New Delhi, in farmers' field (AICRP 1983). Vegetable legumes French bean, pea and cowpea - have been extensively studied. Effective nodulating strains have been isolated and developed after field testing. French been strains have been isolated and developed after field testing. French been strains have been modified from time to time to suit salinity, alkalinity and acidity of soil (Bhardwaj, 1975; Suryanarayana and Pramod Kumar, 1985; Tomar *et al* 1985; Yadav and Sanoria, 1994; Wange *et al* 1996). In iron mine waste soil also rhizobium inoculation could improve nodulation in cowpea. (Ravtray *et al,* 1996). Mutant strains of *Rhizobium leguminosarum* bv phaseoli superior to wild strains in nodulation have been induced in French bean through UV induction, antibiotic resistance, acid and azide resistance at the CCS Haryana Agricultural University, Hissar. For cowpea effective strains of *Rhizobium* sps. have been identified at University of Agricultural Sciences, Bangalore, Tamil Nadu Agricultural University, Coimbatore and CCS Haryana Agricultural University, Hissar.

Mundra and Bhati (1994) studied the role of ion, manganese and *Rhizobium* inoculation on growth, nodulation and iron manganese ratio and protein content in cowpea. They found that 20 kg/ha of *Rhizobium* inoculum increased the number of nodules 38%. The *Rhizobium* sp. also increased nodulation, nitrogen fixation and biomass yield by slow and fast growing cowpea rhizobia in gaur under different environmental conditions (Mand *et al,* 1992). Wange and Patil (1995) studied the response of pea cultivars of *Rhizobium leguminosarum* inoculation. An increased pod yield of 25-66.7% could be obtained in Bonevillae variety. Singh *et al* (1992) in a field trial in sandy loam soil in U.P. found that *Rhizobium* inoculation and phosphate application (46 kg/ha) gave maximum yield in pea Rachna. Singh and Singh (1992) also found that *Rhizobium* inoculation and sulphur application (0-30 ppm) improved seed yield in pea cv T163.

It is estimated that about one third off the electrons made available for reduction of nitrogen in legume nodules are used instead to reduce hydrogen and this gas just escapes into the soil atmosphere carrying away wasted energy. Certain species of *Rhizobium* contain uptake hydrogenase enzyme (Hup+ strains) that oxidizes much of the hydrogen to water before it escapes. During this oxidation ATP is produced from ADP and Pi. This process therefore conserves energy and are more efficient N fixers. The *Rhizobium* strains possessing this enzyme yield more than the strains that do not possess them. The Hup+ strains have been identified by Dadarwal (1986) to be more competitive than Hup- strains.

Genetic modifications in Rhizobium strains

Efforts have been made to develop genetically modified strains incorporating traits for better competitiveness; Khanuja and Suman (1993) obtained heat shock tolerant mutants in the process of improving rhizobial strains. Sharma *et al* (1997) introduced gene markers (Gus) in Rhizobium strains to monitor the release of genetically engineered strains in the environment. However genetically modified strains have not made any impact on legume production under field conditions.

Rhizobium in Root Nodule

The presence of these bacteria in the root nodule can be seen under high power microscope by crushing a washed nodule between two glass slider. Frank (1877) and Beijernick (1888) discovered *Rhizobium* a free living or symbiotic bacteria of leguminous as well as non-leguminous plants.

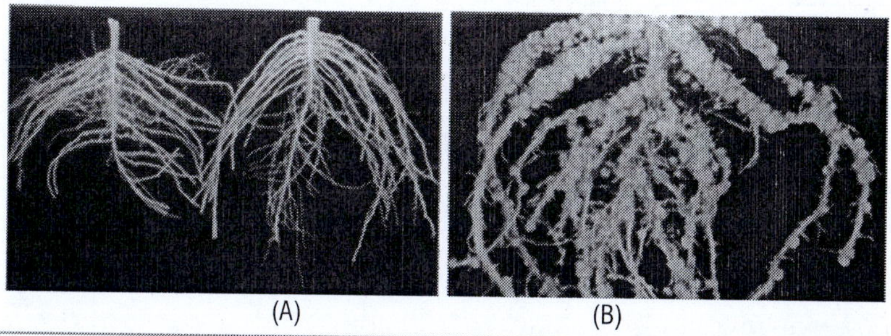

Fig. 5.1: Bare uninoculated roots of soybean plant (A) and nodulated roots of soybean (B) inoculated with *Rhizobium.*

This kind of bacteria, preferentially, infect legumes as compared to non-legumes. Structurally, *Rhizobium* are rod shaped but great variation can be observed during their life cycle. These are coccoid, very small, highly motile and ellipsoidal forms. The bacteroids are usually irregular with X, Y, star and club-shaped forms. They are 0.5 to 0.9 micron by 1.2 to 3.0 micron in size. All the species are non-sporing and non-acid fast. These symbiotic bacteria (*Rhizobium*) are difficult to cultivate in ordinary culture media but they grow on manitol agar.

Rhizobia are susceptible to antibiotics (produced by other microorganisms of soil) and also to the action of bacteriophages. Fungicides, herbicides and other plant protectants are also toxic to *Rhizobia*. Occurrence of different species of *Rhizobium* in leguminous plants can be enumerated as follows:

Agrobacterium (A. radiobacter, A. tumefacions)

It has also been included in the species: *Rhizobium radiobacter*, but this splitting of *Rhizobium* at the generic level into *Agrobacterium* is not accepted by most of the microbiologists.

One more kind of slow growing asymbiotic bacteria of lupin, soybean, cowpea have been identified and placed under a separate genus *Phytomyxa* (*P. japonicum*). In short, the root nodule genera can be delimited into two genera: *Rhizobium* and *Phytomyxa*: the former comprises of three species: *R. leguminosarum*, *R. meliloti* and *R. radiobacter*, while the latter bears only one species *P. japonicum*.

According to Elken (1969) *Rhizobium* can also be classified on the basis of DNA composition, for example guanine and cytosine composition of *R. leguminosarum* and *R. meliloti* is 56.6 to 63.1 per cent while G+C composition of *P. japonicum* is 62.8 to 65.5 per cent (Delay and Rassel, 1965).

Structure and Function of Nodule

The outermost layer of the nodule constitutes the bacteroid zone which is enclosed by several layers of cortical cells. The rate of nitrogen fixation of nodule is directly proportional to the volume of the effective nodule. Sometimes ineffective strains produce ineffective nodules. These nodules are relatively small and contain poorly developed tissue associated with morphological abnormalities.

Formation of Nodule in Leguminous Plants

Biological nitrogen fixation is the reduction of atmospheric nitrogen to ammonia by microorganisms in the soil. It involves a highly specialised and intricately evolved interactions between soil microorganisms and higher plants for harnessing the atmospheric nitrogen. It is a fascinating biological phenomenon studied extensively to provide low-cost nitrogen and improve crop productivity. The nitrogen fixing organisms associated with horticultural crops are the Rhizobium species, living in symbiotic relationship with the leguminous plants and free living fixers belonging to the *Azotobacter* family and the *Azospirilla* living in association with the root system of crop plants. Roots of most of the leguminous plants possess circular out-growth or swelling called nodules or tubercules, which are formed after infection of certain types of nitrogen fixing bacteria particularly the different species of *Rhizobium*.

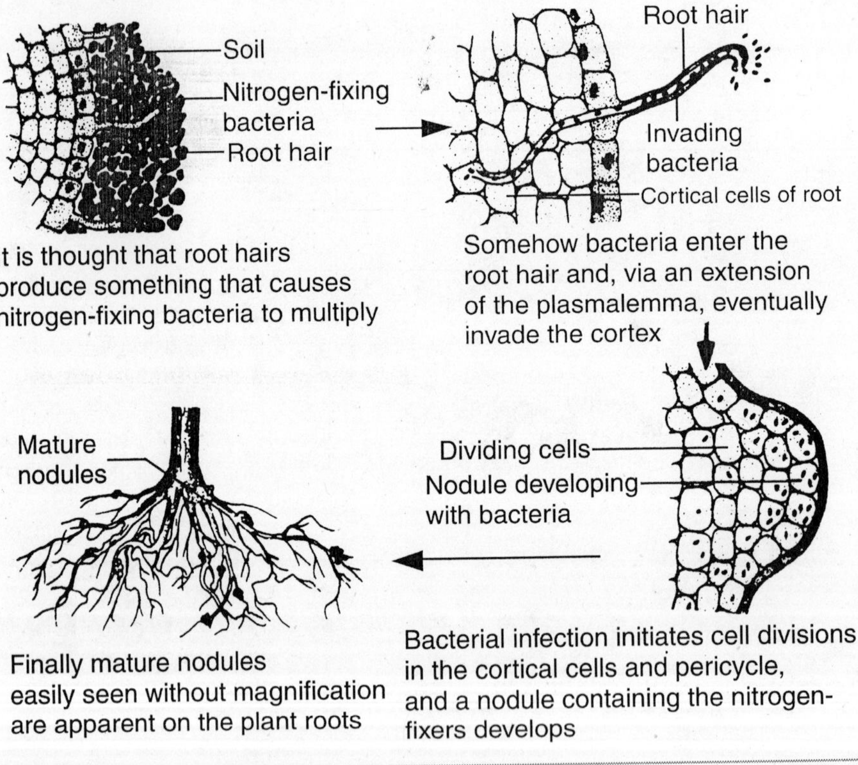

FIG. 5.2: Nodule formation by *Rhizobium* on legume plants. The root hairs are the site of infection. Note the change in the root hair. A characteristic deformation is the curling of the root hair tip to produce a Shepherd's crook. The nodule can be seen on the root.

These bacteria have the capacity to fix free atmospheric nitrogen of soil in nodule.

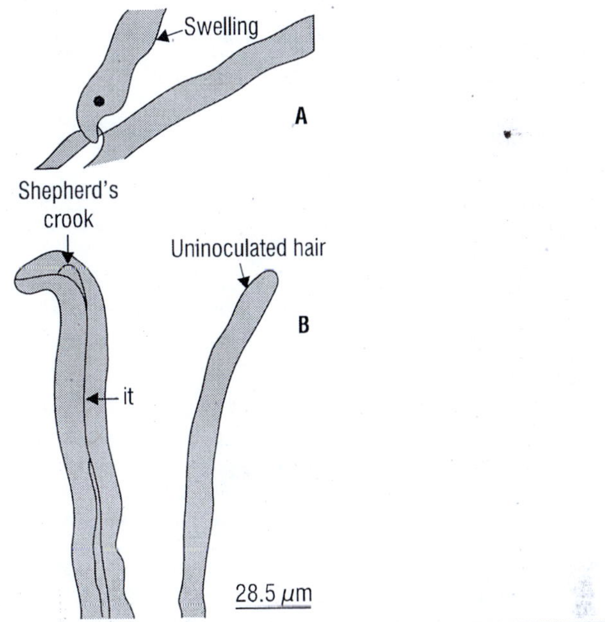

Fig. 5.3: The early host responses to infection leading to the formation of N_2-fixing root nodules. (A) A root hair two hours after inoculation, showing curling and swelling. (B) A root hair (left) tow days after inoculation, showing a shepherd's crook and an infection thread (it). A thin uninoculated root hair is shown on the right.

They infect or get their enterance in root through the soft-hair or other epidermal cells by damaging them. In the first stage of the infection, bacteria grow very profusely at the tip of the root hair and form a long filament in the root hair- called *infection thread*. This thread reaches the endodermis and pericycle area through cortex tissue. Cells of this area (cortex) go on dividing and form a young nodule. The newly formed young nodule pushes the overlying parenchyma and the epidermal tissue towards outside and produces a small swelling on the surface of the root. Anatomically, a nodule is made of thin walled parenchymatous cells which is filled up with the nitrogen fixing organisms. The size and shape of the root nodule varies according to plants in which occur. It is not necessary that all the bacteria, which infect the root, produce nodule.

According to Wipf and Cooper (1939) the root nodule always contains double number of chromosomes as against of normal somatic tissue. If the root lacks such cells (cells with double chromosome number), there will be no formation of root nodule, even after the formation of infection thread. Those plants which do not bear root-nodules will never be able to fix atmospheric nitrogen in the plants. It has been observed by Allen and Allen (1947) that out of 429 genera of family Leguminosae about 179 genera containing 949 species form nodules.

Leghaemoglobin

The effective nodules are usually larger and pink in colour. This is due to the presence of red coloured leghaemoglobin (leg indicating its presence in legume root nodule). This pigment is similar to haemoglobin of blood and is found in nodules between bacteroids and membrane envelops, enclosing them. Chemically, it is haem-protein and it contains a haem meoiety attached to a peptide chain and represents the globin part of the molecule. The molecular weight of leghaemoglobin is 16,000 - 17,000 daltons. The amount of leghaemoglobin in nodules has the direct relationship between amount of atmospheric nitrogen fixed by legumes. The functions of leghaemoglobin are as follows:

1. It represents an active site of nitrogen absorption and reduction.
2. It acts as a specific electron carrier in nitrogen fixation.
3. It regulates the oxygen-supply in the nodule.
4. It acts as an oxygen-carrier.

According to Scholander (1960), leghaemoglobin acts as a biological-valve in regulating the supply of oxygen to bacteroids in nodules.

Site of Nitrogen Fixation in Nodules

According to Bergersen (1969) the membrane envelops surrounding the groups of bacteroids are the site of nitrogen fixation, but in recent years it has been proved that bacteroids are the seat of nitrogen fixation.

Leguminous Plants / *Rhizobiaceae* Symbiosis

Symbiotic N_2 fixation in agriculture can be attributed mainly to legumes – the plants in the *Leguminosae*. It is estimated that *Leguminosae* contains more than 200 genera, and 20,000 species, which ranges from small plants such as the clover to the large plants such as *Acacia* species. Approximately 90% of them can fix nitrogen from the atmosphere with *Rhizobiaceae*, either with *Azorhizobium*, *Bradyrhizobium*, *Rhizobium* or with *Sinorhizobium* in root nodules. The important agricultural legumes can be divided into three groups: crops that grown for their commodities (e.g. grain legumes); forage legumes that may be grazed or harvested for animal fodder; and plants or shrubs in agroforestry systems. Worldwide approximately 1.5 million km^2 of land are cultivated with grain legume plants, mainly in *Glycine max*, *Phaseolus vulgaris*, *Pisum sativum*, *Arachis hypogaea*, *Cajanus cajan* etc.. The annual harvest of grain legumes are about 200 million tonnes, which provides the plant protein source for human and animal consumption, or vegetable oil and other raw materials. The areas partially covered by the forage legumes mainly in *Trifolium*, *Lotus*, *Medicago*, *Stylosanthes*, *Macroptilium* and *Mimosa* etc., is even much larger, about 30 million km^2 of grassland in the five continents. The third group consists of the genera *Acacia*, *Albizia*, *Alnus*, *Leucaena*, *Robinia* etc., which are mostly used as timber, fuelwood or craftwood, and in the pharmaceutical industry as antibacterial and antifungal agents, or food additives and any other functions.

There are five genera of the *Rhizobiaceae*: *Azorhizobium*, *Bradyrhizobium*, *Rhizobium*, *Sinorhizobium* and *Photorhizobium*. All members of *Rhizobiaceae* are characterised by a gram-negative cell wall structure. Cells are generally rod-shaped, non spore-forming and motile with differently arranged flagella. All rhizobia are aerobic bacteria that persist saprophytically in the soil until they infect a root hair cell. The enzymes from the bacteria degrade part of the cell wall and allow bacteria entry into the root-hair cell itself, which lead the root hair to produce a threadlike structure called the infection thread. The bacteria multiply extensively inside the thread, which extends inwardly and penetrates through and between the cortex cells. In the inner cortex cells the bacteria are released into the cytoplasm and stimulate some cells (especially tetraploid cells) to divide. Each enlarged, non-motile bacterium is referred to as a bacteroid. These result in a proliferation of tissues, eventually forming a mature root nodule. A typical root nodule cell contains several thousand bacteroids. The nodule contains a protein called leghemoglobin, which gives legume nodules a pink color due to its prosthetic hem group. Leghemoglobin is thought to help transport O_2 into the bacteroids, which is essential for bacteroid respiration. Nitrogen fixation in root nodules occurs directly within the bacteroids. The host plant provides bacteroids with carbohydrates, which they oxidise and from which they obtain energy. These carbohydrates are first formed in leaves during photosynthesis and then are translocated through phloem to the root nodules. Sucrose is the most abundant carbohydrate translocated, at least in legumes. Some of electrons and ATP obtained during oxidation by the bacteroids are used to reduce N_2 to NH_4^+, which is catalysed by nitrogenase.

Isolation and Identification

Leguminous plants are carefully uprooted and their roots are washed in running water to remove soil particles. Healthy intact and pink nodules preferably on tap root are selected and cut out with a portion of the root.

Table 5.5: New system of classification of root nodule bacteria

Genus	Species	Biovars	Host legumes
Rhizobium	Leguminosarum	a. Viceae	Vicia
		b. Trifolli	Trifolium
		c. Phaseoli	Phaseolus
	meliloti	-	Medicago
	loti	-	Lotus
	fredii	-	Glycine
Bradyrhizobium	Japonicum	-	Glycine
Bradyrhizobium	Species	-	Cicer, Cajanus, Vigna, etc.

The nodules are surface sterilized by immersion in acidified 0.1% $HgCl_2$ for 5 minutes and repeated washing with sterile water. The nodules are then dipped in 70% ethanol for 30-60 seconds followed by several washings with sterile water. Surface sterilized nodules are individually crushed in sterile water with a sterile glass rod and the resulting suspension is streaked on yeast extract mannitol agar (YEMA) plates containing congo red poured in 24 hours earlier to ascertain the sterility of plates and culture media. Upon incubation at 28-30°C for 3 to 10 days depending on the nature of *Rhizobium* (slow or fast growing), the colonies of bacteria normally appear as white, translucent, glistening and elevated with entire margin. *Agrobacterium,* the most common contaminant of *Rhizobium* takes up the colour of Congo-red and appears red in colour.

There are several presumptive tests which are carried out to identify *Rhizobium* isolate. Rhizobia grow very poorly on glucose peptone agar medium, does not grow on Hofers alkalie broth (pH 11.0) and does not utilize lactose. But the best way of identifying rhizobia is to conduct plant infectivity tests and see if effective nodules are produced. A plant test is done under controlled and sterile conditions with assemblies ranging from small tubes to pots. Small test tubes are used for small seeded legumes like clover, berseem, moong and urd, whereas Leonard jar assembly or small pots are used for testing bold seeded legumes like chickpea, arhar and french bean. Agar-agar, vermiculite or sand can be used as the substrate for growing plants. Surface sterilized seeds of specific host are sown in appropriate assembly and inoculated with the bacterial isolate under test. The plants are watered at regular intervals with any N-free nutrient solution. After 5-6 weeks the plants are examined for nodulation, nitrogenase activity and dry biomass. Thus, an efficient strain for a particular crop can be selected. Such strains are then subjected to pot and field trials to assess their potential as inoculants.

After selecting efficient strains, it is necessary to lyophilize them to maintain their efficiency since subculturing at monthly intervals on artificial medium may lead to deterioration of the activity of organisms.

MASS PRODUCTION

The Rhizobial Inoculant Industry has come a long way since its inception in 1895 when Nobble and Hiltner introduced a laboratory culture of Rhizobia under the name 'Nitragin'. The last 100 years have witnessed a number of developments towards the better exploitation of this technique. Carrier-based rhizobial inoculants have almost completely replaced liquid cultures and agar slope cultures which were commonly used a few years ago.

For preparation of carrier-based inoculants the rhizobial broth culture must attain a minimum population of 10° cells/ml. It is then blended with the finely powdered (100-200 mesh sieves) and sterilized carrier. The sterilization of carrier is carried out in autoclaves at 15 lb psi for 3-4 hrs. continuously or 1 hr each on three successive days. Blending is done by mixing of broth culture with sterilized carrier. In trays either manually (with hands washed and smeared with alcohol) or by using a mechanical mixer. The broth is added equivalent to one-third of water holding capacity of the carrier.

After mixing the carrier with the broth culture, raw-blended carrier is kept for 24 hours for curing. During this time the rhizobia get acclimatized with the carrier.

After curing, the inoculant is packed in polyethylene bags. The polyethylene bags are sealed by an electric sealer leaving about two-third vacant space to give proper aeration to the inoculant. Directions for use of inoculant are printed on the packet.

CARRIERS FOR *RHIZOBIUM* INOCULANTS

The medium in which rhizobia are allowed to multiply is an important factor in rhizobial culture preparation. The carrier is used for this purpose. The carrier is a crucial ingredient in the formulation of inoculant because the shelf life of the products largely depends on it. A variety of carriers are in use at present while some are under testing. Some materials whose usefulness as carrier has been established and used as carriers. Even though typical peat is not available in large quantities in our country, the peat like material available in Nilgiris has been found to be a good carrier. Lignite available from Neyveli Lignite mines in South India is another carrier material widely used. The characteristics of good carrier are:

1. It should be indigenously available and cheap
2. Should be non-toxic to the Biofertilizer bacteria
3. Should have around neutral pH
4. Should have an organic matter content around 40%
5. Should have high water holding capacity and surface area.

METHOD OF INOCULATION

A 10% sugar solution or 10% pharmaceutical grade of gum arabic is prepared to serve as sticker for *Rhizobium* cells seeds. This solution is sprinkled on the seeds required for one hectare and the seeds are spread on a polyethylene sheet and mixed uniformly. The content of the inoculant packet are sprinkled over sticker coated seeds and mixed thoroughly.

Table 5.6: Carriers used for *Rhizobium* inoculation in India (Tilak, 1993).

Carrier	Availability	Location	Water holding capacity (%)	Organic matter (%)
Peat	5.5 million tonnes (total)	Ootacamund (Nilgiri Hills)	52-120	21-56
Lignite	3 million tonnes (annually)	Neyveli	92	28
Farmyard manure	Plenty	Everywhere	40-50	30
Charcoal	Plenty	Everywhere	90	77

If required more water is added but only in sufficient amount to coat the seeds fully with the inoculant. The seeds are spread uniformly for drying on gunny bag or cement floor in shade avoiding direct sunlight. After drying, the sowing is done immediately.

In order to protect the rhizobi from adverse situation of soil conditions (acidity, alkalinity, hazard of pesticides, etc.), pelleting method of seed inoculation is recommended. Finely ground calcium carbonate (300 mesh sieve) is added to freshly inoculated wet seeds in a container and mixed rapidly for 2 minutes and until seeds are uniformly coated. Coated seeds are allowed to harden for an hour by spreading on a clean surface. Besides lime, other pelleting agents like gypsum, bentonite, rock phosphate, charcoal etc. have also been recommended.

CROP RESPONSE

The largest contribution of BNF to agriculture is derived from the symbiosis between legumes and species of *Rhizobium*. The *Rhizobium* – legume association can fix upto 50-100 kg N/ha in one crop season and in certain situations leave substantial N for the following crop. The quantity of N fixed varies with the kind of legumes and agro-climatic conditions. Crops like groundnut, chickpea. Cowpea, green gram and soybean can leave being 20-30 kg N/ha for succeeding crops, while crops like Lucerne, pasture legumes and green manuring crops like sunnhemp, *dhaincha* add about 50 kg/ha.

The field response to *Rhizobium* inoculation is quite variable. The average response to *Rhizobium* inoculation ranged from 66 kg/ha at Hisar to 152 kg/ha at Badanpur in chickpea. In general, most types of legumes are reported to respond to *Rhizobium* inoculation but not at all locations. Studies carried out in All India Coordinated Agronomic Research Project taking 5 legumes (moongbean, urdbean, soybean, cowpea, groundnut) indicated significant response at 7 out of 17 locations. Residual effect of *Rhizobium* inoculation on succeeding cereal crop was noticed in 4 cases only. Average response to *Rhizobium* inoculation ranged from 6% in soybean to 18% in cowpea. A residual response of 1 to 3% was noticed on succeeding wheat crop.

FACTOR AFFECTING CROP RESPONSE

Although legume inoculation is a long established practice to maintain soil-health, it has not become popular largely because of its erratic response. Some of the common reasons of poor or no response to inoculation are:

1. **Failure of seed to germinate:** This is due to the use of poor quality seed and lack of sufficient moisture for germination.
2. **Failure of seedlings to become nodulated:** This may be the result of poor quality inoculant, adverse soil conditions, inoculation with wrong inoculant, compatibility of rhizobia with pesticides, etc.
3. **Ineffective nodulation:** This situation may result from the presence of a large native population of rhizobia which may compete with rhizobia applied for nodule formation and makes it ineffective in fixing N with introduced legume.
4. **Nutritional deficiencies:** Nodule formation is severely affected by nutritional deficiencies particularly phosphorus, calcium and molybdenum because of their role in protein synthesis, cell wall development and in nutrition of rhizobia, respectively.
5. **Plant competition:** Failure of the legume plant to promptly nodulate and establish is frequently due to competition for moisture, nutrients, light and space.

Confirming Nodulation

The benefit from the N fixation of Nitrogen fixing species (NFS) plantings, the plants must actually be nodulated and fixing N effectively. These are separately controlled functions. Conflicting reports of nodulation and the influence of environmental factors can result in confusion over whether or not the plants do fix nitrogen. In some woody species, nodules maybe large (that is, 5 cm to 7 cm) or black and woody, not resembling nodules of better-known legumes at all (de Faria et al., 1987). Sprent (1986) reports nodulelike structures that are not nodules because they do not contain rhizobia and have no nitrogenase activity. Such structures can be confused easily by workers without access to appropriate laboratory facilities. The simple test of opening nodules to look for hemoglobin (as indicated by a pink color) also can be misleading because this pink color may be found in outgrowths on roots of plants inhibited by nematodes (Sprent and Sutherland, 1990).

The most certain means of confirming nodulation is to carefully evaluate reports of nodulation using well-established principles. Allen and Allen (1981) list four essential requirements for confirming and reporting nodulation.

1. The host plant should be correctly identified by a competent botanist.
2. Whole nodule specimens should be accompanied by records of their size, shape, location on the root system, relative abundance, color, texture, plant habit and habitat, and the date of collection. Rapid transfer of specimens to plastic bags or dry vials is recommended.
3. Rhizobia should be isolated from the nodule as much as possible, using Koch's postulates and irreproachable bacteriological procedures.
4. Plant trials with the isolate should be conducted for nodule evaluation, using aseptic plant-growth equipment and seeds that are surface, sterilized.

These steps generally will be taken only by microbiologists with an interest and the facilities to conduct this type of procedure. For nonmicrobiologists, perhaps the most thorough means of confirming nodulation is to arrange for a microbiologist to assess nodulation. Nodules to be tested should be placed in a small vial over a desiccant such as silica get for transport to the laboratory. For NFS workers who do not have access to a micro-biology laboratory, the next best alternative is to evaluate nodulation in the field.

FIELD METHODS FOR IDENTIFICATION OF PLANT ACTUALLY FIXING NITROGEN

Simple surveys of plant roots can provide an estimate of the extent of nodulation, although this can be both difficult and time consuming for large plants. The eastest place to examine roots is in the nursery, where seedlings can begin to nodulate as early as 3 to 4 weeks after sowing. Chen and Rolfe (1988) even found nodules on *L. leucocephala* seedlings 12 days after inoculation. Sprent and Sutherland (1990) suggest that in humid areas young seedlings may be carefully dug up and examined; but in drier areas, young plants may not be the best subjects. In very dry areas in well-aerated soils, nodules may be at depths of several meters (Jenkins et al., 1988). This also may be true in light soils in humid areas. For example, effective, nodules were observed at a depth o f1 m on *paraserianthes falcataria* in a very humid site on the island of Oahu in Hawaii. Examination of roots to similar soil depths in the field, particularly in well-acerated soils, will help ensure identification of most nodulation sites.

Examination of roots for the presence of "effective" nodules can be done by carefully noting the color of nodule tissue. The best time to examine nodules in grain and fodder legumes in the field is at their peak of flowering. Although little is known of the periodicity of noudlation for NFSs, the most reliable time to examine them is likely during periods of least stress (that is, when moisture or temperature are least limiting for growth, such as the beginning of the monsoon or rainy season). Roots should be dug up with a shovel or fork to prevent nodules from being stripped from the roots. The nodules that are found need to be removed from the roots, sliced open and examined for color. Those with pink, red, orange or brown color are probably actively fixing nitrogen. Green nodules are older and senescent; white to pale green are ineffective and not fixing N. One exception is the green but otherwise normal nodules produced by effective rhizobia on molybdenum-deficient plants (FAO, 1984).

Leaf color is sometimes used as a rough indicator of nitrogen fixation. In a pot study of *Gliricidia sepium* seedlings, green-leaves seedlings produced higher nodule weights than pale-leaved seedlings (Atta-Krah, 1987a). In the green-leaves group, there also were positive correlations between nodule numbers, shoot and root fresh weights. While chlorosis need not relate to the lack of nitrogen fixation, it can be useful indicator to follow-up with closer examination. Interpreting leaf color is not a reliable method, however. The need for caution is demonstrated by the nonnitrogen fixing legume *Cassia siamea*, which nearly always produces dark-green leaves but does not nodulate.

Some NFSs such as *Sesbania rostrata* (Dreyfus et al., 1985), *Aeschynomene* spp. (Alzard and Duhoux, 1988) and *Casuarina cunninghamiana* (Dommergues et al., 1990) form stem nodules above ground. While aerial nodules are not common, it is likely that other NFS species also nodulate on the stem or on adventitious roots, particularly in high-rainfall areas. Dommergues et al. (1990) note that the aerial nodules of *Casuarina cunninghamiana*, although abundant, are not very conspicuous due to their flat, irregular shape and barklike color.

THE NEED TO INOCULATE

Nodulation will not occur if appropriate rhizobia of *Frankia* are absent from the soil in which NFSs will be planted. Inoculation, the process of adding inoculant to seed or soil, can introduce enough bacteria to infect the roots. Determining whether or not to inoculate should be based on whether there will be a response to inoculation under specific field conditions.

To determine the response to inoculation, two approaches can be taken. The first is to use crop history as a guide. The second is to conduct a field test. Crop history suggests that if the NFS species has never before been planted on the site, inoculation is probably necessary. An example from two soils in Nigeria demonstrates the importance of inoculating NFSs on "new" planting sites (Table 5.6). Most-probable-number analysis for these and other soils in Nigeria indicated that leucaena rhizobia were few or absent in soils without prior leucaena cultivation. This same pattern is true for other species, particularly *Robinia pseudoacacia*, which seems to be very specific in its rhizobial requirements (Turk and Keyser, 1992).

Crop history, however, can be a very imprecise means of predicting nodulation and response to nodulation. Akkermans and Houwers (1979) planted uninoculated *Alnus cordata* and *A. glutinosa* in soil with no *Frankia*. Under these conditions, 100% of the *A. cordata* plants nodulated, while nodulation varied from 56% to 100% among the *A. glutinosa* seedlots. The nodules in this case originated from infections by *Frankia* in the nursery soil.

Table 5.7: Effect of inoculation and fertilization of *Leucaena leucocephala* on two soils in Nigeria.

Treatments	Nodule number per plant	Nodules from inoculant strain (%)	Nodule dry weight (mg/plant)	Shoot dry weight (g/plant)	Plant N (kg/ha)	Total P (kg/ha)
Fashola grassland site, derived savanna						
Uninoculated	36	69	179	26	82	5.1
150 kg N/ha	36	69	87	72	232	15.2
Rhizobium						
IRc 1050	40	78	485	79	220	15.0
Rhizobium						
IRc 1045	15	94	174	68	209	12.8
IITA transition forest savanna site						
Uninoculated	0	0	0	51	174	14.6
150 kg N/ha	0	0	0	130	445	36.0
Rhizobium						
IRc 1050	17	100	198	103	398	26.9
Rhizobium						
IRc 1045	34	100	277	121	448	35.96
LSD* (5%)						
Fashola	11		23	12	53	1.9
IITA	12		25	22	66	0.6

Adapted from Sanginga et al., 1987, *Least significant difference.

The second and more reliable approach to determining response to inoculation is to conduct a field test to determine the need for inoculation. Three treatments are needed for this experiment (FAO, 1984):

1. inoculated plants from seeds inoculated with the best inoculant available;
2. noninoculated plants receiving no fertilizer treatment; and
3. noninoculated plants furnished with fertilizer nitrogen

The density of indigenous rhizobia has been identified as the primary determinant of response to inoculation in grain and forage legumes (Singleton and Tavares, 1986; Thies et al., 1991). Turk (1991) found a similar relationship with six legume plant species in four soils. High populations of indigenous rhizobia can dramatically reduce the response to inoculation. This factor, plus the level of combined soil nitrogen, can greatly influence the economic return to an investment in inoculation.

SELECTING AND HANDLING INOCULANTS

The following suggestions for selecting and storing inoculants largely come from an excellent practical handbook on the use of legume inoculants (FAO, 1984):

1. Different legumes require different inoculants. Be certain to have the right one. Table 5.7 demonstrates the effectiveness of *Rhizobium* and *Bradyrhizobium* strains on seven NFSs.
2. Be certain that the inoculant is of the high quality and is still fresh. High-quality inoculant should contain only large numbers of rhizobia or *Frankia* capable of producing nodules and fixing large amounts of N on the host (e.g., greater than 1×10^7 per ml). They must be free of other bacteria that may harm the rhizobia or host NFS.
3. Keep the inoculant in a cool place, preferably in a refrigerator, until ready to use. High temperatures will harm bacteria, but inoculant stored at temperatures of 4°C or lower will remain effective for several months. Fig. 5.1 shows the effects of high temperature on rhizobia survial.
4. Inoculate seeds just before planting because rhizobia die quickly on the seed.
5. When inoculating seeds, avoid making them too wet. Use only enough water for good adhesion of the inoculant to the seed.
6. Chemicals, fungicides or insecticides applied to seeds or soil may be toxic to the bacteria in the inoculant. If seeds are chemically treated, use a larger amount of inoculant and plant immediately after inoculating.
7. Store inoculated seed in a cool place protected from hot sunshine and excessive drying.
8. Leftover inoculant may be kept safely for short periods if it is in a tightly closed container to prevent drying.
9. Inoculated seed should not be eaten or fed to animals.

INOCULATION

Inoculation – the introduction of appropriate bacteria to the rooting zone of nitrogen-fixing plants – can supply enough rhizobia or *Frankia* to meet a plant's requirements. Seed or soil can be inoculated, although inoculation of seed or roots generally is preferred due to grater ease and lower costs. Seedlings, such as those in dibble tubes, can be inoculated with an inoculum/peat mixture suspended in water.

Methods of application

Methods like coating seeds with biofertilizer, drenching biofertilizer suspension on crop rows, if not applied at the time of sowing, are based on the principle to place the nitrogen fixing organisms on seed coat or at closet proximity of roots in rhizosphere. Bio-inoculants are sprinkled on moist seeds so as to load each seed with a minimum of 1000 viable cells. Most of the producers now have stared having adhesives mixed at the time of manufacturing and earlier methods of sprinkling *Rhizobium* treated gur/molasses solution is not required. Soaking the seed in water followed by culture inoculation gives better results because or removal of water soluble seed coat toxic agents.

Quantity of the biofertilizer applied in a unit area depends on the number of viable cells present in g^{-1} of biofertilizer and size of seeds. Usually 200 g *Rhizobium* biofertilizers are used for treating 10 –12 kg of pulse seeds. On an average, 500 g lignite/charcoal based biofertilizer is sufficient to treat the seeds required for sowing one hectare. Under normal conditions 10^3-10^6 when seeds are sown in adverse conditions like drought and acidity caused by fertilizers.

Biofertilizer contains living organisms and all due care and precautions should be taken to save them from extremes of temperature and desiccation. Treated seeds must be sown immediately and should not be mixed with acid forming chemical fertilizers. Recommended pesticides on pulses have no adverse effect on nodulation and yield. Soybean seeds when treated with fungicides followed by *Rhizobium* treatment or *vica versa* did not cause significant yield difference (Gupta *et al.*, 1988). Out of four herbicides tested Exadiazona 1.0 kg (active ingredient) ha^{-1} significantly increased the nodule number and yield of soybean (Khare *et al.*, 1989).

Inoclation methods should be chosen to match both locally available resources and the environment into which the inoculated seed will be planted and the type of seed to be inoculated. A summary of inoculation methods for different soil or planting conditions and seed sizes is found in Table 5.8; additional details on inoculation methods are given in supplement A of part 3.

Table 5.8: Recommended methods of inoculation

Soil or planting conditions	Recommended method of inoculation
Small-seeded NFSs	
Normal-good temperature, moisture, pH and seedbed	Slurry using peat-based inoculant, 30 ml slurry/kg of seed.
Normal-but with heavy infestation of native ineffective rhizobia	Inoculant pellet. Apply 45 g inoculant/kg seed. Mix 15 g inoculant with 22 ml of gum arabic solution (45%) and apply to seed. Use the remaining 30 g inoculant to dry the seed immediately after applying the slurry.
Acid soils, rough, poor seedbed	Lime pellet. Inoculant slurry prepared with gum arabic or synthetic cellulose, 22 ml/kg seed. Apply double dose of inoculant.
Seeds treated with toxic chemicals, other soil and planting conditions good	Plant without inoculation. Use soil inoculant after seedling emerge.
Large-seeded NFSs	
Normal-good temperature, moisture, pH and seedbed	Slurry using peat-based inoculant, 30 ml slurry/kg of seed.
Seeds treated with toxic chemicals, other soil and planting conditions good	Use soil inoculant. Apply in seed furrow or nursery soil with seed at sowing time.

Soil or planting conditions	Recommended method of inoculation
Hot, dry soil	Use soil of granular inoculant. Apply inoculant 2.5 cm to 5 cm below seed for deeper if moist soil is deeper.
Fragile seeds	Use soil of granular inoculant. Apply in furrow or nursery soil with seed.
Fields of nonnodulated NFSs	
Small seeds sown by broadcasting	Sow granular inoculant at a depth of 5 cm.
Large or small seeds in rows	Sow granular or liquid inoculant in rows beside plants at a depth of 5 cm to 7 cm.
Seedlings or cuttings	Prepare a slurry by mixing the inoculant with water: one volume inoculant to two volumes of water. Dip seedlings of cuttings in slurry and allow to soak for 1 minute or longer. Plant as soon as possible.

Adapted from: FAO, 1984

BIOFERTILIZERS AND THEIR USAGE IN FIELD CROPS

Biofertilizer are applied in different ways in the field crop production.

1. seed treatment or seed inoculation
2. seedling dip
3. soil application.

Seed Treatment

Each packet (200 g) of inoculant is mixed with 200 ml of rice gruel or jaggery solution. The seeds required for one ha are mixed in the slurry so as to have uniform coating of the inoculant over the seeds and then shade-dried for 30 minutes. The treated seeds should be used within 24 hours. One packet of inoculant is sufficient to treat 10 kg seeds. Rhizobium, Azospirillum, Azatobacter and Phosphobacteria are applied as seed treatment.

Seedling Root Dip

This method is used for transplanted crops. Five packets (1.0 kg) of the inoculant are required for one ha and mixed with 40 litres of water. The root portion of the seedlings is dipped in the solution for 5 to 10 minutes and then transplanted. Azospirillum is used for seedling root dip particularly for rice.

Soil Application

Ten packets (2.0 kg) of the inoculant are required for one ha of main field. The inoculant is mixed with 20 kg of well dried, powdered FYM and broadcast in the field just before transplanting / sowing.

OPTIMIZING NITROGEN FIXATION IN NFS

Although there remain many gaps in our knowledge of how to manage NFSs for optimum N fixation, the best information we have suggests we can improve N fixation through management. The following are site characteristics or management properties that are likely to result in improved N fixation for most, but not all, NFS species.

1. Well-aerated soils that are not waterlogged or drought-stressed
2. Soil moisture content of 60% to 75% of the soil water-holding capacity
3. Optimum temperatures for plant species growth
4. Mature but not senescent plants
5. Soil pH above 5.5
6. Intermediate spacings
7. Dual inoculation with *Rhizobium* or *Frankia* and mycorrhiza
8. Starter N, P and K applied in the nursery
9. Other soil nutrients such as Ca, Mn and Mo above limiting levels

EFFECTIVE AND INEFFECTIVE STRAINS

Soil is the best medium for microorganisms to grow and flourish. It contains effective and ineffective strains. The relative capacity of the symbiotic system, once established to fix molecular nitrogen, is known as it's effectiveness. Many *Rhizobium* strains are highly effective while others are largely or completely ineffective. A strain that does not permit fixation at a rate sufficient to meet the demands of the host is either partially effective or totally ineffective. Consequently, mere presence of nodules is not a guarantee that leguminous crop is being benefited with fixed nitrogen.

Ineffective *Rhizobium* strains produce a greater number of nodules than effective stains but the nodules are smaller in size and tend to be ore widely distributed over the root system. On the other hand, effective *Rhizobium* strains cause larger few nodules congregated on main root. Effective nodules are distinctly red in colour because of the presence of iron containing substance known as leghemoglobin. In contrast, the nodules produced by ineffective strains have neither leghemoglobin. In contrast, the nodules produced by ineffective strains have neither leghemoglobin nor red pigmentation. Consequently, it is possible by visual examination of splited nodule for its redness to ascertain the relative capacity for active nitrogen fixation. At field level, soybean, chickpea and pigeonpea have shown positive correlation in yield with nodule weight and not with nodule number. Plant nodulated by ineffective strains showed lower dry matter yield, grain yield and nitrogen content, irrespective of the number of nodules than those nodulated by effective *Rhizobium* strains. The yield increase obtained from the sue of biofertilizer relates not only to the nitrogen enrichment but also to the antagonistic effect on harmful organisms in rhizosphere.

ESTABLISHMENT OF INOCULATED *RHIZOBIUM*

The extent of establishment of inoculated rhizobia in soil through biofertilizer is dependent on the type of legumes, effectiveness of strains, inorganic or mineralisable nitrogen content of the soil, level of available soil phosphorus and potassium, pH and presence of suable forms of secondary nutrients. Influence on nitrogen fixation due to unfavourable climatic and seasonal variations is generally more on physiology of host than the symbiotic association.

Nodulation as well as nitrogen fixation are also dependent on the carbohydrate: nitrogen ratio within the host tissues. A widening of the ratio stimulates the resulting from very high light intensities retards the process. retardation of nitrogen fixation in nitrogen rich soil or consequent on the application of fertilizer nitrogen may be associated with narrowing of the carbohydrate: nitrogen ratio status in host. The influence of phosphorus and potassium on nitrogen fixation and nodulation is associated with the vigour and well being of the host rather than the specific stimulation of symbiosis.

Soil pH affects microsymbiont, macrosymbiont and their symbiotic interaction. Infection does not occur much below pH 5.0 as it will be inhibited by the high hydrogen ion concentration as well as the presence of high iron or aluminium and low molybdenum ions. Molybdenum plays an important role in electron transfer. Soils poor in molybdenum, give less response to biofertilizer application.

Rhizobium species are susceptible to attack by bacteriophages and, therefore, bacteriophage resistant/tolerant *Rhizobium* strains should be used preparing biofertilizer.

PRODUCTION OF *RHIZOBIUM* BIOFERTILIZER

Pulse and oilseed legumes occupy about 30 million hectares in India and to cover the whole area with *Rhizobium* biofertilizer, there is a need to produce 15,000 tonnes considering the rat of application at 500 g ha^{-1}. for a realistic coverage of 25 per cent of this area, the annual requirement is 3750 tonnes of biofertilizer (Verma and Battacharyya, 1992). The present production level is 1000-1200 tonnes including *Rhizobium, Azotobacter* and *Azospirillum* containing biofertilizers, though the total production capacity of the existing 60 production units is 2685 tonnes. There were nearly 30 functional production units with output of 1000 tonnes in 1991.

Biofertilizer production technology includes isolation, selection of suitable strains, preparation of mother or seed culture, inoculant production, carrier preparation and their mixing, followed by curing, packaging, storage and dispatch. The requirement of raw materials for the production of one tonne of *Rhizobium* biofertilizer has been worked out and presented in following Table 5.9.

Production of biofertilizer is very technical and complex as it involves microbial lives, though it looks simple like mixing broth and carrier material. This feeling has attracted pseudotechnocrats in business of biofertilizer production with the consequence that biofertilizer of poor/substandard quality are increasingly becoming available in the market. Use of such product has eroded the faith of farmers on the efficiency of this proven low input and important technology. The recently held National Seminar on Organic Farming at Indore, M.P. (Sept. 1992), attended by more than 250 delegates took stock of the situation and strongly recommended for the enactment of "Quality Control Act" for products of biological origin and its strict application on similar lines as they exist in the case of fertilizers and insecticides. The Seminar held the view that 'Biofertilizer Quality Act' is essential to safeguard the interest of farmers and to instill in them confidence on the efficiency of biological routes of soil fertility management.

Table 5.9: Blue Print of 1 tonne *Rhizobium* production, a general idea (with 40%

Requirements

1. **Raw Material**
   a) 4 litre seed culture containing tested efficient crop specific *Rhizobium* strain providing 10^6-10^7 rhizobia per ml culture.
   b) 396 litre yeast Extract Manitol Broth (Sterilised)
   c) 590 kg carrier material like charcoal/lignite, passed through 100-200 mesh, sterilized for 4 h on first day and for on h on subsequent two days at 121.6°C in 5 cm layers in autoclave. Lime 10-15 per cent added in lignite to bring up the pH to neutral.
   d) 10 kg gum acacia aqueous solution (sterilized)
   e) 5000 low density polypropylene packets of 50-75 micron, capacity to contain 200 g inoculant
   f) 125 corrugated 5 ply boxes each having capacity to contain 40 culture packets.
2. **Equipments**
   a) Autoclave
   b) Rotary shaker or fermenter
   c) BOD Incubator
   d) Mechanical Mixer
   e) Sealer
   f) Steel trays
3. **Glassware and chemicals**
   a) As per requirement.

Selection of Strains

Selection and choice of appropriate strains of rhizobia for biofertilizer production is as important as the use of quality seed for crop production. Effective strains are isolated from health, characteristic regular shaped pink nodules. More than 70 per cent of the area under pulses and oil seeds in India falls under rainfed/dryland farming zones. Inoculated rhizobia has to face extremes of low moisture regime in soils of such areas and, therefore, the strains for biofertilizer production shall be from root nodules of plants growing vigorously with a good stand in soils with low moisture regime. Other characteristics of efficient strains are good competitiveness with native *Rhizobium* strains, capacity of withstand extremes of soil and weather conditions, better host infection ability, nodule forming and nitrogen fixing capacity, bacteriophage resistance and ability to exhibit long shelf life in carrier material and seeds and tolerance to pesticides, fertilizers and other agricultural chemicals. It is equally important to maintain the selected strains without any contamination. The use of 200 ppm actidione to curb fungal and 100 IU/ml penicilline to avoid bacterial contamination at the time of isolation has been suggested (Bhattacharyya and Singh, 1992).

Both multistrains and monostrain bioinoculants are produced in the country. But majority of the production units in India prefer to concentrate on monostrain biofertilizer production. Biofertilizer are produced for individual leguminous species and not for group of legumes included in cross inoculation groups. The use of multistrain inoculants for different hosts or a single is likely to dominate over other strains (Date, 1974). The practice of using wide spectrum strain of cowpea miscellany (most of Indian pulses are indicated) is generally based on results obtained in green house trials where effective cross-infections within a range of pulses have been demonstrated (Subba Rao, 1988). When multistrain biofertilizer is produced, individual strains should be grown separately in conical flasks. Adequate care should be taken that all strains have equal growth rates to avoid absence of any strain due to its inability to withstand the competition in carrier material. Many times infection specificity has been observed at variety level which is suggestive of a need for detailed studies on selection of strains for individual pulse species, agroclimate zonewise.

Established effective strains can be also obtained from National Facility of *Rhizobium* Germplasm Collection, Division of Microbiology, IARL, New Delhi 110 012; ICRISAT, Patancheru, Andhra Pradesh 502324; Institute to Microbial Technology, Chandigarh; National Biofertilizer Development Centre, CGO II, Kamala Nehru Nagar, Ghaziabad 201002 and Production Centres of State Agricultural Universities. Some established promising strains are given in Table 5.10. For International strains, one may request to one of the eight Microbiological Resource Centres (MIRCENS). Some of the locations are Bangkok (Thailand) Brisbane (Australia), Cairo (Egypt), NifTal Hawaii (USA), Porto Alegre (Brazil) and Guatemala (Central America).

Table 5.10: Some promising *Rhizobium* strains

Host legumes	Promising Rhizobium strains	
Groundnut	NC-92 IGR-6 IGR-40	Niftal 1000 Niftal 968 Nitragin 8 All
Pigeonpea	CC-1 F-4 BDN-A2	P-241 Niftal-569 Nitragin-21A16
Chickpea	F-6 F-75 F-76 H-44 H-45	Nitragin-27A3 Nitragin-27A8
Soybean	SB-102 SB-103 SB-119 SB-120 UASB-29 Pantnagar	Niftal-102 Nitragin-61A148 11TA-2127
Lentil	L-6 L-1-87	Niftal-640 C-4201
Pea	RL(P) 114	Nitragin-128A12
Greengram	M-10 GMBSI	UPLB-M6 THA-301

Host legumes	Promising Rhizobium strains	
	MO-5	Nitragin-176A22
Blackgram	DU-2 M-11 KU-1 BDN-F DC-28 DC-2 DC-6 FB-77	THA-301 Niftal-442 UPLB M-6 Niftal-174 Niftal-189 Nitragin-176A22

*Principal Investigator (Microbiology), AICRIP (Except Groundnut and Soybean), ** USDA Rhizobium Culture Collection Catalogue, Baltsville, MD.

Carrier Materials and their Sterilization

Medium matrix on which microbial inoculants grow to a reasonable higher number with long shelf life are called carrier materials. These include peat, lignite, charcoal, bunt rice husk, vermiculite, soil, FYM, coir dust, press mud and polymers. Peat is the most suitable carrier material. Most of the production units of developed countries are using peat. Peat of a good quality (more than 75% carbon) is a rare commodity in India. The available low grade peat retards the keeping quality. Nilgiri peat is of poor quality (below 50% carbon) and does not provide a good medium for the growth of rhizobia. According to the availability and cost at production site, choice is only with lignite and charcoal in India. As per the suitability the order is peat >lignite >charcoal >FYM, soil, rice husk, coir dust, oxalic acid industry waste, press mud and others. Lignite can be obtained from Palana (Rajasthan), Neyveli (Tamil Nadu) and Singrauli (M.P.).

Though charcoal is used by production units, quality standards like fineness, extent of pyrolysis, water holding capacity, density, carbon and volatile compounds, ash content and active surface area are not properly adhered to, resulting in the production of inadequate number of rhizobia with low shelf life. Few physical properties of common carrier materials are compared in Table 5.11. Good quality carrier material must have good moisture holding capacity, be free from toxic substances, sterilisable, and have readily adjustable pH to 6.5 –7.0. Under Indian conditions where extremes of soil and weather conditions prevail, there is yet no suitable carrier material identified capable of supporting the *Rhizobium* growth even for six months and biofertilizer producers feel the urgent need of research for suitable carrier material.

The carrier materials are sundried to five per cent moisture and pulverized to fineness of 100-200 mesh. Coarser materials form balls on mixing with broth and do not adhere on the seeds at the time of inoculation. Lignite powder begin acidic (pH 5.5) in nature, lime is mixed to bring up the pH to 6.7 – 7.0. Using one kg lime is required for nine kg of lignite. Charcoal powder being neutral in pH does not require addition of lime

Better growth of *Rhizobium* is obtained in sterile carrier. Best method is Gamma irradiation of sterilization for carrier materials. While using autoclave, lime mixed lignite or plain charcoal powder is filled upto two third capacity of steel trays (size depending on the size of horizontal autoclave) and is sterilized at 15 PSI for 3-4 hours continuously or one hour each for three successive days. At the National biofertilizer Development Centre, Ghaziabad, it was experienced that single pass, even for 3-4 hours, often causes temperature tolerant actinomycetes spores to germinate and usually second or third autoclaving for one hour at 121°C on alternate days was essential for achieving desired sterilization, causing time constraint at peak production period. Use of ethylene oxide though is fast, is associated with persistent residual gas and unfavourable growth environment.

Table 5.11: Chemical and physical characteristics of peat, lignite and charcoal

Parameters	Peat (Indian)	Lignite	Charcoal
A. Chemical*			
Appearance	Dark brown	Brownish black	Black
PH	Acidic, needs	Acidic, needs	Neutral
Odour	Nil	Nil	Nil
Carbon (%)	54.20	72.79	84.00
Hydrogen (%)	5.70	4.74	2.30
Oxygen (%)	31.60	19.60	10.70
Nitrogen (%)	1.50	0.98	0.01
Ash (%)	7.00	7.00	3.00
B. Physical **			
Water hoding	149.3	198.9	200.00
Capacity (%)			
Moisture (%)	40-50	40-50	20-30
Bulk density (g/cm^3)	1.02	1.08	0.43
Particle density (g/cm^3)	2.18	1.66	1.62
Porosity (%)	53.16	34.79	73.39
Total surface area	646.60	556.40	870.90
C. Availability ***			
Quantity	5.5 million tonnes	3 million tonnes	Plenty
	Nilgri valley Tamil Nadu	a. Neyveli (Tamil Nadu) b. Singrauli (M.P.) c. Palana (Rajasthan)	Everywhere

Source: * Arora, B.K. (1984), ** Tilak, K.V.B.R. & Subbarao, N.S. (1978), *** Subba Rao, N.S. (1982)

Gamma irradiation is capable of penetrating and reaching through out the product under treatment, killing microorganisms. It is known as one of the perfect and established methods of sterilization. High sterilization efficiency, non/low residual effect on sterilized material, simplicity in operation with control of only one parameter i.e. time of exposure, possibility of sterilizing plastic cans and bags, safety against heat, being cold process and rapidity are additional benefits in favour of Gamma irradiation over the methods of sterilization. In most of the developed countries like United Kingdom, Australia, France, etc. carrier materials in pouches are sterilized through Gamma irradiation. Sterilization of carrier materials is a very important step in biofertilizer production to keep up and improve the standard of quality and shelf life. The recommended dose of Gamma irradiation for sterilization of carrier materials in low density polyethylene (LDPE) packet is 5 mega rads or 50 killo Grey (Bhattacharyya and Singh, 1992). At present, this facility is available in BARC, Trombay which may not be able to meet the requirement of existing 60 production units in the country. There is a need of establishing a National Facility for Gamma irradiation, specifically for sterilization of carrier materials.

Suitable Nutrient Broth

Most commonly accepted suitable medium for rhizobia is Yeast Extract Mannitol (K_2HPO_4-0.5g, $MgSo_4$ $7H_2O$-0.2g, NaCl – 10.0g, Mannitol – 10.0 g Mannitol – 10.0 g, Yeast Extract –1.0 and distilled water – 1000 ml). Mannitol is the single expensive item which regulates the production cost of biofertilizers. Recently, the use of cheaper carbon and energy source like 5 g molasses or 5 g sucrose per litre in place of 10 g mannitol per litre (Bhattacharyya and Singh, 1992) and soybean protein extract in place of yeast extract as organic nitrogen and growth factor source (Wadoux, 1991) has been suggested for commercial production of *Rhizobium* in fermenter. These nutrient substitutes have been reported to support a population of 10^9 – 10^{12} cells ml^{-1}. Distilled water requirement in production unit is enormous and so as the cost and, therefore, ordinary tap water is used for preparation and also as diluent of broth, before mixing with carrier material. Adequate care should be taken that water used is free of excess mineral salts or chloride residues. Demineralised water is used in production units under the Government of India.

Suitable Packing Material

Three types of polymer bags are used for dispensing required quantity (about 200 g) of biofertilizer. LDPE packets are used universally, as they can be readily sealed, allow high gas exchange with relative low moisture loss (Thompson, 1991). If production technology is based n carrier material sterilization through autoclave, polypropylene or high density polypropylene (HDPE) bags should be used. LDPE bags cannot withstand the temperature of sterilization (121°C). In case sterilization through Gamma irradiation is in practive, LDPE bags should be used as polypropylene is unstable while exposed to irradiation. Small pin holes in bags are both unnecessary and harmful as they cause increased moisture loss and allow easy entry of contaminants. Recommended specifications for LDPE are 50-75 micron thick, flexible sheet pouch type and 15-25 cm in size. Each packet should have information regarding name of product, crop for which intended, name and address of manufacturer, type of carrier, batch or code number, date of manufacture, date of expiry, net quantity, the area (0.4 ha) mutant for instructions for storing as: **STORE IN COOL PLACE AWAY FROM DIRECT SUN AND HEAT.**

Suitable Adhesive Material

Wall paper glue (5%) gum arabic (4%) carboxy methyl cellulose (4%), honey (10%), powder milk (10%), sugar (10%) etc. are used as sticker and partly as nutrient source until infection process is completed, for rhizobia (Rennie, 1991). Inclusion of adhesives within the prepared inoculation is practical, convenient and recommended (Day, 1991). Good quality adhesives do not set, form balls on impregnation, or lose glueing properties during sterilization. In India "gum arabic" also called gum acacia powder, is used as adhesive. Acacia has many species like *A. niotica* (Babul), *A. auriculiformis (Bengal Babul)* etc. Gum arabic is obtained from *Acacia arabica* and some times from *Acacia Senegal.* Unavailability of genuine gum arabic is a major problem. Adulteration is usually with *Azadirachta indica* (Neem) which does not withstand autoclaving.

Fermentation

Starter cultures of selected strains are obtained after ascertaining their performance in green house and at field levels. The tested strain which is kept in tubes is transferred to 250 ml conical flasks placed on rotary shaker and the contents incubated for 3-7 days depending whether they are fast growing or slow growing. The contents of these flasks usually attain a load of 10^5 – 10^6 cells ml^{-1} called seed (500 or 1000 ml) on shaker for charging five litre capacity conical flasks or fermenter. Broth is prepared either by swirling on rotary shaker or in fermenter.

Rotary Shaker Method

Use of conical flasks on rotary shaker allows visual observations of the contents, which will help in checking the contamination and maintaining the quality. Aeration is dependent on rotation, but increased agitation does not proportionately increase the rate of growth. No contamination should be allowed because even a single contaminant which has a mean generation time of few minutes reduces the *Rhzobium* count considerably during a period of 5 days in the case of slow growing species like *R. japonicum* specific for soybean.

In the National and Regional Biofertilizer Development Centres 5 litre conical flasks are being successfully used. Difficulty in operation is experienced while using the rotary shaker of a holding capacity of more than 80 litres broth. Broth, though is being produced by using rotary shaker in most of the State Government Production Units, has limitations like low rate of production (maximum of 500 packets per rotary shaker per day) and as such, large scale demand is not possible to be met considering limited time available for sowing crops. It requires quality control tests of individual unit which is time consuming and there is continuous irritating sound in the shaker room.

Bioreactor

Commercial biofertilizer production through locally fabricated fermenters (bioreactors) is being successfully carried on by many production units. In these units rotary shaker method has given way to fermenter method of broth preparation. Broth production rate through fermenter is 7-10 times more than that through rotary shaker. Good fermenter has growth vessel with provision of air inlet connected with filter to obtain sterile air, inlet ports for inoculant and antifoaming agents, ample drawing port, product harvest system and a convenient system for *in situ* sterilization for a fermenter of larger capacity.

In fermenter aeration is regulated by forcing sterile air through spargers. Aeration and pH are the two important factors which regulate the multiplication rate of *Rhizobium*. Growth is not increased by violent aeration or agitation. Oxygen partial pressure of 0.15 atmosphere is optimum for respiration which can be achieved by introducing 5 litres of air per hour.

For *Rhizobium* broth production, a battery of small fermenters of capacity less than 100 litres in batch fermentation system is considered superior over larger fermenters because of increased chance of contamination accompanied with increased fermentation time in larger units. A chance contamination can be reduced by decreasing the fermentation period.

Studies on optimization of parameters of production technology are in progress at the National Biofertilizer Development Centre, Ghaziabad, with UNDP aid and at the Indian Institute of Technology, New Delhi, with DBT support. Preliminary experiments have shown that by using fermenter, in a medium containing carbon 5 g l^{-1} through molasses or sucrose, the use of 10% inoculum load (24 hours old culture with 10^{-10}*Rhizobium* cells ml^{-1} in 24 hours.

In the United Kingdom, sterilization system of biofertilizer production technology is used where carrier in pouch is sterilized by Gamma irradiation. Fermenter is emptied through medium dispenser and sterile rubber pipe fitted with disposable needles. The bag containing sterilized carrier material is quickly wiped with alcohol and the *Rhizobium* broth (80 ml per 200 g carrier material) injected straight from fermenter into the sealed bags. The injection holes are sealed with a piece of adhesive tape. The inoculated packets are blended by using automatic blender and incubated for a week and after labelling, the packets are ready for sale (Day, 1991).

While using rotary shaker, broth prepared in conical flasks are mixed with sterile carrier material 1:3 or 2:1 depending on water holding capacity of carrier material. Verma and Rawat (1991) have suggested that 60 per cent of the water holding capacity of carrier material should be saturated with broth. Thoroughly mixed broth and carrier material is spread in thin layers in trays, covered and incubated for 48-72 hours at 25°C. Aggregates are broken up by intimate mixing and finally pulverized. The homogenous material is passed through 250 mesh sieve and packed in LDPE bags.

Final product must conform with the quality standards set by the Bureau of Indian Standards (earlier ISI) vide standard IS: 8268 – 1976. The salient parameters mentioned are that the *Rhizobium* inoculant should be carrier based and pH of the carrier should be between 6.0 – 7.5. It shall contain a minimum of 10^8 viable cells g^{-1} of carrier on dry mass basis within 15 days of manufacture and 10^7 within 15 days before expiry date marked on the packet when stored at 25-30°C. It should have maximum of six months expiry period. *Rhizobium* inoculant should result in 50 per cent or more dry matter production compared to control.

FRANKIACEAE SYMBIOSIS

Some of the actinorrhizal plants like Casuarina and Alnus are found to have nodules infected by Frankia. In has been reported that N fixed by *Frankia* in Alnus ranges from 50 to 150 kg/ha/year. The potentiality of these microorganisms has been demonstrated in increasing the biomass yield of *Casuarina* and *Alnus*.

About 200 plant species from 25 genera, 8 families and 7 orders, all non-leguminous angiosperms, have been found to form nodule symbioses with N_2-fixing *actinomycetes* belonging to the genus of *Frankiae*. All of these species are perennial dicots and, with few exceptions, woody shrubs or plants. They are named actinorhizal plants from *actino* in *actinomycete*, and from *rhiza* in the Greek word for root. Actinorhizal plants can be found both on every continent except Antarctica and in most climatic zones. They typically grow on disturbed marginal soils and are pioneer species early in successional plant community development, such as *Dryas* species in arctic tundra; *Casuarina, Hippophae, Myrica* and *Elaeagnus* species in coastal dunes; *Alnus* and *Myrica* species in riparian; *Alnus* and *Dryas* species in glacial till; *Casuarina, Purshia, CeanothusCercocarpus, Comptonia* and *Cowania* species in chaparral and xeric; *Alnus* species in alpine; *Alnus, Casuarina, Coriaria* and *Shepherdia* species in forest. Globally, especially in wherever indigenous legumes are absent or rare, actinorhizal plants have potential applications in soil amelioration and reforestation, as fuelwood, timber and pulp, and as windbreak or even for addressing **pyrodenitrification**.

The symbiont *Frankia* is a gram-positive, filamentous bacterium belonging to the family *Frankiaceae* within the order *Actinomycetes*. Speciation in *Frankia* is not yet clear and within the genus different isolates are classified. Almost all of *Frankia* are characterised by three structural forms of hyphae, sporangia and vesicles. The hyphae are branched with a diameter of 0.5 to 1.5 µm and the mature vesicle is spherical with a diameter of 2 to 4 µm. Both the hyphae and the mature vesicles are septate. When actinorhizal plants, such as *Alnus* and *Casuarina* seedlings are excavated from soil containing *Frankia*, numerous small, multilobed, coralloid-type, amber or whitish nodules are found on their root systems. The oldest and biggest nodules are close to the stem base and the youngest are on the distal parts of the root system. Some *Casuarina* nodules may reach a size of around 5 – 10 mm in diameter and a weight of about 1 g in dry matter. It has been reported that *Frankia* produced sporangia and vesicles as soon as the microorganism escaped from the mother nodule. This indicates that the formation of *Frankia* structures inside the nodule might be under host control. In general, the nitrogen fixation rates of *Frankia*–actinorhizal plant symbioses are comparable to that of *Rhizobia*–legume symbioses.

AZOSPIRILLUM BIOFERTILIZER

Azospirillum have been found more effective towards contributing to increased yields in sorghum, pearl millet, *kodo, kutaki,* safflower, maize, mustard and wheat. In general, C_4 plants like corn, sorghum, panicum and setaria showed better response to inoculation than C_3 crops. The optimum temperature for *Azospirillum* growth is 32-35°C. This might be the reason for better performance of inoculated crops in summer under irrigated condition. *Azospirillum* biofertilizer consumption in Tamil Nadu ahs increased in the last three years because of consistently good performance of rice crop. Tamil Nadu followed by Gujarat is the largest producer and consumer of *Azospirillum* biofertilizer in the country.

Soil nitrogen affects the response to inoculation. In multi-location experiments with pearl millet, maximum of 20 per cent increase in grain yield due to individual inoculation of *Azospirillum* and *Azotobacter* and 27.2 per cent and 18.18 per cent increase in plant nitrogen uptake due to inoculation of *Azospirillum* and *Azotobacter* respectively (Wani, 1992) were observed at zero level of nitrogen application. Increased nitrogen dose retarded the performance of inoculum. The reported nitrogen assimilation of only 5 kg^{-1} in pearl millet seems to be too small a quantity of cause increase in yield on inoculation. It is believed that the increase in yield is due mainly to release of growth promoting agents and protection from diseases than is from the contribution of nitrogen fixation. The presence of *Azospirillum* internally in root system has not been yet conclusively proved.

Tilak and Subba Rao (1987) have reported highest increased grain yield (33.67%) and plant biomass yield (32.71%) by the application of coinoculatns *Azospirillum brasilense* and *Glomus fasiculatum* on sorghum.

Major Species

There are four major species of *Azospirillum* viz. *A. lipoferum, A. brasilense, A. amazonense* and *A. seropedicae*. Of the four species, first two species are commonly found. The first species (*A. lipoferum*) is commonly found in the roots of C^4 plants while the second one (*A. brasilense*) is found in C^3 plants in tropical and subtropical regions.

Isolation

Half centimeter long root pieces are surface sterilized in 70% alcohol for 3-5 seconds and washed repeatedly in phosphate buffer pH 7.0). These roots pieces are then placed in semi-soild nitrogen-free sodium malate medium. The tubes are incubated at 35°C for 48 hours. Characteristic growth of *Azospirillum* is indicated by the formation of white pellicle 2-4 mm below the surface of the medium. The isolates thus obtained are purified in two steps. Firstly, the isolates are enriched by repeated transfer into a semi-solid sodium malate medium in small screw cap tubes. These cultures are then subjected to dilution upto 10^{10}. From each dilution 0.1 ml is inoculated into a fresh tube of sodium malate medium. The least dilution which gives good growth is found to be pure and used for further studies.

Mass Multiplication

For mass multiplication of *Azospirillum*, the organism is allowed to grow in flasks containing malic-acid medium augmented with 1% ammonium chloride and incubated at 35°C for 3 days. When there is good growth, the broth culture is mixed with carrier and the carrier-based culture is then packed in small polyethylene packets. Soil and FYM in the ratio of 1:1 sterilized for 3 hours consecutively for 3 days were found to be the best carrier wherein bacterium was able to survive upto 6 months with counts of 10^6 cells/g of carrier material.

Method of Inoculation

The most suitable method used for carrier based culture is the seed inoculation. The technique adopted for inoculating the seed with *Azospirillum* culture is same as that described in the case of *Rhizobium.* In case of transplanted rice, the dipping of roots of rice seedlings in the slurry of inoculant is practiced.

Conditions Needed for Successful Inoculation

Seed inoculation with effective strains should be done in shade and subsequently the sowing should be done on the same day. The field should be well saturated during rice transplanting and after transplanting, the flooding should be done only after 5-6 days. For upland crops, moisture at field capacity should be maintained at the time of seeding so that added inoculum may get congenial condition for its growth and development.

Crop Response

The *Azospirillum* inoculation has been found beneficial in a variety of crops including cereals, forage and other crops. The crop response, however, varies greatly types of crop and variety, location, season, soil fertility level, native microorganisms interaction, etc. Crop response to *Azospirillum* inoculation is attributed to its ability to fix atmospheric nitrogen and to produce phytohormones.

A number of field experiment conducted on various crops to find out crop response to *Azospirillum* inoculation observed a wide variation in increase in yield (0-64%) over control. Significant increae in yield was found at about 50% of the locations. On an average, an increase of about 2-3 q/ha in crop yields and an economy of 20-30 kg N/ha equivalent is possible by *Azospirillum* inoculation. Among various field crops, sorghum and millets respond better than others under rainfed conditions. In Tamil Nadu, rice responds very well to *Azospirillum* inoculation.

AZOTOBACTER BIOFERTILIZER

Out of three different typical species of *Azotobacter*, *A. chroococcum* and *A. beijerinckii* are the most commonly occurring species. The latter is found commonly in acid soils while *A. chroococcum* in neutral and alkaline soils. Aged cells of *A. chroococcum* are characterized by the presence of insoluble black brownpigment (melamin) found as a result of oxidation of tyrosinase, a copper containing enzyme.

Table 5.12: Response of pearlmillet to inoculation with *Azospirillumbrasilense* on grain yield (q/ha^{-1})

Treatment	Centers				
	Kanpur	Hyderabad	Parbhani	Delhi	All India
Uninoculated	12.8	12.5	14.5	9.8	12.4
A. brasilense	15.5*	12.8	17.7	11.3	14.3
10 kg N ha^{-1} (urea)	16.3	14.3	19.5	11.8	15.5
10 kg N ha^{-1} + *A. brasilense*	19.1*	14.5	22.8*	14.8*	17.8
20 kg N ha^{-1} (urea)	17.5	14.8	22.5	12.5	16.4
20 kg N ha^{-1} + *A. brasilense*	18.9	17.0	23.2	15.8	18.7
40 kg N ha^{-1} (urea)	20.1	18.5	24.3	18.5	20.4
CD 5%	2.5	1.5	2.9	3.0	2.2

The figures in brackets represent the soil pH values. * Significant increase over corresponding control.
Source: (Tilak and Subba Rao, 1987).

Isolation

Azotobacter spp. are isolated by soil dilution plating method. Soil sample of 10 g is transferred in 100 ml of sterilized water and serial dilution of the suspension are prepared by further dilutions using sterile distilled water. Petriplates and pipettes are sterilized. Anyone of the nitrogen free agar media specific for *Azotobacter* is prepared and poured into petriplates. One ml aliquot of appropriate dilution are spread evenly over cooled and set agar medium in petriplates. The plates are incubated at 30°C in an incubator. After 3-4 days of incubation, soft, milky and mucoid colonies of *Azotobacter* are formed on agar plates. Direct isolation of *Azotobacter* has been practiced by Russian Scientists. Lumps of soils are spread out on a nitrogen free agar medium and the plates are incubated at 28°C. After three days of inoculation *Azotobacter* colonies develop on the agar plates.

Another method is to prepare a paste of soil in nitrogen free liquid medium. This paste is incorporated in a layer of washed charcoal placed in a petriplate. Soil is aerated by burying a glass tube in it. After incubation of the plates at 28°C for 3-4 days, small *Azotobacter* colonies, if present, develop on the soil surface. This method is popularly known as 'soil plate' method.

Mass Production

Mass production of *Azotobacter* for preparing inoculating in India is done either by growing the culture in large flasks on a rotary shaker or in batch fermentors depending on the quantity needed. All precaution and sterilization production in the case of *Rhizobium* are also observed in case of *Azotobacter*.

Method of Application

A slurry of carrier based culture is made with minimum amount of water and the seeds are inoculated by mixing with slurry. The seeds are dried in the shade and then sown. For transplanted crops, seedlings are dipped in the slurry for certain time (20-30 minutes) and planted immediately. For sugarcane, many packets of culture are recommended at regular intervals in the early stages of growth. Second and subsequent inoculation are made by pouring slurry near root zone. The inoculant is also mixed in FYM and broadcasted for the root zone.

Crop Response

The results from a large number of experiments conducted in last four decades, have shown a positive response to *Azotobacter* application on a wide range of crops like cereals, millets, vegetables, cotton, sugarcane etc. *Azotobacter* inoculation increased the crop yields by 0-30% (Table 5.13). The beneficial effect of *Azotobacter* is attributed to its N_2 fixing capacity and also the ability to produce growth promoting substances. The actual contribution under field conditions has been doubled due to its low population in soil. The intensive research efforts are required to develop the *Azotobacter* strains which can multiply in soil and establish high productions.

Table 5.13: Effect of *Azotobacter* inoculation on crop yields

Crop Increase	Fertilizer supplementation (kg/ha)			% yield over control
	N	P	K	
Wheat	120	-	-	11
	-	-	-	10-30
Rice	120	60	60	23
Maize	-	-	-	34
Sorghum	-	-	-	15-20
Pearl millet	-	-	-	0-27
Onion	-	50	100	22
Tomato	-	-	-	2-24
Cotton (fibre) Irrigated	63	30	-	10-20
Non-irrigated	-	-	-	11-16
Sugarcane	-	-	-	24

Source: Venkataraman and Tilak (1990)

Methods of Application of Azospirillum and Azotobacter

Application of *Azospirillum* and *Azotobacter* through seed coating is similar to that of *Rhizobium* biofertilizer. In Tamil Nadu the use of *Azospirillum* in paddy has gained popularity among farmers. Considering the benefits of *Azospirillum* inoculation, the State Department of Agriculture has recommended the reduction of nitrogen by 25 per cent in those fields receiving *Azospirillum* treatment. *Azospirillum* inoculation has been recommended for paddy, sorghum and sunflower. The methods of application and doses are as follows:

1. Transplanted paddy and sorghum: Three packets (600 g) of *Azospirillum* biofertilizer are mixed in sufficient water wherein seeds required for one hectare are soaked over night before sowing in the nursery area. Root portion of the seedlings ready for transplanting are dipped for 15-30 minutes in bacterial suspension prepared from 5 packets (1000 g) of

Azospirillum biofertilizer and transplanted immediately.

2. Direct seeding paddy: Ten packet (2kg) of *Azospirillum* are mixed with seed before sowing. Little quantity of rice 'Kanji' can be added for sticking the biofertilizer to the seeds.
3. Rainfed sorghum and sunflower: Seeds required for one ha are treated with 3 packets (600 g) of *Azospirillum* inoculants and in the main field 10 packets (2kg) of *Azospirillum* biofertilizer ha^{-1} mixed with 25 kg farm yard manure and 25 kg soil are applied.
4. Sugarcane: 25 packets (5.0 kg) of *Azospirillum* ha^{-1} are applied in two split doses at 30 and 60 days after planting. For uniform spread, the culture can be mixed with about 500 kg of powdered FYM applied along the base of the clumps.

A detailed study conducted by Sugarcane Breeding Institute, Coimbatore (TN) on four different soil types at four locations four seasons during 1982-1984 has shown that inoculation of *Azospirillum* with 180 kg N ha^{-1} could give an yield of 106.4 t ha^{-1} which compared well with what was obtained at the level at 240 kg ha^{-1} of inorganic nitrogen (Anonymous 1987) indicating a saving of 60 kg N ha^{-1}. In on-farm trials (106), *Azospirillum* performed better in well drained clay soils in increasing the sugarcane yield and improving sucrose per cent in juice (Table 5.14). Banger *et al.*, (1992) on comparing the application methods through sett inoculation and soil inoculation found *Azotobacter* was performing better than *Azospirillum* under water stress condition. The yield due to *Azotobacter* along with 225 kg N ha^{-1} was found equal to that of 300 kg N ha^{-1} alone, thus saved 75 kg N ha^{-1} (Table 5.14).

Table 5.14: Response of sugarcane to *Azospirillum* and *Azotobacter* inoculation **(a)** with two doses of nitrogen (study conducted on four different soil types at found locations and four seasons during 1982-84)

Treatment	Cane yield t ha^{-1}	Sucrose % juice
120 kg N ha^{-1} + *Azospirillium*	93.2	16.12
120 kg N ha^{-1} + *Azotobacter*	80.8	16.22
180 kg N ha^{-1} + *Azospirillum*	106.4	16.40
180 kg N ha^{-1} + *Azotobacter*	93.8	16.13
240 kg N ha^{-1} Urea alone	99.6	16.24

(b) In various types of soils and varieties through 106 on-farm trials in taemer's fields in Tamil Nadu (1983-'85)

		Cane yield		Sucrose	
Soils	**Treatment**	**t ha^{-1}**	**Increases over control**	**% juice**	**% increase over control**
Clay	*Azospirillum*	148.5	13.3	16.50	13.0
(well drained)	*Azotobacter*	112.0	--	14.95	2.3
	Control	131.0	--	14.60	--
Clay loam	*Azospirillum*	116.9	9.5	17.10	10.3
	Azotobacter	110.2	31	17.05	10.0
	Control	106.8	--	15.50	--
Loamy	*Azospirillum*	110.9	4.8	16.77	9.3

		Cane yield		Sucrose	
Soils	Treatment	t ha^{-1}	Increases over control	% juice	% increase over control
	Azotobacter	120.6	13.9	17.37	13.2
	Control	105.8	--	15.34	--
Sandy	*Azospirillum*	105.8	--	15.34	--
	Azotobacter	105.2			

Complied from Anonymous (1987).

Table 5.15: Effect of biofertilizers on yield and quality of sugarcane (Co 7318)

Treatment	Cane yield (t ha^{-1})	% increase over control	Sugar (t ha^{-1})	Net return Rs ha^{-1}
300 kg N ha^{-1} (Recommended N)	92.50	38.8	11.52	23,075
*Azospirillum** + 75% N	85.40	28.1	10.30	20,418
*Azospirillum** + 50% N	76.38	14.6	9.63	17,008
*Azospirillum*** + 75% N	87.81	31.7	10.85	21,298
*Azospirillum*** + 50% N	77.26	15.9	9.63	17,304
*Azotobacter*** + 75% N	94.60	36.8	11.94	23,947
*Azotobacter*** + 50% N	89.07	28.8	11.36	21,910
*Azotobacter*** + No	66.63		8.59	13,455
*Azotobacter*** + No	69.14		8.89	14,377
CD at 5%	13.47		1.84	--

Source: Bangar, et al. (1992) *Sett inoculation ** Soil application

BLUE-GREEN ALGAE INOCULANT

Blue green algae are referred to as paddy organisms because of their abundance in the paddy fields. Many species belonging to the genera, *Tolypothrix, Nostoc, Schizothrix, Calothrix, Anoboenopsis* and *Plectonema* are abundant in tropical conditions. Most of N fixing BGA are filamenters, consisting of chain of vegetative cells including specialized cells called heterocyst which function as a micronodule for synthesis and N fixing machinery. BGA, *Anabaena azollae*, also occurs in symbiotic relation with the aquatic fern, *Azolla*.

In addition to N fixation, BGA also synthesis and liberate growth promoting substances such as auxin and amino compounds which stimulate the growth of rice.

As early as 1939, Dr. P. K. De suggested that fertility of tropical rice fields was largely due to activity of the N-fixing BGA. Blue-green algae are photosynthetic nitrogen fixers i.e. they use energy derived from photosynthesis to fix atmospheric nitrogen. They are also called cyanobacteria and are free living organisms. In general BGA show dominance over other classes of algae in India. The utilization of blue-green algae as a biofertilizer for rice is very promising. A judicious use of these algae could provide to the country's entire rice acreage as much nitrogen as obtained from 15-17 lakh tonnes of urea. Methods have been developed for mass production of algal biofertilizer and rice growers in many parts of the world have already started using these algae. Recent researches have shown that algae also help to reduce soil alkalinity and this opens up possibilities for bioreclamation of such inhospitable environments.

Blue green algae (BGA) when cultured in small pit and used as an inoculum in paddy fields at the rate of 12-15 kg ha^{-1} resulted in nitrogen economy to rice crop of the order of 20-25 kg N ha^{-1} with yield increase of 5-14 per cent.

DISTRIBUTION

The principal genera consisting of maximum number of nitrogen fixing species are *Aulosira, Nostoc, Anabaena, Tolypothrix, Scytonema, Plectonema,* and *Cylindrospermum.* Contrary to general belief that BGA are present invariably in tropical rice soils, the distribution of soils harbouring N fixing BGA varies from as low as 4% to as high as 80%. Species of *Nostoc* and *Anabaena* were ubiquitous in all regions in varying production and formed the major constituent of the algal biotypes in 9 out of 16 states. An all India survey has indicated that out of 2213 soil samples examined, only 33% were found to contain useful N fixing BGA. This indicates the need for inoculation of efficient strain of BGA.

ISOLATION

BGA are generally covered with mucilage and it is easy to locate them as colonies on rice fields. Isolation of BGA involves the transfer of an algae from its natural habitat to medium where it can be grown without any contamination.

MASS PRODUCTION

Currently, the technology for mass production of soil-based algal inoculants is well developed in active use in India. Production involves a simple process with a few modifications to enable the production of algae both at small and large scale. The four common methods of algal production are:

1. Tropical method
2. Pit method
3. Field method
4. Nursery cum algae production method

Based on the natural ecology of these algae, a simple rural oriented open-air method of producing them in bulk has been developed at the IARI, New Delhi. The basic principle is to grow them in natural sunlight under conditions simulating those in the rice field. The starter culture – a soil based mixture of *Aulosira, Tolyporthrix, Nostoc, Anabaena* and *Plectonema* is supplied by the Indian Agricultural Research Institute, New Delhi. The details of multiplication from the starter culture are as follows:

Shallow tray (15 cm × 7.5 cm × 22.5 cm) of galvanized iron sheet, or brick and mortar or pits lined with polythene sheets are prepared. The size can be increased if more materials is to be produced. About 10 kg soil is placed and mixed with 200 g superphosphate. The trays are then filled with water (5-15 cm) depending upon the local conditions and rate of evaporation; the pH of the soil should be around neutral. After soil settles down saw-dust and the starter culture are sprinkled on the surface of the standing water. The whole assembly is allowed to expose to sun. In hot summer months, the growth of the algae will be rapid and in about a week a thick algal mat will be formed on the surface of the soil and sometimes even is added intermittently. When the algal growth becomes sufficiently thick, addition of water should be discontinued and the water is allowed to dry up in the sun. The dried algal flakes are collected from the surface and stored in bags for future use in the fields. The trays are again filled with water and a small amount of the dry algal flakes is added, as further inoculum. The process is continued as above. Once the soil in the tray is exhausted (usually 3-4 harvests), fresh soil is put and mixed with superphosphate and process is repeated as before. To prevent the breeding of insects, application of Folidol (0.001 ppm) or Malathion (0.00075 ppm) or Carbofuran (3% granules) is recommended.

METHODS OF FIELD APPLICATION

Algae are applied @ 10 kg/ha over the standing water in the field one week after transplantation. The field is kept water-logged at least for a couple of days immediately after algal application.

Recommended pest control measures and other management practices generally do not interfere with the establishment activity of these algae in the fields.

The sun-dried algal material can be stored for long and used in the field. The algal material should not be stored in direct contact with chemical fertilizers or other agricultural chemicals.

Crop Response

The results of the field trials conducted in India have indicated about 10-15 per cent increase in rice yields with BGA application. The response of BGA application is better in dry season than in wet season and long duration rice varieties show greater response compared to short duration ones. Use of mixed BGA inoculum is superior to unialgal inoculation. Among different strains, agronomic potentially of *Aulosira* is well established in Indian condition. Grain yield measurements suggest that algalization produces a cumulative residual effect. Comparison of N fertilizer and BGA indicates that BGA inoculation is equivalent to the application of 20-30 kg N/ha. Besides nitrogen, these algae are known to provide growth promoting substances.

Investigations have shown that HYVs of rice respond to BGA inoculation even in presence of moderate level of N application. The observed BGA effect in presence of fertilizer may be due to supply of growth promoting substances to the crop as well as to the N-fixed by the organism.

Table 5.16: Performance of BGA on farmers' fields

Treatments	Vellore (11 trials)	Grain yield (kg/ha) Trichy (8 trials)	Cuddalore (10 trials)
$N_{50}P_{25}K_{25}$	4235	4557	3789
$N_{50}P_{25}K_{25}$+BGA	4455	4735	3884
$N_{75}P_{37.5}K_{37.5}$	4400	4485	3953
$N_{75}P_{37.5}K_{37.5}$+BGA	4565	5117	4156

Factors Affecting Crop Response

The factors like the shortage of P in soil, the presence of combined nitrogen in flood water, presence of predators of BGA and use of herbicides in rice often limit the growth and activity of cyanobacteria. The non-availability of BGA inoculants at large scale due to scarcity of water and high temperature during June in North West India, and cubmber-some method of multiplication are the two major hinderances in their wider adoptability. The development of BGA which multiply fast and could be inoculated directly in the field may go a long way in the popularization of its use.

Blue Green Algae (BGA) and Azolla

Blue green algae (Cyanobacteria) are photosynthetic nitrogen fixing organisms owing to the presence of nitrogen fixing cells (heterocysts). The algae are congenial to rice ecosystem and are used as biofertilizer in rice crop. The predominant genera in Indian soils are *Anabaena, Nostoc, Aulosoria, Calothrix and Tolypothrix.* The fresh water fern, *Azollapinnata* contains specialized leaf cavities inhabited by the blue green algae, *Anabaena azollae.* The *Azolla – Anabaena* association in unique because it exists through out the vegetative and reproductive phases of the fern.

The various studies conducted in the country have shown that BGA applied @ 10 kg ha^{-1} a week after transplanting of paddy contributed 20-30 kg N h^{-1} $season^{-1}$ and 5-14% increase in grain yield. In the elaborate on-farm trials conducted by the Indira Gandhi Krishi Vishwa Vidyalaya, Raipur, (M.P.) during 1981-91, the efficiency of BGA was fond decreasing with increasing dose of nitrogen with maximum increase in yield recorded in combination with 20 kg ha^{-1} (Adil and Katre, 1992). Large scale on-farm trials conducted in Tamil Nadu (118) and Madhya Pradesh (56) had indicated increase in yield at 0 level nitrogen in M.P. and at 100 kg N ha^{-1} T.N. by BGA application (Table 5.17).

Table 5.17: Average paddy yield (q ha-1) due to BGA (10 kg ha-1) inoculation on-farm trials conducted in two major rice growing stats

Treatment	Madhya Pradesh* (56 trials)		Tamil Nadu (118 trials)	
	Gain	% increase	Grain	% increase
ON: 40 P: 15K	24.1			
ON: 40 P: 15K + BGA	28.2	17.0		
100 N: 50P: 50K			5.11	
100 N: 50P: 50K + BGA			54.8	7.2

Sources:* Anonymous, (1978) **Venkataraman, G.S. and Shanmugasundaram, S. (1992)

Inoculated BGA on establishment and under good conditions of growth add upto 5 t ha^{-1} organic matter in addition to improvement of rice root respiration, uptake of phosphorus and reduction in sulphide injury. The nitrogen assimilated by BGA, on degradation is utilized by rice plant. During decomposition 10-15 per cent biological nitrogen forms complex with humic acid-type polymers (Verma and Martin, 1976), which is indicative of the residual effect of BGA on the subsequent crops. Positive influences of algal biofertlizers on rice fields have also been observed in Egypt, the Phillippines, former U.S.S.R, Sweden, Nepal, Sri Lanka, China and Burma.

MASS PRODUCTION AND APPLICATIONS

Azolla can be conveniently grown round the year. However, the multiplication of *Azolla* is rapid in winter season than in summer. The occurrence of standing water and application of phosphorus are essential requirement for rapid spread of *Azolla.* In India fresh weight of *A. pinnata* increased 2-6 fold in a week.

Two methods of *Azolla* application have been advocated. One is as a green manure by amending it in fields prior to rice cultivation and the other is dual cropping with rice when the fern is grown along with principal crop for sometimes. The rate of inoculation is usually higher for multiplication plots than for dual cropping. The inoculum of 0.1-0.3 kg/square metre is recommended for multiplication plots and 0.5-1.0 kg/square metre (0.5-1.0 t/ha) for dual cropping.

In green manure crop, the fields are prepared for rice cultivation, flooded with water followed by seeding with *Azolla.* After 5 to 10 days, the water of field is drained off and *Azolla* mat is ploughed into the soil. The process may be repeated for two crops of *Azolla*. In dual cropping *Azolla* is grown simultaneously with rice seedlings after transplanting. Normally, it is inoculated after one or two weeks of transplanting and field is flooded with water. When *Azolla* forms a mat on water surface, water is drained off and *Azolla* mat is incorporated into the soil. The *Azolla* mat required repeated incorporation to avoid chocking of rice plants due to *Azolla* mat.

CROP RESPONSE

Studies carried out in India have shown that *Azolla* green manuring produces 10-12 t/ha of green matter containing 20-40 kg N, whereas its dual cropping provides 20-30 kg N/ha. The supply of N can be further increased by raising more than one crop of *Azolla,* first as green manure and then as a dual crop along with rice. *Azolla* green manuring followed by dual cropping is equivalent to that from 60 kg N/ha as fertilizer. Combined use of *Azolla* and N fertilizer give better results. Besides other management practices, grain yield response varies greatly with *Azolla* species and strains. On an average, the increase in grain yield due to *Azolla* green manuring ranges from 0.5-2.0 t/ha, whereas dual cropping produces additional yield of 0.5-1.5 t/ha over control.

FACTORS AFFECTING RESPONSE

The growth of *Azolla* and crop yield are significantly influenced by water availability, fertilizer application, rate and time of inoculation, method of rice planting and insects, pests and diseases. Water is prerequisite for multiplication of *Azolla*. *Azolla* cannot withstand desiccation and cannot be used in upland rice. Water depth of 5-10 cm is maintained during *Azolla* multiplication. *Azolla* requires at least an application of 20-30 kg P2O5/ha. Split application of fertilizer P helps in producing higher *Azolla* biomass. Soluble P sources are found to be more effective than rock phosphate.

The amount of inoculum significantly affects the growth rate of *Azolla*. Under poor growing conditions and in areas of excessive algal growth, higher inoculum rate should be used.

Azolla inoculation in early stages of rice growth is preferred because a delay reduces the fern growth due to increased shading of rice. Temperature higher than 34°C + 1°C during day time is detrimental for *Azolla* growth. *Azolla* is often attacked by insects and the entire crop is damaged within a few days, if they are not controlled timely. Pests like *Pyralia* and snails (*limnae acuminata*) and disease particularly black rot caused by *Rhizoctonia solani* are serious ones affecting *Azolla* multiplication. To control pest damage, insecticide application like Carbofuran is usually recommended.

BGA Production Technology

BGA production technology is simple and can be practiced on farmers field incurring negligible cost, although commercially the cost of 10 kg BGA is Rs. 20.00. BGA production and utilization is most popular among the farmers of Tamil Nadu, but is not catching up in other states like Uttar Pradesh, Madhya Pradesh, West Bengal, Orissa etc. The experience is that growth rate on farmers fields is very low even when sun shine is ample. Added to it, there is a problem of contamination of green algae and poor competitiveness of inoculated BGA. Farmers also do not get the BGA established in rice fields on inoculation. There is also need for devising appropriate testing kits which can be easily handled by extension workers and farmers to determine whether the algae growing in the fields are BGA or nitrogen consuming green algae.

Acidic or highly alkaline soil conditions, direct contact with acid fertilizers, poor competitiveness due to slow growth rate, low tolerance to pesticides, predators and cyanophages and low availability of phosphorus and trace elements, are factors responsible for poor establishment of applied BGA in rice fields. Sundried BGA inoculant can be stored for a long time under dry condition and out of direct contact with chemical fertilizers without any quality deterioration. Application of BGA at three consecutive season helps in establishing, if available phosphorus level is not low in the soil.

The methods recommended are simple, rural oriented, open air method and inexpensive. Laboratory or sophisticated equipments are not required as in the case of *Rhizobium* biofertilizer.

Trough Method

Troughs of 2m × 1m x 22m capacity each and made up of galvanized iron sheet are filled with 5-10 cm water and placed under sun light. A mixture of 8-10 kg soil and 200 g super phosphate is sprinkled uniformly on the water. Next day BGA mother culture obtained from BGA subcentres (60 BGA subcentres are in operation supported by Govt. of India) is added left under sun. In 7 to 10 days a thick mat of BGA appears. Water is allowed to evaporate and dried algal flakes are collected.

Pit Method

Shallow pits of similar dimension as of trough are made, lined with polythelene sheet to prevent drainage. Other procedures are the same as the trough method.

Mass Multiplication of BGA in the Field

Field production of algae is adopted for multiplication of the material on a commercial scale. In the wetland, about 40 m^2 area is demarked and puddled to facilitate water stagnation. The area is encircled with strong earthen bunds of about 15 cm height. Super phosphate is applied at 2 kg for 40 m^2 area. About 5 kg of composite algal culture is required for 40 m^2 area. To control predators like daphnids, snails, and mosquitoes, carbofuran 3 G is applied at 200g. In about 7-10 days period a thick mat of algae develops in clay soils. In loamy soils, about 3-4 weeks are required for full growth of algae. Once the algae have attained full growth with complete coverage, algal flakes are collected and stored in sacks for further use.

During summer months (April - June), the average yield of algae ranges from 20 to 30 kg/40m^2. Continuous harvest of algae may be made from the same area by reflooding the plot and applying super phosphate and pesticides. Addition of further inoculum is not necessary for subsequent production.

DUAL CROPPING OF *AZOLLA* WITH RICE

Out of six species, *AzollaPinnata* is more common and abundantly found in natural ponds (Verma 1992). It can double its biomass in 3-5 days and assimilate 60-80 kg ha^{-1}. It can be grown in bunded rice fields and incorporated into the soil while field is puddled for transplanting paddy seedlings. The other method includes growing *Azolla* amidst standing paddy crop as dual crop. The influence of *Azolla* on paddy yields as affected by different methods of application is presented IN Table 5.18.

There is no technology available for using *Azolla* as dry inoculum as in the case of *Rhizobium* or other bacterial (including blue green algae) biofertilizers. *Azolla* multiplies through vegetative propagation. Huge quantities of inoculum (10-12 t ha^{-1} as green manuring or 11 t ha^{-1} as dual cropping) are required and difficulty in transport during rainy season is the major constraint experienced in its adoption as biofertilizer.

Table 5.18: Influence of Azolla as affected by methods of application on yield rice (IR-36)

Treatment	Grain yield q ha^{-1}	% increase due to *Azolla*
30 kg N ha^{-1}	33 (28)	17.8
60 kg N ha^{-1}	38 (28)	35.7
Azolla dual unincorporated	33 (28)	17.8
Azolla dual incorporated	35 (28)	25.0
Azolla basal (green manuring)	32 (27)	18.5
Azolla basal + dual cropping (once)	35 (27)	29.6
Azolla basal + dual cropping (twice)	38 (27)	40.7

Figures in parenthesis are grain yield due to no nitrogen. Complied from – Singh P.K. and Singh, D.P. (1992).

AZOLLA-ANABAENA SYMBIOSIS

A small floating water fern, Azollais commonly seen in low land fields and in shallow fresh water bodies. This fern harbours a blue-green algae, *Anabaena* andAzollae. The *Azolla-Anabaena* association is a live floating nitrogen factory using energy from photosynthesis to fix atmospheric nitrogen amounting to 100-150 kg N/ha/year from about 40-60 tonnes of biomass. The studies carried out under AICARP showed that incorporation of 5 t ha-1 *Azolla* gave rice yields equivalent to about 30 kg N ha-1 at several locations. Reports of its current use as a biofertilizer for rice China, Vietnam, India, Indonesia, Thailand and other East and South Asian countries are available. An integrated system of rice Azollafish has been developed in China.

Azollais widely distributed in the rice growing tracts of tropics and temperate zones and grows on the irrigated rice fields. There are six species of Azolla: *A. caroliana, A. nilotica, A. filiculoides, A mexicana, A. microphylla and A, pinnata*. The common species occurring in India is *A. pinnata*. A heat tolerant species *A. microphylla* is introduced to India from the Latin America. *A. filiculoides* is a temperate species got from Europe. New crosses between *A. microphylla* and *A. filiculoides* are also available. Several of hybrids have high nutritive value (4.0 – 4.5 % N on dry weight basis) and grow fast. The hybrids can also withstand both heat and cold.

Azolla mutants with a high degree of tolerance to salinity have been developed at International Energy Agencies, Vienna. Scientists in China and U.S. are now working on introducing a *Bacillusthuringiensis* (Bt) toxin in *Azolla* to fight against insect pests.

Multiplication of Azolla

Propagation

The water fern is mainly propagated by vegetative means. It will yield on an average about 1.5 kg per square meter in a week. Under normal conditions, the fern doubles itself in three days. Multiplying *Azolla* through sporocarp technique has also been developed in many research institutions in the world.

Azolla can be multiplied in small cement cisterns or galvanized iron trays of convenient size. *Azolla* can also be multiplied in the field nursery.

Cement Cisterns

Azolla can be multiplied in cement cisterns of 2 × 1 x 0.3 m. A layer of 7.5-10.0 cm soil is provided at the bottom of the cisterns. About 5.0 – 7.5 cm depth of water is maintained in the cisterns. Super phosphate @ 8-12 g and carbofuran at 2.0 g are applied. About 200 g of fresh *Azolla* is inoculated to each cistern. In about 12-15 days, about 2.0 –2.2 kg of *Azolla* can be harvested. After each harvest, super phosphate is applied to facilitate fast multiplication. Every time, 200g of *Azolla* is inoculated for further multiplication.

Mass Multiplication in Nursery Plots

The field selected for *Azolla* nursery should be nearer to shade, thoroughly ploughed and inoculated. The field is divided into plots of 40 m^2 area (20 × 2 m) by providing suitable bunds and irrigation channels. Water is maintained to a depth of 10 cm. About 10 kg of fresh cattle dung is mixed with 20 litres of water and sprinkled uniformly throughout the plots. Fresh *Azolla* is inoculated at 8 kg/plot and super phosphate (100g) is applied twice at an interval of 4 days. Carbofuran is added at 100g per plot, 7days after inoculation to control insect pests. About 50-75 kg of fresh *Azolla* can be harvested 15 days after inoculation. *Azolla* can be again inoculated with the addition of cattle dung, super phosphate and carbofuran granules.

Enahnced Crop Production

Nitrogen fixing microorganisms convert about 139-170 million tonnes of nitrogen every year into fertilizer nitrogen for which the energy bill is paid by Nature. It is at least several folds greater than the inputs of nitrogen from chemical fertilizer factories to the tune of 65 million tonnes per year. The fact that the total world biological nitrogen fixation is three times that of industrially produced nitrogen, demonstrates the significance of biological nitrogen fixation in agricultural and natural nitrogen cycles.

Nitrogen fixing systems offer an economically attractive and ecologically sound means of reducing external inputs and improving internal resources. Symbiotic systems such as that of *Azolla* and *Anabaena* complex and that of leguminous green manures with *Rhizobium* and *Azorhizobium* associations can be of particular value to wet land rice crop and renewable biological source supplementing inorganic N for cereals.

In TNAU significant contributions have been made on *Azolla* biofertilizer particularly on the selection of new strains of *Azolla microphylla,* which is highly tolerant to high temperature and also tolerant to salinity with higher biomass producing capacity (25t/ha). *Azolla* hybrid has been developed by sexual hybridization. This hybrid is also highly adaptive and has higher N_2 fixing capacity in soil ecosystem. The nitrogen fixation by free-living blue green algae (cyanobacteria) also contributes significantly to the replenishment of soil N. Recently, several attempts have been made to improve the associations between rice and nitrogen fixing soil bacteria and possible nitrogen fixing rice endophytes to achieve a higher contribution of biological nitrogen fixation to rice crop. In these soils, sustainability depends on symbiotic and non-symbiotic BNF system.

Nitrogen fixing free-living cyanobacteria play a vital role in the maintenance of soil fertility and sustainability. The large scale production of soil based BGA inoculum developed at Tirur by TNAU during 1977 is a crude technology, which is two decades old, and hence has to be updated. Immobilized cyanobacteria inoculum is a recent development in the biofertilizer technology, which was developed in the biofertilizer technology, which was developed by TNAU, Coimbatore with the active collaboration of the Division of Life Sciences, King's College, London, and the Department of Biochemistry and Molecular Biology, University of Seville, Spain, under an EEC project.

Immobilization of nitrogen fixing cyanobacteria under rice field conditions can be done with solid matrices such as polyurethane foam, polyvinyl foam, sugar cane waste, paper waste, and rice husk. The immobilized cyanobacteria could continuously excrete the ammonia extracellularly which could be used for continuous supply of nitrogen to rice crop. The immobilized cyanobacterial system in solid matrix when inoculated in rice field is like a mini *in situ* fertilizer factory which produces ammonia continuously. The accumulation of ammoniacal N in the flood water as a result of immobilized cyanobacterial inoculation will reflect on the growth and yield of rice.

5.19: Field application methods of biofertilizer in different crops

Crops	Recommended biofertilizers for nitrogen	Method of application	Quantity required/ha of each biofertilizers	Time of application
1. AGRICULTURAL CROPS				
A. FIELD CROPS				
a) PULSE				
1. Chickpea, Pea G. Nut Soybean, Beans	Rhizobium	Seed Treatment	1.0 – 2.0 kg	Sowing Time
2. Lentil, Lucern, Barseem, Green Gram, Black Gram, Cowpea, Pegion Pea, Guar	Rhizobium	Seed Treatment	0.4 – 0.6 kg.	Sowing Time
b) CEREALS				
1. Wheat, Oat, Barley	Azotobacter	Seed Treatment	1.0 – 2.0 kg.	Sowing Time
2. Rice	Azospirillum	Seedling Treatment	2.0 – 3. 0 kg.	Transplantation Time
	BGA	Broadcasting	10 kg.	After one week of Transplantation
	Azolla	Dual Crops	500 1000 kg.	do
c) OIL SEEDS				
1. Mustard, Seasmum, Kusum, Linseeds	Azotobacter	Seed Treatment	0.2 kg.	Sowing Time
2. Sunflower, Castor	Azotobacter	Seed	0.5 – 0.8 kg.	Sowing Time

Crops	Recommended biofertilizers for nitrogen	Method of application	Quantity required/ha of each biofertilizers	Time of application
		Treatment		
d) MILLETS				
1. Pearl millets, Finger millets, Kodo millet	Azospirillum	Seed Treatment	0.2 kg.	Sowing Time
2. Maize & Sorghum	Azospirillum / Azotobacter	Seed Treatment	0.4 – 0.6 kg	Sowing Time
e) FORAGE CROPS & GRASSES				
1. Baramuda grass, Sudan grass, Napier grass, Guinea grass, Paragrass, Rhodo grass, Star grass etc.	Azospirillum	Seed Treatment	0.8 – 1.0 kg.	Sowing Time
f) FIBRE CROPS				
1. Cotton	Azotobacter	Seed Treatment	0.6 – 1.0 kg.	Sowing Time
2. Jute	Azotobacter	Seed Treatment	0.8 – 1.0 kg	Sowing Time
3. Sunhemp etc.	Rhizobium	Seed Treatment	0.8 – 1.0 kg	Sowing Time
g) SUGAR CROPS				
1. Sugarcane	Acetobacter / Azotobacter / Azospirillum	Soil/Sets Treatment	4.0 – 5.0 kg.	Sowing Time
2. Sugarbeats	Azotobacter	Seed Treatment	2.0 – 3.0 kg.	Sowing Time
B. HORTICULTURAL CROPS				
a) VEGETABLE CROPS				
1. Potato	Azotobacter	Soil/Tuber Treatment	4.0 – 5.0 kg.	Sowing Time
2. Reddish, Spinach, Ladyfinger	Azotobacter / Azospirillum	Seed Treatment	0.4 – 0.6 kg.	Sowing Time \
3. Turnip, Carrot, Perwal Cucerbits Crops	Azotobacter	Seed Treatment	0.2 kg.	Sowing Time
4. Onion, Brinjal, Knotknot, Cauliflower, Cabbage, Tomato, Chillies	Azotobacter / Azospirillum	Seeding Treatment	1.5 – 2.0 kg.	Transplantation Time
b) FLOWERS				
1. Merigold and all other seasonal plantation (annuls, perianuals, Herbs/shrubs) flower as well as omamental plants.	Azotobacter / Azospirillum	Seeding Treatment	1.5 – 2.0 kg.	Transplantation Time
2. Buds (Gladiolus,	Azotobacter	Buds/Tuber	3.0 – 4.0 kg.	Sowing/Transplantation

Crops	Recommended biofertilizers for nitrogen	Method of application	Quantity required/ha of each biofertilizers	Time of application
Garlic, Zinger, Turmeric, Arbi, banana) Tubers & Sets of Rose Plants etc.		Set Treatment		Time
c) OTHER MISC. PLANTATION CROPS				
1. Tobacco	Azotobacter	Seedling Treatment	2.0 – 3.0 kg.	Transplantation Time
2. Tea, Coffee	Azotobacter	Soil Treatment	3.0 – 4.0 kg	Sowing Time
3. Rubber, Coconuts	Azotobacter	Soil Treatment	20 - 30 gm/Plant	Transplantation Time
II. AGRO-FORESTYR/FRUIT PLANTS				
A. FIELD CROPS				
1. All fruit/agro-forestry (herb, shrubs, annuals & as usual perianual) plants grown for fuel wood, fodder, fruits, gum, spices, medicines saps, sugars leaves, flowers, nuts and seeds purpose	Azotobacter	Soil Treatment	8 – 10 gm/Plant	
12 – 15 gm/Plant				
20 – 30 gm/Plant	Plantation Time			
At nursery stage				
After Transplantation as usual				
2. Leguminous plants/Plants	Rhizobium	Soil Treatment	20 – 30 gm/Plant	

Note: 1. In addition to Nitrogen fixation biofertilizers. Same quantity of phosphatic solubilizing biofertilizers (PSB) may also be used to provide phosphorous through soil treatment for all crops as well as by other method of biofertilizer application as recommended above.
2. @ 1 kg/acre of potash mobilizing bacteria recommended for all crops.
Source: "Millennium Biofertilizer Guide", Complied & Edited by Rajnish Jain, Delhi

CHAPTER 6

Mechanism and Estimation of Nitrogen Fixation

The universal presence of nitrogen in all living matter explains the close relationship that exists between environmental chemistry and biological systems. Although much of the work studying the biological transformation of nitrogen has been conducted on soil ecosystems, it is much the same for those conditions occurring in aquatic ecosystems.

The supply of nitrogen in the atmosphere is vast. However, it exists as an inert gas, a form unusable to life. In order for nature to utilize this immense reservoir of free nitrogen it has to be combined with other elements. The relationship that exists between the various forms of nitrogen compounds and the changes that can occur in nature are illustrated by a schematic diagram of the nitrogen cycle.

This diagram shows how nitrogen in its free form is converted by electrical discharge to nitrates. These nitrates along with nitrogen fixing bacteria and algae form plant protein which in turn provides protein to animals and humans to carry on their life processes. Within the animal body, the protein is used for the growth and repair of tissue and also as a source of energy.Waste products produced by the life processes are formed and excreted as fecal matter and urine. The fecal matter, along with the protein from dead animals and plants, is converted by bacteria to ammonia; the urine is hydrolyzed rapidly to ammonium carbonate. The ammonia produced can be used to produce additional plant protein or be oxidized to form nitrites and nitrates, thus completing the nitrogen cycle.

The form in which nitrogen appears is indicative of the stage of stabilization of the organic matter. Illustrates the various nitrogen forms and how they vary with time. Initially most of the nitrogen will appear in the form of organic nitrogen and ammonia. As time goes on, the organic nitrogen is converted to ammonia. Under aerobic conditions, the ammonia will ultimately be converted to nitrites and nitrates.

Nitrogen in liquid waste can exist in 4 forms, all of which are of interest to the agricultural sanitary engineer. The 4 forms are organic nitrogen, ammonia nitrogen, nitrite nitrogen, and nitrate nitrogen. The total of these 4 forms constitute total nitrogen.

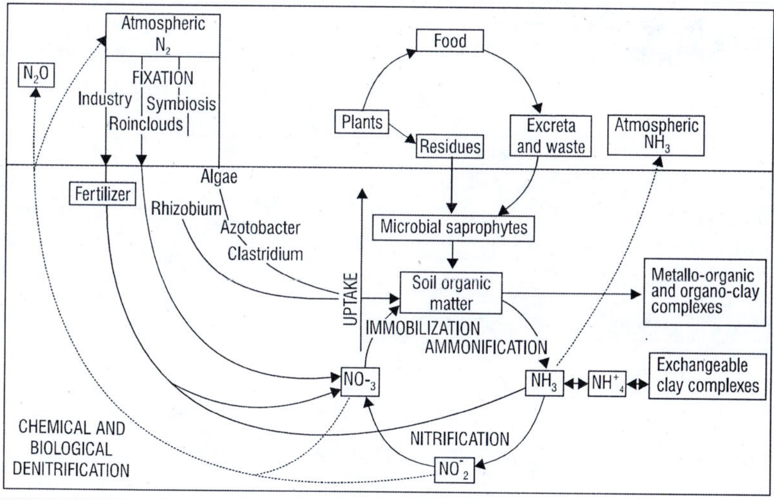

Fig. 6.1. The chemical conversion in the nitrogen cycle. The main activities are indicated together with some of the organisms concerned.

Table 6.1: Forms of nitrogen

Organic Nitrogen	All nitrogen present In organic compounds is considered to be organic nitrogen. The nitrogen-containing organic compounds are derivations of ammonia, the oxidation of which forms ammonia nitrogen.
Ammonia Nitrogen	The ammonia nitrogen is a result of bacterial decomposition of organic matter. Fresh sewage is generally high in organic nitrogen and low in ammonia nitrogen. The sum of organic and ammonia nitrogen should remain constant for the same liquid wastes, unless ammonia is allowed to escape to the atmosphere because of septic action. The total concentration of the two serves as a valuable index for evaluating the strength of liquid waste and for determining the type of treatment process to select.
Nitrite Nitrogen	Nitrite nitrogen is formed by bacterial oxidation of ammonia nitrogen. It is not present in fresh wastes but appears after bacterial activity has taken place. The presence of nitrite nitrogen indicate that the waste has undergone partial decomposition and is unstable. Nitrites can either be reduced back to ammonia or oxidized to nitrates.
Nitrate Nitrogen	Nitrate nitrogen is formed by the oxidation of nitrites and represents the most stable form of nitrogen. It is an indication of stability and is a determination of the completeness of the biological decomposition process.

Gaseous nitrogen is the most abundant element of atmosphere (78.08% v/v), and seems to have a highly complex nutrient cycle in the terrestrial and aquatic ecosystems. There is an interrelationship between different elements required for plant growth in environment. These all are affected by microbes. Atmospheric nitrogen is chemically inert. Therefore, it can be used, as such by most of the living organisms. The common source of nitrogen in the nature is the nitrate ions (NO_3). The nitrate ion is often absorbed by the plants as a mineral raw material from the environment. The absorbed nitrate ions are then converted into amino groups ($-NH_2$). The conversion of NO_3 to NH_2 group is only possible by plants and not by animals. Minerals accumulation in soil, dissolution of rock by water, addition of fertilizers in soil are also the sources of NO_3 in the soil.

Apart from these sources, atmospheric nitrogen and oxygen combine together in presence of lightening which come down with the help of rain-water to the ground, and serve as NO_3 source on the earth. Absorbed nitrogen (in the form of NO_3) incorporates into organic structures of plants which remain intact until death. Animals also obtain nitrogen by eating plants. Finally, both plants and animals die or decay. Through gradual decay of all nitrogen of dead plants and animals, it gets concentrated into ammonia (NH_3). The so forming NH_3 is then utilized as a source of nutrition for nitrifying bacteria. Usually, these bacteria are of two kinds:

- Those bacteria which absorb ammonia and convert it into nitrite ions i.e.,

$NH_3 \rightarrow NO_2$

- Those bacteria which absorb nitrite ions and convert it into nitrate i.e.,

$NO_2 \rightarrow NO_3$

Thus, the environment again gets NO_3 by the activities of these nitrifying bacteria. The available NO_3 in the environment acts upon denitrifying bacteria which converts it into molecular, atmospheric nitrogen. The free molecular nitrogen is then, utilized by nitrogen fixing organisms in which few species of bacteria and blue green algae have been included. They are often present in soil and water. They absorb nitrogen and incorporate it into ammonia and protein. The process is called **Nitrogen fixation**.

Nitrogen is the mineral nutrient most needed by plants and it often limits plant growth. Some plant species have formed mutualistic symbioses with nitrogen fixing *prokaryotes* and some *eukaryotes*. Those organisms that can directly utilise atmospheric nitrogen as a nitrogen source, are called *"diazotrophs"*, and belong to the kingdoms *eubacteria* and *archaebacteria*. They are able to live independently on soil N for growth. Inside the root nodules, the bacteria on the other hand are supplied with carbon from the host and are sheltered from competition with other organisms. The most important mutualistic symbioses are root nodule symbioses, mainly formed by members of the genera *Rhizobium*, *Bradyrhizobium* and *Azorhizobium* on legumes, and by the genus *Frankia* on some non-legumes. The reduction of atmospheric dinitrogen (N_2) to ammonium (NH_4^+) which occurs in nitrogen fixing organisms is catalysed by an enzyme complex called *nitrogenase*. A minimum 16 ATP and $8e^-$ are needed for a reduction of one molecule of N_2. It is estimated that nitrogen fixation could range from one to a few kg ha^{-1} $year^{-1}$ in lichens and free living bacteria, and up to a hundred or possibly a few hundred kg ha^{-1} $year^{-1}$ in legumes and actinorhizal plants.

Nitrogen fixation has always been a subject of great importance in biology, but since two decades the recent biotechnological development has lent its practical urgency. Nitrogenous fertilizers have become more and more expensive in terms of fossil energy, and hard cash; the population pressure have increased the need for high protein plant food. Biotechnologists are actively engaged in improving the efficiency of microorganisms to fix atmospheric nitrogen in fields. This will improve the functioning of the nitrogen cycle in biosphere which will ultimatly increase plant productivity. The programme of improvement also includes the creation of nitrogen-fixing symbiotic association other than those existing between leguminous plants and bacteria of the genus *Rhizobium.* This will help to improve agricultural production.

DIAZOTROPHIC MICROORGANISMS

The fixation of atmospheric nitrogen by prokaryotic organisms is known as diazotrophy. The phenomenon has also been observed in eukaryotic unicellular green alga, isolated from hot spring, by Yanada and Sakaguchi (1980).

Biological Nitrogen Fixation

This process is brought about by nitrogen-fixing microorganisms. These are soil inhibitants e.g., bacteria, blue green algae, molds, etc. Those microorganisms which utilize gaseous nitrogen directly and independently in the soil are known as asymbiotic or non-symbiotic nitrogen fixing organisms and the process by which they brought about is referred as non-symbiotic nitrogen fixation. Those microorganisms which utilize gaseous nitrogen indirectly through the mediation of other living organisms in the soil are called symbiotic nitrogen fixing bacteria and the process brought about by them is referred as symbiotic nitrogen fixation.

Asymbiotic Nitrogen Fixation
Microorganisms

Bacteria: As described earlier, soil contains free-living bacteria that are capable of fixing molecular nitrogen into its compounds. This process was first recognized in bacterium *Clostridium pasturianum* which is a Gram-positive, spore forming rod-shaped bacterium. After seven years of the isolation of *C. pasturianum*, two more important free-living nitrogen-fixing bacteria were isolated by Beijernick (1901). These newly identified bacteria were *Azatobacter chrooccum* and *A. agile*.

Azatobacter chrooccum is usually found in field-soils. They are short, thick, rod-shaped with rounded ends. Some larger ovoid forms are almost yeast-like in appearance. The cells are arranged singly or in end to end pairs. They measure 2 to 3 nm by 6 nm in size. They appear-brown in old cultures. They are Gram-negative and non acid fast. They are motile and the motility is due to the presence of polar flagellum. Physiologically, they are aerobic and remain active in between 25°C. They fix atmospheric nitrogen only after utilizing carbohydrate (dextrose, maltose, lactate, etc.) and give carbon dioxide as a by product. *A. glile* also actively fixes atmospheric nitrogen and produces carbon dioxide.

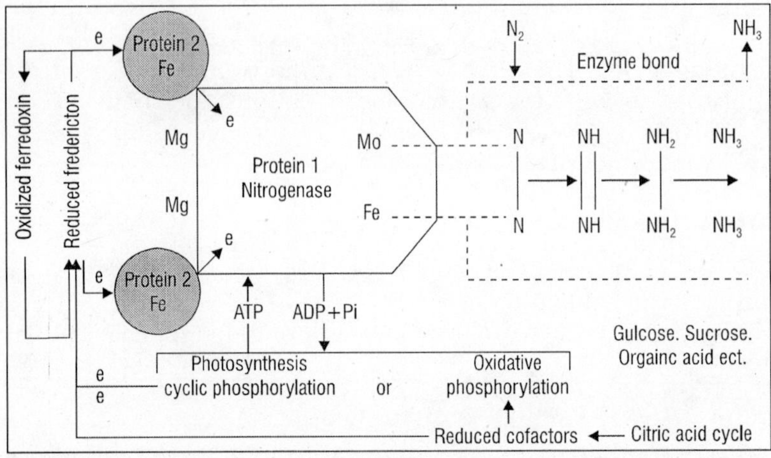

Fig. 6.2: A generalised scheme for the action of nitrogenase (incorporating results from studies on various organisms).

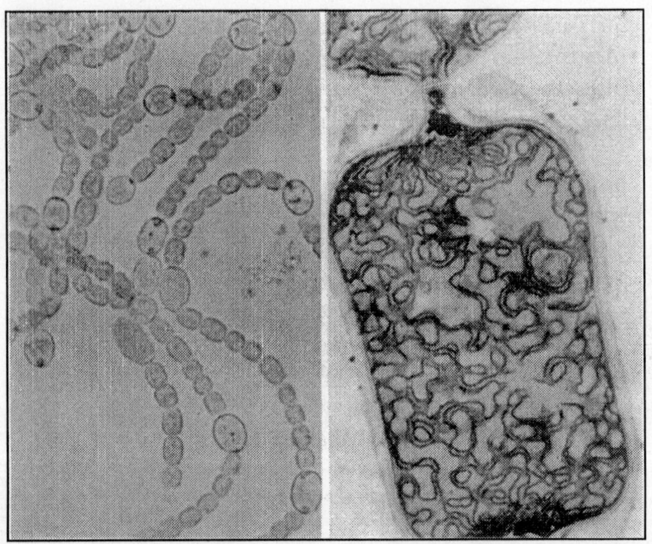

Fig. 6.3. Ultrastructure of the heterocyst of cyanobacteria *Anabaena* (a) filaments of *A. azollae* showing enlarged heterocyst in different stages of development, (b) cross section through a heterocyst of a *A. cylindrica* showing the contorted thylakoid membranes.

Apart from Azatobacter, other non-symbiotic bacteria like Rhodospirillum pneumoiniae, Rhodopseudomonas, Chlorobium, Diplococcus pneumoiniae, Azatobacter aerogenes, Micrococcus sulfurens have also been known to fix atmospheric nitrogen asymbiotically.

The ability to fix nitrogen by microorganisms was confirmed by means of the technique known as *acetylene reduction* to *ethylene* by diazotrophic microorganisms. The mechanism of this conversion is controlled by an enzymatic complex (or nitrogenous enzyme), which can reduce gaseous nitrogen to ammonia.

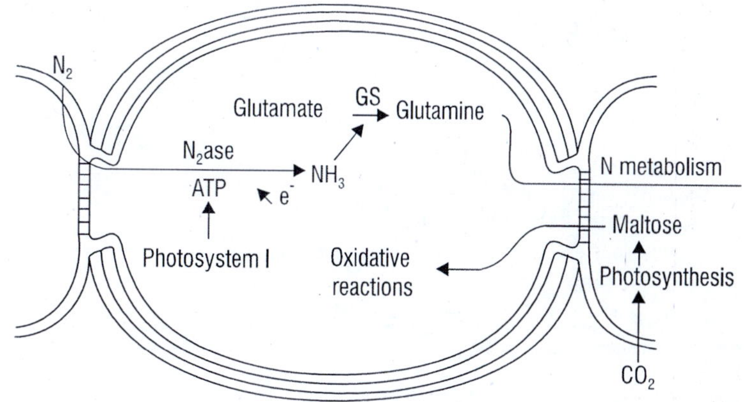

Fig. 6.4: Nitrogen and carbon flow associated with nitrogen fixation in the heterocyst of blue green alga. N2ase = the nitrogenase enzyme as GS = glutamine synthetase, oxidative reactions including the oxidative pentose phosphate pathway. Polar nodule are not shown.

The enzyme nitrogenase is sensitive towards oxygen. There are reports, which reveal that several groups of microorganisms fix nitrogen in presence of minute quantities of oxygen. Such microorganisms are known as **microaerobic fixers**. They oxidize methane. The examples are bacteria, Spirilla (*Aquaspirillum and Azosprillium*), *Xanthobacter autorophicus* and *S. flavus*, *Thiobacillus ferrooxidans and* non-heterocystous microaerobic cyanobacteria (*Oscillatoria, Plectonema*). Certain coliform bacteria also fix nitrogen only in aerobic conditions (e.g., *Citrobactor, Enterobacter, Erwinia*). A comprehensive lists of microorganisms have been presented in the book, but in near future they may need revision in light of findings of biochemical and physiological researches. It is established that oxygen not only inhibits the activity of nitrogenase enzyme but also regulates its biosynthesis, so that in conditions that are physiologically hardly favourable to nitrogen fixation, the enzyme is not synthesized.

Mechanism of Asymbiotic Nitrogen Fixation

The knowledge of asymbiotic nitrogen fixation is still insufficient. It was previously believed that *Azatobacter* fixes atmospheric nitrogen with the help of enzyme system called azatase and nitrogenase is only one of the compounds of this system. Beijerinck (1901) indicated that fixation of atmospheric nitrogen took place at the nitrous acid stage, but it was not proved. According to one more concept, nitrogen is fixed through the stage of hydrolysis to hydroxylamine or to ammonium nitrate, which is produced from nitrogen and water as shown below:

$N_2 + 2H_2O \rightarrow NH_4NO_2$

But this concept has not been proved. The hydrolysis of nitrogen is very difficult under normal conditions.

It was further proposed that *Azatobacter* fixes nitrogen after its reduction by the active hydrogen. However, this hypothesis was also not proved.

At last the mechanism of nitrogen fixation was described on the physiological basis. It was proved that the protoplasm of *Azatobacter* produces special catalysts which take part in fixing free nitrogen. The catalysts contain carboxyl and amino group as important parts, which do not take any active part in fixing atmospheric nitrogen. Phenyl hydrogen and hydroxylamine are also present in the composition of the catalysts. It was presumed that molecular nitrogen reacts with catalysts of nitrogen fixation through oxygen of carboxyl group and produces hydrazine as the first product of the reaction. The hydrazine, is then, further reduced by active hydrogen and is transformed into amino acids. They finally enter the proteins of the cell protoplasm. This incorporation of amino acid into protoplasms helps into the growth of a bacterium.

REQUIREMENTS OF NITROGEN FIXATION

1. Oxygen and carboxyl group of the catalysts play an important role in nitrogen fixation.
2. A very small amount of chemical energy (ATP) is used up during nitrogen fixation. About 1 kilo calories fix 2 mg of atmospheric nitrogen. It shows the relationship between respiration and nitrogen fixation.

Some strict anaerobic bacteria are more sensitive to oxygen. *Clostridium pasteurianum* is an obligate anaerobe, therefore, the mechanism of nitrogen fixation of such organism is similar to that *Azatobacter*, because they produce large amount of hydrogen during their metabolic reaction. The various possible pathways of nitrogen fixation have been shown in different illustrations.

SYMBIOTIC NITROGEN FIXATION

Agriculturists have noted that a relatively large group of plants- the legumes, are capable of fixing atmospheric nitrogen through a symbiotic association with soil bacteria called *Rhizobium*. These bacteria or nodule bacteria, usually dwell in the roots of various beans, peas, lupines, cloves and other leguminous plants.

Hellriegal and Wilfarth (1888) first of all demonstrated that an increase in nitrogen content in soil is due to the presence of small tumor-like outgrowth on the roots of leguminous plants. It was found that in absence of root nodules, growth of succeding plant-crop was retarded. The results of various experiments reveal that root nodule forming plants can feed upon atmospheric nitrogen and this capacity of plants can be obserbed only in those plants which are growing in non-sterilized or bacteria containing soils. These symbiotic bacteria e.g., *Rhizobium* utilize the free atmospheric nitrogen and synthesize it into new nitrogenous compounds, which are utilized by plants for their growth; and bacteria get their food from these plants. Such a mutual beneficial association of bacteria and plants is referred to symbiosis.

Symbiosis was originally defined by Bary in 1879. Symbiosis is the living together of any two organisms. Mutulism is a relationship where both partners benefit, though it greads into those where the advantage or disadvantage to one of the partner may not always be obvious. Commensal relationships are usually rather loose association of the benefit of another without any effect on itself.

MECHANISM OF NITROGEN FIXATION IN LEGUMES

The mechanism of nitrogen fixation in legumes has not yet been understood. However, the basic prerequisite for nitrogen fixation in extracts of nodule-bacteriods are similar to those that have been described for asymbiotic bacteria such as *Azatobacter*. It has not been well established that how in both asymbiotic and symbiotic system ^{15}N is converted into NH_3 and then to amino acids, because the intermediate products between ^{15}N to NH_3 have not been determined.

A good number of theories have been proposed by various workers. Two of them are important, which are as follows:

Theory of Virtanen

Nitrogen fixation in root appears immediately after nodule formation. Virtanen (1948) proposed the following scheme of nitrogen fixation in root nodules:

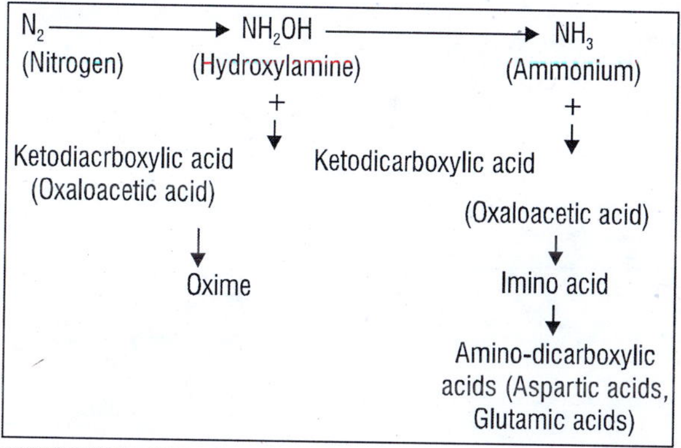

Fig. 6.5: Mechanism of nitrogen fixation (Virtanen, 1948).

According to him young plants fix more amount of nitrogen than the old plants. A great part of nitrogen is converted into L-aspartic acid and L-glutamic acid. Apart from these, α-alanine is also present in the nodule which is produced from L-aspartic acid by decarboxylation. In addition to this, small amount of Oxime-N and Nitrite-N are also present.

Theory of Burris and Wilson

According to this theory hydroxylamine is the central compound of nitrogen fixation, from which ammonia is formed through reduction or it can also directly react to the keto acid. The proposed scheme of Burris and Wilson (1945) is as follows:

Both these theories reveal the formation of amino dicarboxylicacid as the primary products in biological nitrogen fixation

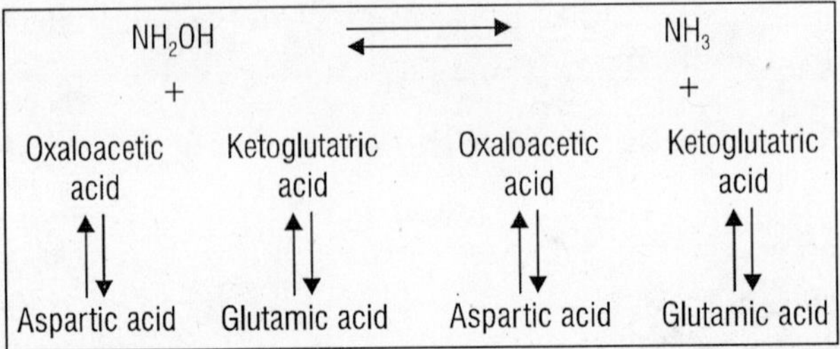

Fig. 6.6: Theory of Burris and Wilson.

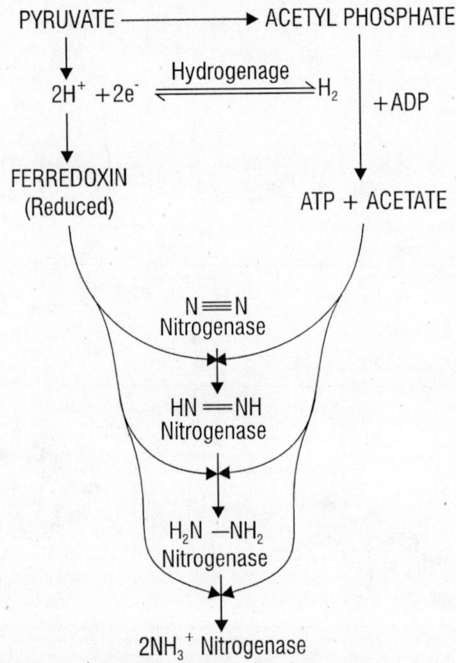

Fig. 6.7: Scheme proposed for mechanism of nitrogen fixation.

NITROGEN CONVERTERS IN THE SOIL

The oxidation of ammonia to nitrate in the soil is one of the important phases in nitrogen cycle. This is brought about by two groups of bacteria- *Nitrosomonas* and *Nitrobacter*. The required energy for the growth of these bacteria is obtained through the oxidation of ammonia or nitrate. Winogradsky (1891) first isolated these autotrophic bacteria and described the conversion of ammonia to nitrite, and nitrite to nitrate. This is called **nitrification**. The conversion of nitrate to nitrous oxide and nitrogen gas by various kinds of soil organisms is called **denitrification**. The process releases nitrogen gas to the atmosphere and completes the complex nitrogen cycle in nature.

CYANOBACTERIA AND NITROGEN FIXATION

Frank (1889) first reported the ability of nitrogen fixation by blue green algae. According to Drewes (1928) *Nostoc punctiforme* and *Anabaena variabilis* have the ability to fix nitrogen in the soil. The principal genera of this group known to fix nitrogen are *Nostoc* spp. *(N. cummune, N. punctiforme, N. muscorum); Anabaena* spp. (*A. ambigua, A. cylindrica, A. fertilissima. A. circularis*, etc.), *Cylindrospermum* spp. (*C. gorakhporense, C. licheniforme)*- all of them belong to the family Nostocaceae. A few of members of the family Microchaetaceae and Rivulariaceae like *Aulosira fertilissima* and *Calothrix brevissima* respectively are known to fix atmospheric nitrogen in rice fields. Significantly most of the free living nitrogen fixing members of this group produce**heterocyst**which is considered to be the site of nitrogen fixation (Kulasooriya, 1972).

The free living diazotrophic cyanobacteria are the largest contributors of the process of biological nitrogen fixation, the second most important biological process on this planet. Their nitrogen fixing ability was first related to the presence of specialized non-oxygen evolving cells called heterocysts in which nitrogenase; a highly oxygen-sensitive enzyme is produced from oxygen.

NITROGEN FIXING FUNGI

The fungi also independently play an important role in fixing free nitrogen in soil. The fungi that are responsible for this function are of species of *Mactosporium commune, Cladosporium herbarum, Phoma, Alternaria tenuiis, Rhodotorula* spp. etc. The occurrence of these fungi has been noted in forest-land.

Biochemical Aspects of Diazotrophy

An American company (DuPont) in 1960 started the fundamental research on the biochemistry of nitrogen fixation specially on nitrogenase activity in cell free extracts of *Clostridium pasteurianum.* The experiments were conducted in absence of air because of the very great oxygen sensitivity of nitrogenase. The techniques like chromatography, electrophoresis and fractionation under anaerobic conditions, helped the researchers to isolate nitrogenase from *Clostridium pasteurianum.* Thereafter, nitrogenase was easily isolated from other anaerobic nitrogen fixers while this extraction was more difficult with *Azotobacter vinelandii.* The so isolated enzyme has the property to accept electorn from sodium dithionite and was oxygen-tolerant and evolved hydrogen in the absence of nitrogen. In 1981, nitrogenase in the form of crude preparations was isolated from about 30 microbes and out of which about 6 of them were purified.

Biochemically, nitrogenase is a binary enzyme and consists of two brown metalloproteins, of which the joint activity is essential to reduce nitrogen to ammonia.

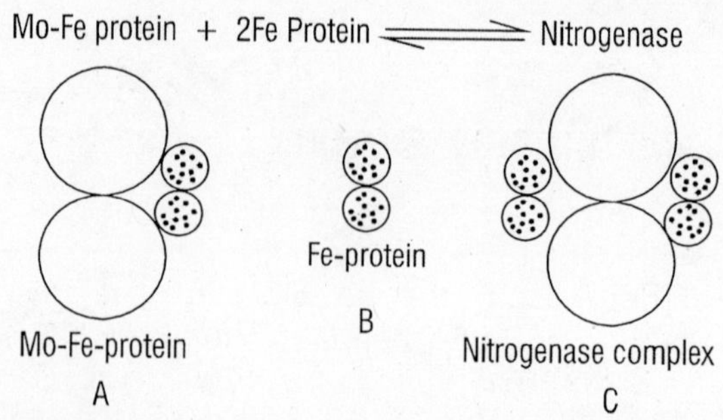

Fig. 6.8: Nitrogenase enzyme.

During nitrogen conversion, nitrogenase binds to molybdoferroprotein (MoFe-protein), which was named **disnitrogenase**. When nitrogen binds at the molybdenum level in a separable fragment of protein, called **FeMoco**. The two component proteins of the enzyme nitrogenase are quickly and irreversibly destroyed by exposure to oxygen.

Nitrogenase is also able to reduce other substrates in the place of nitrogen, including hydrogen ions and, in this case it evolves hydrogen.

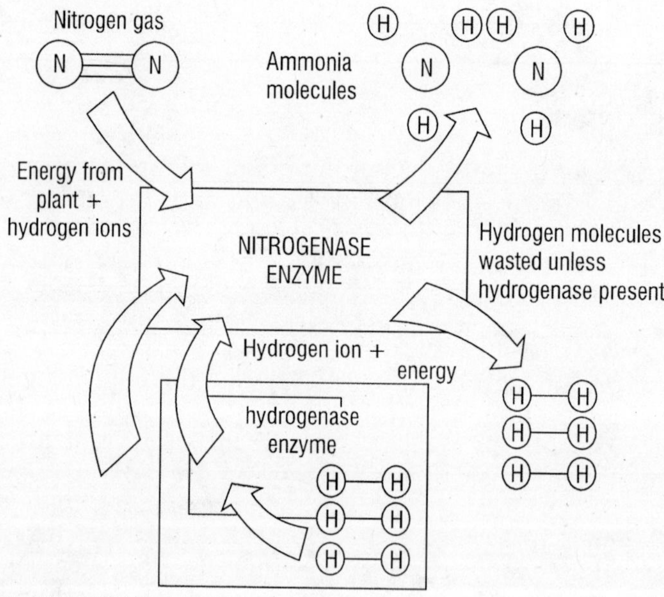

Fig. 6.9: Representation of the mechanism of nitrogen fixation and the action of hydrogenase.

NITROGENASE PRODUCING *E. COLI* CELLS

Because of the close relationship between *E. coli* and *Klebsiella pneumoniae, E. coli* is specially qualified for the investigation of the nitrogenase system. With *E. coli* C-M74 cells, high nitrogenase activity can be attained in culture medium, supplemented with 50mg/l ammonium sulfate, 1% aspartic acid, and 0.1% yeast extract.

During reduction of nitrogen to ammonia, electrons are required which are transferred to nitrogenase by a specific transport proteins; the *Ferrodoxins* or the *Flavodoxins.* ATP serves as energy source. The consumption of ATP molecules is 12 to 15 per nitrogen molecule. The reason for this high consumption of ATP has not been entirely elucidated.

As evident from the reactions that hydrogen is produced during nitrogen fixation. Certain nitrogen-fixing aerobic bacteria possess hydrogenase enzyme which has the ability to recycle hydrogen produced by nitrogenase in vitro. Therefore, the re-utilization of this hydrogen generates more ATP (by oxidative phosphorylation) and consequently improves the efficiency of nitrogen fixation.

*Rhizobium*is also associated with soybean plants which contain hydrogen uptake (or *hup*) genes which have the ability to recycle hydrogen and back into the nitrogenase system and fix nitrogen. Through this mechanism the energy is harvested by plants, which is otherwise lost.

The nitrogenase enzyme of bacteria fix nitrogen by converting nitrogen gas into ammonia using hydrogen ions and energy by host plant. During the process there is a production of energy rich hydrogen gas. Some bacteria possess hydrogenase enzyme which can capture hydrogen gas and converts it into hydrogen ions and release energy which can feed back into nitrogenase enzyme.

Genetics of Free-living and Symbiotic Diazotrophs

Organization of Nitrogen-fixation Genes

In molecular terms, very little is known of the means by which the prokaryote specifically recognizes and invades its respective host-plant. However, there has been a growing interest in approaching this problem by identifying and analysing the relevant symbiotic genes and their gene products. Because of the agricultural importance of the *Rhizobium*-legume interaction, the most detailed studies have been conducted on this system. Also, because of bacteria are simpler than plants. Our knowledge of the symbiotic genes of *Rhizobium* is better than that of the plant genes. There have been several recent advances in the analysis of the *Rhizobium* genes that are involved in nodulation, host-range specificity and nitrogen fixation. By dissecting the structure, organization and regulation of these genes, a better understanding of their functions may be forthcoming.

In the fast-growing *Rhizobium* species, which nodulate temperate legumes such as clover, alfalfa, peas and field beans, the genes for nodulation and nitrogen fixation are clustered on large symbiotic plasmids. One such plasmid, pRL1JI, was isolated from a strain of *R. leguminosarum* (which nodulates peas) and in this case the genes that determined the ability to nodulate peas were located between two groups of genes required for nitrogen fixation. The study of the nitrogen-fixation (*nif*) genes in *Rhizobium* has been greatly facilitated by the fact that at least some of them are very similar to the corresponding *nif* genes in the free-living nitrogen-fixing bacterium - *Klebsiella pneumoniae.*

In *Klebsiella,* 17 *nif* genes have been identified which are variously involved in *nif* gene regulation, the synthesis of the iron-molybdenum cofactor (FeMoCo). Using DNA sequence comparisons, *Rhizobium* species have been shown to contain the *nifHD* and *K* genes (nitrogenase proteins) *nifA* (regulation) and *nifB* (synthesis of FeMoCo). Further studies may well demonstrate similarities between other *Rhizobium* and *Klebsiella nif* genes.

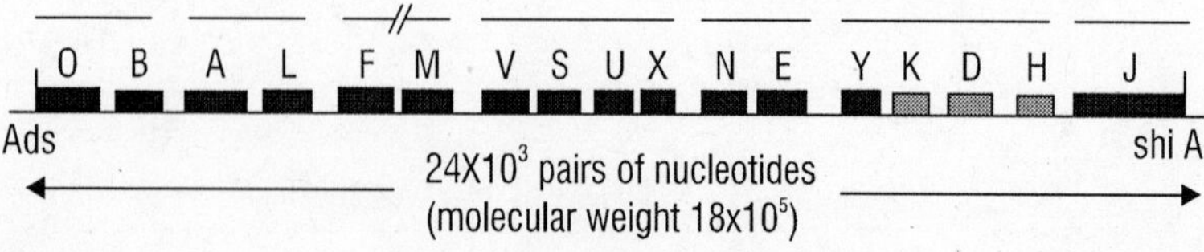

Fig. 6.10: Map of the *nif* gene cluster of the *Klebsiella pneumoniae* chromosome. The capital letters denote the genes (cistrons), the *nif* cluster being located between the genes for biosynthesis of histidine (his) and those for shikimic acid uptake (shi A). The arrow indicate the separate operations and the direction of their transcription. In 1981 the status of F and M as one or two distinct operons was not clear. The size of most of the genes, however, were known.

The study of nitrogen fixation genes was first carried out in *Klebsiella pneumoniae* (strain M5al); and spectaular progress achieved in 1970s. Although the first report came from the mutants of nitrogen fixing bacteria, which was isolated and biochemically characterized as early in 1959. The genetic research by Streicher et al. (1971) of U.S.A. and Dixon of Agric. Research Council, Unit of Nitrogen fixation (U.K.) added more knowledge in this field. Dixon and Postgate (1971) successfully transferred the nitrogen fixation genes, called *nif,* to *Klebsiella pneumoniae.* Later on they could able to transfer these genes to *Escherichia coli.* This work encouraged the researchers of this field specially for *Rhizobium* and the presence of *nif* genes in this microorganism. More than a dozen genes, termed the *nif* genes are involved in the assembly of nitrogen-fixation apparatus. The *nif* genes are not found in a free state but are clustered in one region (they are not scattered throughout the bacterial DNA). Therefore, it is much easier to stretch of DNA in a *Rhizobium* chromosome and insert the whole batch into another organism. This has also been done in *Agrobacterium tumefaciens* (Ti plasmids). In 1981, the number of *nif* genes was 17 and they were organized in eight transcription units (or operons), one of which was monocistronic. Brill (1980) also included two other genes: *nif*-R between L and F, and *nif*-W between F and M.

Nod-Genes for Nodulation

Concerning the genes (called *nod* genes) for nodulation and host-range specificity, an important observation is that, despite the morphological complexity of the infection process, only a few genes on the symbiotic plasmid are required for nodulation. In the *R. leguminosarum* symbiotic plasmid pRL1JI, less than 10kb appears to be required for nodulation and for the determination of host-range specificity for peas. The reason has been analysed by DNA sequencing (and thus identifying genes as long open reading frames), isolation and characterization of mutants and by studying the regulation of the *nod* genes.

The organiztion of *nif*-genes of other free-living diazotrophs from which *nif*-mutants were isolated (*Azatobacter vinelandii, Azospirillum lipoferum, Clostridium pasteurianum, Rhodopseudomonas capsulata* and a few cyanobacteria) was as like to that of *Klebsiella pneumoniae.*

Regulation of Nitrogen-fixation Genes

The regulation of *nif* genes is rather complex. Ammonia and nitrogenous compounds (like nitrates and amino acids) inhibit the expression of nitrogenase activity.

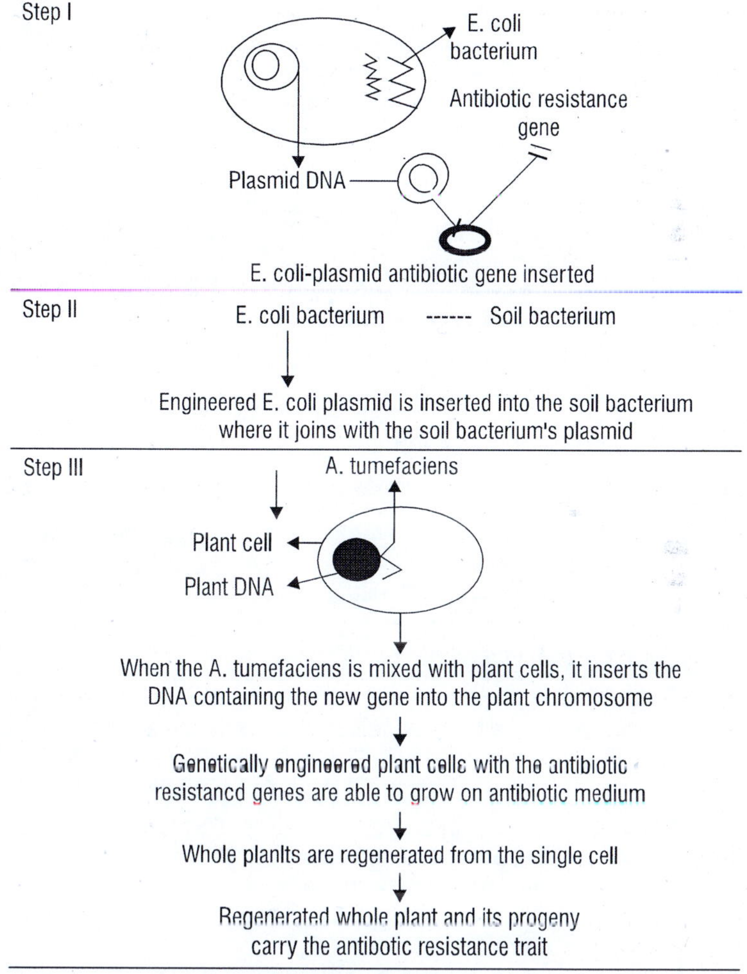

Fig. 6.11: Schematic presentation of plant genetic engineering with Ti plasmids from *Agrobacterium tumefaciens.*

Excess of aeration in the culture medium also inhibits the expression of *nif*-genes. For induction of *nif*-genes, the presence of nitrogen is not necessary, because these genes can express themselves in cells cultivated in an argon atmosphere. In fact, the *nif*-genes expression is the result of a de-repression rather than that of an induction

Eady et al. (1978), Roberts et al. (1978) reported that all the products of *nif*-genes are not present in the cultures carried out in the presence of ammonia. When ammonia is added to a completely derepressed cultures, a rapid repression of the *nif*-phenotype occurs and nitrogenase activity as well as the enzymic proteins disappear within 45 minutes.

The presence of ammonia in the medium plays very important role in the synthesis of various biomolecules, e.g., in presence of ammonia, glutamine synthetase reacts with ATP and seems to lose the enzymic and regulatory functions of *nif*-genes.

The presence of oxygen in the culture medium represses the biosyntheis of nitrogenase enzyme in *Klebsiella pneumonia.* Similar results have also been observed in aerobic and anaerobic bacteria. However, the mechanism of oxygen repression seems to be independent from that of ammonia in such microorganisms.

Molybdate does not have any direct effect on the regulation of b-*nif*-genes, however, molybdoprotein is probably involved in the regulation of at least one of the *Klebsiella nif* operons (*nif* YKDH). The presence of molybdenum transport and storage proteins is known in *Clostridium pasteurianum* and other diazotrophs.

It is now established, after number of current reports that the regulation of the *nif*-genes of *Klebsiella pneumoniae* is carried out by both *nif* A and *nif* L genes as well as by genes that are remote from the *nif* cluster (*gln* genes). However, there are other genes which are involved in the expression of *nif*-genes, they are as follows:

nar D = which is involved in molybdenum processing.
unc and a gene in his operon = which influence ATP supply.
nim gene = gene of the uncertain function near tip.

Genetics of Symbiotic Diazotrophs

Much works has been done on the fast growing nitrogen-fixing bacteria-*Rhizobium.* The genetic map of *Rhizobium* is now established. The studies reveal that *R.leguniosarum, R. phaseoli* and *R. trifolii* have very similar gene organization.

The genes for the component protein subunits of nitrogenase (molybdoferroprotein and ferroprotein) are formed by a single operon in *Rhizobium leguminisarum.* Similar situation has been reported in *Azatobactor vinelandii* and in *Klebsiella pneumoniae.*

TRANSFER OF *NIF*-GENES TO MICROORGANISMS

As mentioned earlier that *nif*-genes are not found in a free state but are clustered in one region (they are scattered throughout the bacterial DNA). Therefore, it is much easy to cut out the relavent stretch of DNA in a *Rhizobium* chromosome and insert the whole batch into another organism.

The construction of self-transmissible plasmids (Ti-plasmids) bearing *nif*-genes made it possible to shift the nitrogen-fixing capacity to non-fixing species or to non-fixing mutant of diazotrophic species like *E. coli (E. coli* C-M74), *Salmonella typhimurium, Serratia marcescens, Pseudomonas flurescence,* and *nif*-mutant of *Azatobacter vinelandii.*

It is quite easy to transfer *nif*-genes from one species to another, or form one microbial genus to another under laboratory conditions. This may be done by three methods of gene-transfer: conjugation, transduction and transformation. Therefore, the transfer of *nif*-genes between the plasmid and bacterial chromosomes can take place under laboratory as well as under natural conditions.

TRANSFER OF *NIF*-GENES AND DEVELOPMENT OF NEW NITROGEN-FIXING PLANTS

It is easy to transfer grouped genes andthe *nif*-gene cluster. Plant viruses could be used as vectors for this purpose. The cauliflower mosaic virus is a deoxyribovirus to which it is possible to join certain *nif*-genes which could become integrated into the plant genome or which could be transcribed into the plant cell at the same time as those of the virus when the latter's replication takes place.

*Agrobacterium tumefaciens*is a pathogenic bacterium, which contains a large plasmid, Ti and is responsible for the induction of tumours in plant tissues on the site of the infection by the bacterium. These plasmids are of 150 to 230 kilobases and can able to develop tumours, therefore, they are called Ti or **tumour inducing plasmids**. Ti -plasmids are introduced by either transformation or by conjugation. The DNA of Ti-plasmids have two following regions:

1. The larger region contains information that makes it possible for the bacterium to catabolize derivatives of the basic amino acids, or opines, synthesized by the tumour cells.
2. It corresponds to the T-DNA and is incorporated into the genone of the plant cell. This T-DNA includes the ability of utilization of the latter as well the transformation of its metabolism in order to produce opines. The molecular weight of T-DNA is 15 x 10^6 daltons.

The development of tumours in plant tissue by Ti -plasmid of *Agrobacterium tumefaciens* is brougnt about by the transfer of a segment of bacterial DNA (from a prokaryotic) and its integration into the genome of eukaryotic cells. This introduction of Ti -plasmid was successfully made by Schell and his coworkers. Out of the seven genes of T-DNA of Ti -plasmid, they found that five of them are responsible for tumorization, that is in blocking cell differentiation. It is of interest to mention that *Agrobacterium tumefaciens* infects only dicotyledonous plants, however, it is not easy to introduce Ti plasmids into the cells of monocotyledonous plants like cereals.

Kemp (1981) used nucleic acid recombination and cloning techniques to transfer genes to plant cells. He has transferred genes for the bean protein-**phaseolin** to sunflower plants. This transfer was followed by the synthesis of mRNA of the storage protein in the sunflower cells, made tumoral by the T-DNA of *Agrobacterium tumefaciens* which was used as a vector for the gene.

The development of plants which can able to fix atmospheric nitrogen independently of symbiotic microbes remains a difficult and complex task. However, in the present situation, it is possible to develop and gain better knowledge of the factors that induce nitrogen-fixing symbiosis, with a view of improving their efficiency

ESTIMATES OF NITROGEN FIXATION

Both direct and indirect methods have been employed to estimate nitrogen fixation associated with various non-legumes in pot culture and field studies. Direct methods include nitrogen balance studies and ^{15}N based techniques such as ^{15}N isotope dilution and ^{15}N incorporation while acetylene reduction assay (ARA) is the widely used indirect techniques.

^{15}N BASED METHODS

15N isotope dilution technique provides the most accurate estimates of the contribution of biologically fixed nitrogen to the nitrogen nutrition of legumes and non-legumes alike. Nitrogen fixation associated with sugarcane, cereals and actinorhizal plants has been estimated by this technique with and without addition of inoculum.

Ventura and Watanabe (1983) reported that a substantial proportion of plant nitrogen (32-35%) was derived from associative fixation in uninoculated rice plants under flooded condition in the Philippines. Yoo *et al.* (1986) observed contribution of 11-19 percent nitrogen by biological fixation when rice was inoculated with *Klebsiella oxytoca.* de Freitas *et al.* (1984) found that biological fixation contributed 22-26 percent of plant nitrogen was derived from biological fixation in forage grasses. Biological fixation contributed 53 to 65 percent of plant nitrogen when *Casuarina equisetifolia* was inoculated with *Frankia* (Sougoufura *et al.*, 1990). Boddey *et al.* (1991) reported that certain sugarcane varieties were capable of obtaining upto 80 percent plant nitrogen from biological fixation.

Nitrogen fixation associated with tropical grasses, sugarcane, rice and sorghum has also been estimated by ^{15}N incorporation technique.

NITROGEN BALANCE STUDIES

Non-symbiotic N_2 fixation to the extent of 18-20 kg N ha^{-1} $year^{-1}$ was reported from long term nitrogen balance studies at Rothamsted, England (Jenkinson, 1977). Substantial positive balances for nitrogen have been reported in pot experiments with sorghum, pearl millet, finger millet and napier bajra at ICRISAT, Hyderabad (Dart and Wani, 1982). App *et al.* (1980) reported positive N_2 balance for flooded soils in pots planted to rice. There was relatively higher level of nitrogen fixation associated with grasses in the tropics when compared to that in temperate countries. Dobereiner *et al.* (1972) reported fixation of 90 kg N ha^{-1} $year^{-1}$ by *Paspalum notatum.* Substantial nitrogen contribution by *Azolla-Anabaena* associations, blue green algae and photosynthetic bacteria to the total nitrogen input of rice has also been reported (Venkataraman, 1975; Watanabe, 1981).

ACETYLENE REDUCTION ASSAY

This is a simple and sensitive assay for studying the nitrogenase activity of microbial cultures and plants in pot culture and field experiments. Nitrogenase activity has been reported in many grasses by the excised root assay when the roots were exposed to acetylene after pre-incubation for 8-18 h under reduced oxygen tension (Dobereiner and Day, 1975). Soil root cores removed from the field at harvest have also been used for measuring nitrogenase activity of both grasses and grain crops (Day *et al.*, 1975). Wani *et al.* (1984) developed a non-destructive intact plant assay for pot grown plants and it can also be used for screening tube-grown seedlings or bacterial strains in association with plants for their nitrogenase activity. The method has been well adapted for agronomic studies and plant breeding programmes on N_2 fixation in the field. Observations recorded in the IRRI have established that a quantity of nitrogen equal to 50 percent of the nitrogen removed by the crop was fixed in the standard pot, planted with rice where blue-green algae were spontaneously growing. Alimango and Yoshida (1977) reported nitrogenase activity showing a fixation of 18-33 kg N_2 ha^{-1} in Alby province in the Philippines when assayed by ARA method. The ARA was attributed mostly to blue green algae and *Azolla* and no activity was observed when soil was covered with black cloth.

Table 6.1: ^{15}N dilution estimates of N_2 fixation associated with various crops

Test Plant		Reference Plant	Soil treat-ment	Percentage of plant N derived from fixation	References
Species/cultivar	Inoculum	Species/cultivar			
Oryza sativa L. IR 42	Nil	*Oryza sativa* L. IR 42	U, C	32-35	Ventura and Watanabe, 1983
Oryza sativa C 5444	*Klebsiella oxytoca*	*Oryza sative* C 5444	H, C	11-19	Yoo *et al.*, 1986
Triticum aestivum L. Fielder	*Azospirillum brasilense*	*Triticum aestivum* L. Fielder	U	5-11	Kucey, 1988
Saccharum Sp. NA 5679, CB 4176	Nil	*Glycine max* (non-nodulating)	--	22-26	de Freitas *et al.*, 1984
Sorghum sudanense	Nil	*Triticum aestivum* L. (Ticona 2)	--	16-39	Vose *et al.*, 1982
Panicum maximum (11 ecotypes)	*A. lipoferum* + *A. brasilense* + *A. amazonense*	*Brachiaria radicans* (IRI 442)	U	24-39	Miranda and Boddey, 1987

Reference Plants are not inoculated
U - unsterilised, H-heat sterilized, C – Covered with black lid
(Adopted from Chalk, 1991).

QUANTIFYING NITROGEN FIXATION IN NFSS

Quantifying nitrogen fixation by NFSs under field conditions is a very difficult and expensive task; a full description of the methods for evaluating biological N fixation is beyond the scope of this book. For such a discussion, readers should see Bergersen (1980) or Sprent and Sprent (1990). A number of methods are in use, but these methods for quantifying nitrogen fixation do influence the results as demonstrated by Sanginga et al. (1990b) in a study using leucaena and *Faidherbia albida*. The isotope dilution methods they used resulted in large negative estimates of nitrogen fixed by *F. albida*, while theymeasured positive values using the A value method.

Table 6.2: Estimated nitrogen fixation from grain and forage legumes

Species	Nitrogen fixation (kg/ha/yr)
Arachis hypogaea	60-80
Desmodium sp., *Lespedeza* sp.	100-400
Medicago sativa	100-300
Trifolium sp.	100-150
Vicia faba	240-325
Vigna unguiculata	85

Source: NAS, 1979

The most common method used for assessing nitrogen fixation is acetylene reduction, although this method actually measures nitrogenase activity, not nitrogen fixation. It has appeal as a tool because it is sensitive, simple and relatively inexpensive, but it has many potential sources of error. The primary limitations are related to the destructive and short-term nature of the method (Weaver and Frederick, 1982) and to the need to calibrate the rates of ethylene production with the actual rates of N_2 fixation (Peoples et al., 1988). Despite the constraints of using acetylene reduction, it is still a useful tool for evaluating the relative effects of short-term treatment on N_2 fixation and for screening purposes.

N-difference or isotope dilution methods are more costly, but they can provide accurate time-averaged estimates of the proportion of plant N derived from N_2 fixation. It must be noted that while these methods have been successfully used with herbaceous plants or in pot studies with tree seedlings, their application in field studies of NFSs is still limited due to complexity, cost and (or) methodological problems.

It is clear that the choice of methods is restricted by financial, technical and environmental constraints and that methods need to be selected in relation to overall research objectives.

Table 6.3: Advantages and limitations of methods for estimating N2 fixation by legumes in the field.

Advantage	Limitations
Acetylene reduction method	
Relatively inexpensive Simple Rapid	Indirect (must establish conversion factor for N_2 fixation, which may vary) Nitrogenase activity possibly inhibited by acetylene

Advantage	Limitations
Very sensitive	Short-term estimate Considerable error possible due to incomplete recovery or damage of nodules unless using an in situ assay
N-solute method	
Inexpensive Simple Analyses done by simple colorimetric assays Need not be totally destructive May be done on an individual plant basis Assesses plant dependence on atmospheric N_2 and soil N	Indirect (must establish calibration with plant N_2-fixing status) Cannot be used for interspecific comparisons without calibrating each species Calibrations possibly influenced by development stage of some species Cannot be used on some amide-producers Short-term estimate
N difference method	
Direct Relatively simple Adjusts for soil-derived N	Requires suitable non N_2-fixing reference plant Same amount of soil N must be absorbed by legume and reference plant
Isotope dilution techniques	
Direct Can get time-integrated estimate of percentage N fixed	Requires suitable non-N_2-fixing reference plant
^{15}N enrichment method	
Potentially accurate	Expensive Requires addition of ^{15}N-labeled compound Critical that legume and reference plant absorb the same relative amounts of soil N and added N Change in ^{15}N level of plant available soil N possible with change in depth and time Reference plant must derive all its N from soil
Natural ^{15}N abundance method	
Additional N not requires Concentration of 15N in plant-available Soil N relatively constant with depth and time	Relatively expensive Insensitive if amount of ^{15}N in soil nears amount of ^{15}N in air Large field variability possible in some cases May have to allow for isotopic fractionation

Adapted from: Ledgard and Peoples, 1987.

QUANTITIES OF NITROGEN FIXED

Estimates of how much nitrogen NFS species fix vary greatly due to differences in methods used, the conditions under which the trees are growing at the time of the measurement and the degree to which sample data are extrapolated. Thus estimates of the amount of N fixed, such as those included in table 6.4, should be taken with great caution. Some authors have argued that actual N fixation rates can be very low, that is, less than 2 kg N/ha/year (Roughley, 1987), although this estimate was for acacias in arid areas where low rates of N fixation generally are expected. Yet the studies that used the most reliable methods still suggest that annual rates of 50 kg

N/ha to 200 kg N/ha are not unreasonable and are in the same range as values reported for a number of herbaceous legumes

These values should only be used as a very approximate estimate of the quantities of nitrogen fixed by species. The influences on N fixation noted in previous sections and the differences in assessment methods limit the use of these values for any kind of extrapolation or generalization.

FACTORS INFLUENCING BIOLOGICAL NITROGEN FIXATION

The factors that affect nodulation in the field have been extensively studied for agricultural crops but have been little studied for NFSs. Nevertheless, it appears from a number of studies that there are fundamental similarities between NFSs and herbaceous legumes with respect to rhizobial relations. Since symbiotic biological nitrogen fixation is a product of both the host plant and rhizobia or *Frankia,* factors that influence growth and survival of both the host plant and microbial symbiont are important.

Tree growth is dependent upon interactions between genetic potentials and the environment in which the tree grows. Environmental influences such as radiation, temperature, water, nutrient availability, wind, competition, pollution, fire and allelopathy are compounded by biotic influences such as insects and disease and the influence of man. Each of these factors in turn involves its own set of complex interactions. For example, radiation is an important factor in plant growth, but the extent to which radiation influences growth is dependent upon light intensity and quality plus the shade tolerance of the tree and the management of competition and leaf area. In many cases, experimental results do not provide clear pictures of specific environmental influences on plant vigor or nitrogen fixation.

Simply stated, symbiotic nitrogen fixation probably is limited by any factor that reduces tree vigor and health. Fixation of nitrogen is limited by the biological yield of the host plant even if a suitable symbiont (i.e., *Rhizobium* or *Frankia*) and appropriate soils are present because the host plant is both the source of energy for the symbiont and the sink for the fixed nitrogen (App et al., 1980). Dawson and Gordon (1979) demonstrated the influence of photosynthesis, as reflected by leaf area, on nitrogen fixation in *Alnus glutinosa.* Similar effects of the supply of photosynthate on nitrogen fixation have been found for alder and peas (Wheeler and Lawrie, 1976). Many examples are cited in the next section of how nitrogen fixation can be affected, often by influencing both tree and symbiont. Most of what is known about environmental influences on nodulation comes from research on herbaceous crops. Much of this work has focused on the effects of single factors (e.g., combined nitrogen or temperature); very little has been reported on the effects of multiple factors.

An effective symbiotic association between *Rhizobium* and legume can provide most of the nitrogen (N) required for the growth and productivity of the host plant. However, for this to be true, the legume host must be efficiently nodulated by its homologous *Rhizobium.* Inoculant strains/native *Rhizobium* must be capable of survival as saprophytes in soil and be able to compete with other rhizobia in soil for nodulation sites. Further, as plants must develop rhizosphere to support rhizobial growth and also nodules to house the invading rhizobia, it is prudent to assume that any factor that adversely affects plant growth will also cause competition for nodulation. Environmental and biological factors that influence biological nitrogen fixation are discussed here.

Biological Factors

Antagonistic interactions exerted by the soil microbial population against *Rhizobium* have much to do with the degree of resistance achieved by different strains of *Rhizobium. Rhizobium japonicum* USDA 123 declined 1000-fold during first five days of incubation in a non-sterile Miamian soil (Vidor and Miller, 1980). The authors established a direct relationship between the population decline of this strain and the simultaneous population increase of lytic organism, probably a bacteriophage. Similarly, Chatel and Parker (1973 b) reported that a number of problems encountered with pasture establishment in Australia could be explained by anatagonistic interactions between *Rhizobium* and other soil micro-organisms. Specifically, they showed that the second year clover mortality problem in Western Australia was due to such an interaction (Chatel and Parker, 1973 a).

Schwinghamer and Brockwell (1978) showed that lysogenic strains of *R. trifolii* grown in sterile broth and peat culture strongly suppressed the growth of sensitive strains. In other studies it was found that the addition of virulent bacteriophages to paired strains of *R. trifolii* decreased the numbers of sensitive strains on the roots of clover plans white resistant strains were favoured for nodulation (Evans *et al.*, 1979; Barnet, 1980). There are many reports of pleiotrophic effects on symbiotic ability due to phage resistance or sensitivity (Ali *et al.*, 1980; Stacy *et al.*, 1984). This influence may be due to the fact that there is a common site of interaction between phage and bacterium and between bacterium and plant, possibly the lipopolysaccharide (Zajac and Lorkiewcz, 1983).

It is difficult, however, to extrapolate field responses from laboratory studies on antagonistic effects of bacteriophages. In the field, phage *Rhizobium* interactions might be altered by clay particles or organic matter, which could reduce or eliminate antagonistic effects (Roper and Marshall, 1978). Field studies on the role of bacteriophages in persistence and competition of *Rhizobium* in the soil are inconclusive. Bacteriophages that were active against commercial inoculants of *R. trifolii* were isolated from clover fields in New Zealand. Resistant and partially resistant *Rhizobium* strains were isolated from these fields (Patel and Craig, 1984). In other field studies no correlation was found between the number of *R. trifolii* and the number of bacteriophages capable of infecting this species (Lawson and Barnet, 1984). Inoculation of soil with bacteriophages did not affect the total number of rhizobia or the number of nodules formed (Barnet, 1980).

Bacteriocinogeny is the ability of a bacterium to produce bacteriocins, with antibacterial action towards other similar bacteria (Schwinghamer,

1971). Bacteriocin producing strains are common among Rhizobiaceae (Schwinghmaer and Reinhardt, 1963). Two groups of bacteriocins are recognised in *Rhizobium*: (a) low-molecular weight non-phage like substances that diffusion readily in agar and (b) less diffusible high molecular weight phage like compound (Schwinghamer, 1975). Two types of bacteriocins, small and medium (according to lysis zone) were found in field isolates of *R. leguminosarum* (Hirsch, 1979). The genes encoding medium bacteriocin production were present on large plasmids, one of which also carried *nod* genes (Johnson *et al.*, 1978). Recently, Brahmaprakash and Hegde (1990) reported bacteriocin nogeny amongst redgram rhizobia.

Bacteriocinogenic strains of *R. trifolii* suppress the growth of sensitive strains in liquid culture (Schwinghamer and Brockwell, 1978). Hodgson *et al.* (1985) demonstrated the advantage of using a mixed inoculant of highly effective strains (bacteriocin resistant) and a bacteriocinogenic strain. When a bacteriocin resistant strain and a sensitive strain were used to inoculate *Trifolium* in glass tubes, the bacteriocin resistant strain formed less than 10% of nodules. When a bacteriocin producing strain was added to the mixed inoculant the resistant strains formed 75% of the nodules. The results were similar when the experiment was repeated in unsterile soil containing a population of bacteriocin sensitive ineffective population of *R. trifolii.* Use of bacteriocinogenic strains in the inoculant increased the proportion of nodules formed by the bacteriocin resistant effective strain. Schwinghamer and Belkengren (1968) described a *R. leguminosarum* bv. *trifolli.* They suggested that the basis for the nodulation competitiveness of T 24 could be its ability to produce anti-rhizobial compounds. Triplett and Barta (1987) have shown by transposon mutagenesis that the symbiotic competitiveness expressed by T 24 includes expression of two phenotypes, nodulation as well as trifolitoxin production. Both nodulation and trifolitoxin production are necessary for nodulation competitiveness because mutants defective in either of the properties could not inhibit nodulation by an effective trifolitoxin sensitive strain in a mixed culture.

Legumes commonly interact with the fungi belonging to the four genera viz., *Acaulospora, Gigaspora, Glomus* and Sclerocytis to produce vesicular-arbuscular mycorrhizal (VAM) associations. VAM are obligate symbionts that provide usable phosphates to the plant in return for fixed carbon. The legume-*Rhizobium*-VAM; tripartite association is more efficient than the association of any two of these alone (Powell and Bagyaraj, 1984).

Oxygen

An abundant supply of oxygen is necessary for the formation and functioning of nodules, primarily to ensure adequate respiration of the root and nodule tissue, McConnell (1959) determined that nodule formation and nitrogen fixation in *Alnus glutinosa* are considerably reduced at oxygen levels below 0.21 atmospheres. This is consistent with results obtained for soybeans and clover (Bond, 1951a; Ferguson and Bond, 1954).

The oxygen level in soil usually is less than 20% (in part due to high CO_2

concentrations resulting from respiration) and in heavy clay sols can be much less, presenting an important limitation to nodule formation and N fixation. The failure of plants to nodulate in water or in waterlogged soils can be attributed to the lack of oxygen (Lie, 1974). Periods of excessive water supply also can cause a temporary reduction in nitrogen fixation since even a thin layer of water surrounding the nodule will reduce the supply of oxygen (Sprent, 1969). Poorly aerated soils also cannot nitrify ammonium. Under these conditions, the lack of nitrate from nitrification and nitrogen fixation can cause severe nitrogen stress.

Water stress

Soil moisture can be limiting, both when too low or too high. Experiments with alfalfa, soybean and *Acacia mellifera* have shown that the optimum level of soil moisture for n fixation is approximately 60% to 75% of soil water-holding capacity (Fred et al., 1932; Habish, 1970). Moisture contents above or below this level can adversely affect the number of nodules formed and their longevity. Huang et al. (1975) demonstrated that nitrogen fixation can be reduced indirectly by the effects of moisture stress on photosynthesis.

Sprent demonstrated the detrimental effects of water stress on the number of nodules, nodule activity and average nodule weight (Table 6.4) She also showed that acetylene-reducing activity declined by 90% in stressed plants (Sprent, 1976). Under moisture stress conditions, the wilting of lower leaves is usually a good indicator that nodules are operating at suboptimal rates (Sprent, 1972). Under conditions of moisture stress in sandy soils, nodulation in NFSs may occur deeper in the soil (Beadle, 1964).

Table 6.4: Effect of moisture stress on 44-day-old *Phaseolus vulgaris* plants inoculated with two strains of *Rhizobium*.

	***Rhizobium* strain**			
	3601		**3605**	
	Control	**Stressed**	**Control**	**Stressed**
Pmoles C2H4/mg/min	16.45	1.75	37.45	3.15
Nodule number	28.4	8.3	18.50	4.5
Average nodule wt (mg)	1.44	0.95	1.72	1.16
Water content of sand (%DW)	6.54	0.71	6.34	0.81

Source: Sprent, 1976.

Waterlogging also can depress nitrogen fixation, primarily due to oxygen deficiency, although there is large variation among species. The NFS *Alnus glutinosa* growing on permanently waterlogged soils has been found to nodulate primarily in the well-aerated, humid, upper soil horizons, but not on permanently submerged roots (Akkermans and Van Dijk, 1976). In a study of *Seshania sesban* and *S. rostrata*, Ndoye et al. (1990) demonstrated a decrease in the amount of N_2 fixed per plant for *S. sesban*, but not for the stem-nodulated *S. rostrata* when both were grown under waterlogged conditions.

Temperature

The effects of temperature on N fixation in NFSs have not been well studied, but its effects on herbaceous legumes suggest that species differences, plant age and environmental factors influences the degree to which temperature affects N fixation (Gibson, 1976). In the field, temperature effects may be most closely related to development of the host plant. Fyson and Sprent (1982) found that nitrogenase activity began at about the same physiological stage for *Vicia faba* plants grown at two different temperatures.

Different strains of *Rhizobium* also may respond differently to temperature variations. Workign with several tropical grain legumes, Dart et al. (1976) reported little difference in nitrogen fixation between strains at 21°C, but found significant differences at air temperatures of 27°C and 33°C.

Perry et al. (1979) demonstrate the effect of temperature on two non-legume NFSs, *Alnus rubra* and *Cytisus scoparius.* Optimum temperature for nitrogenase activity for *C. scoparius* was approximately 22°C, while the optimum for *A. rubra* was approximately 35°C. At temperatures of 11°C and 21°C, seedlings of the subalpine species *Alnus viridis* fixed significantly more N at the higher temperature (Benecke, 1970). Since growth rates were significantly lower at the lower temperature, however, the total accumulated plant nitrogen as a percentage of total plant growth was greater at the lower temperature. In this case, nitrogen fixation was not the factor limiting plant growth at lower root temperatures.

Age

The age of host NFSs also influences nodulation and nitrogen fixation, but the extent of this influence on NFSs has not been thoroughly studied. Rates of N fixation in *Alnus rubra* from greenhouse and field studies suggest that seedlings can fix as much as 100 kg N/ha during the first year of growth and over 200 kg N/ha/yr in 30-year-old stands (Zavitkovski and Newton, 1968). Sprent (1983) hypothesized a substantial decrease in the amount of N fixed per unit of biomass as trees become older, with an increase in total N fixed per unit area. However, the few studies over time on N fixation in NFSs indicate this may not be entirely correct. In studies of growth and nodulation of *Alnus glutinosa,* Akkermans and van Dijk (1976) concluded that both N fixation and weight of dry nodules per tree increased over time. While this study does not provide biomass data, using reasonable biomass values for 7- and 18-year-old trees (the approximate ages evaluated) and rates on N fixation at those ages, the rates of fixation per unit of biomass do not likely decline.

pH and Soil Nutrients

Low soil pH generally inhibits nitrogen fixation by reducing the development of rhizobia, increasing the numbers of ineffective rhizobia or disrupting the infection process (Sprent and Sprent, 1990; Mengel and Shubert, 1988). Low-pH soils canbe low in phosphorus, calcium, magnesium and molybdenum and may have toxic levels of aluminum and manganese, all of which can affect NFS vigor and health. Plants that depend on nitrogen fixation have been found to be more sensitive to pH than plants of the same species supplied with mineral nitrogen (Andrews, 1976). In general, the critical pH for nodulation of most legumes is above pH 4.5 to 5.5 (Dixon and Wheeler, 1983).

Table 6.5 demonstrated the negative influence of low pH on nodulation, nitrogen fixation and plant growth of several nonleguminous NFSs. Bond (1951b) found that the symbiont of the nonlegume *Myrica gale* was substantially better suited to acid conditions than many legume symbionts, with nodulation occurring at a pH as low as 3.3. In this study, seedlings fertilized with nitrate N showed the best growth at pH 3.3. indicating that poor growth of seedlings at this pH was due to difficulties in nodule formation, not due to the inability of the plant to grow under such strongly acid conditions. Griffiths and McCormick (1984) found nodulation of the nonlegume *Alnus glutinosa* was reduced below pH 5.5.

TABLE 6.5: Influence of age on nitrogen fixed and nodule weight in *Alnus glutinosa.* (Adapted from: Akkermans and van Dijk, 1976)

Species	Nodulation	N fixation and plant growth
Myrica gale	poor at 3.3 good at 5.4-6.3	best at 5.4
Alnus glutinosa	zero at 3.3	best at 4.2-5.4
Casuarinas cunninghamiana	zero to poor at 4-5 good at 6-7	best at 6-7
Hippophae rhamnoides	zero to poor at 4.2 best at 5.4-7.0	best at 5.4-7.0
Shepherdia Canadensis	zero at 5.0 best at 6.0	
Coriaria myrtifolia	zero to poor at 5.0 best at 6-9	best at 7-9

Adapted from: Wheeler and McLaughlin, 1979.

Ahmad and Ng (1981) reported on dramatically reduced leucaena growth with lower pH, total soil nitrogen and concentrations of *Rhizobium.* Given the low concentration of rhizobia present in the soil at lower pH levels, it appears that differences in pH had the greatest influence on plant growth by reducing the number of rhizobia. Umali-Garcia (1988) found that liming of acid soils to reduce the pH to 6.5 improved the performance of the symbiosis in *Acacia mangium.*

In southern Vietnam, *Acacia auriculiformis* grown in acid sulfate soils of pH less than 4.0 nodulated profusely in the undecomposed litter layer (Oi horizon), but not in the highly acid mineral soil (MacDicken, 1991b). This demonstrates the ability of the symbiosis to adapt to strongly acid soils where effective nodulation would not be possible in the mineral soil. A similar example is the stem-nodulating green manure species *Sesbania rostrata.* Stem nodules are less affected by flooding and combined nitrogen than root nodules (Rinaudo et al., 1988).

Nitrogen-fixing plants need the same nutrients as other green plants, but they may need additional quantities of cobalt, iron and mylybdenum to produce leghemoglobin and nitrogenase (Sprent and Spreng, 1990; Becking, 1976). The availability of each of these micronutrients is pH-dependent. Molybdenum availability is lowest in acid soil, while the availability of iron and cobalt is lowest in strongly alkaline soils.

In a pot study of inoculated seedlings of *Acacia dealbata*, the omission of P, K or Ca from the nutrient solution reduced specific nitrogenase activity (Chau et al., 1986). Reinsvold and Pope (1987) found that increased P applications increased nodule number, weight, shoot height, above-ground biomass and acetylene reduction activity in inoculated *Robinia pseudoacacia.*

Combined Soil N

The inhibitory effects of combined soil nitrogen on the nitrogen fixation process have been recognized for over 100 years (Wilson, 1940). Legumes, for example, do not fix enough N to obtain maximum growth response (Harper, 1974) and therefore require soil nitrogen. Yet high levels of combined soil N may reduce nodulation and nodule activity and accelerate senescence (Gibson and Nutman, 1960; Becana and Sprent, 1987). The levels that inhibit N fixation are not necessarily the same for different rhizobial strains (Murphy, 1988), NFS species (George and Singleton, 1989) or clones of the same species (MacKay et al., 1987).

The inhibitory effects of high levels of combined N reported for herbaceous legumes also have been reported for both legume and nonlegume NFSs. Zhang et al. (1990) found a marked decrease in the nodulation rate of *Casuarina cunninghamiana* seedlings as soil N increased. Soil nitrogen in excess of 10 ppm had the greatest effect in reducing the nodulation rate. Zavitkovski and Newton (1968) estimated the relative proportions of fixed and combined soil N at different levels of total soil N with *Alnus rubra*. Reductions of 35% to 45% in number of weight of *Gliricidia sepium* nodules studies (Umali-Garcia, 1990). In the same study, *Acacia auriculiformis* and *Paraserianthes falcataria* showed no negative response to those levels of nitrogen application, but they had virtually no nodulation at 300 kg N/ha.

The response of NFSs, particularly young seedlings, to combined soil N is critical to nursery management of nitrogen fixation. For some species, starter N in nursery soils is detrimental to N fixation and growth. In the case of *Alnus viridis*, seedlings grown under soil N levels of 0, 5, 25, 50 and 100 ppm exhibited maximum growth with no nitrate N added (Benecke, 1970). Nodulated seedlings with no N performed better than unnodulated plans at any level of added nitrogen. However, Hill (1970) found that *Leucaena leucocephala* harvested 10 weeks after planting gave a positive response to applications of 33 kg N/ha and 66 kg N/ha. Plants in both the fertilized and unfertilized plots were well nodulated with pink nodules. The response of seedlings to combined soil N is likely dependent on how rapidly the seedlings nodulate – the longer the period before N fixation begins, the greater the response to applied N or higher levels of combined soil nitrogen.

There also may be differences in response to combined N among clones or varieties of the same species. MacKay et al. (1987) demonstrated a difference in relative performance of two *Alnus glutinoa* clones propagated through in vitro methods. Nitrogen treatments containing from 0.10 mg N/g to 0.68 mg N/g in dry substrate stimulated N fixation and plantlet growth, but one clone (AG-8) was more dependant upon combined N than the other (AG-4). Varietal or clonal differences may be important in determining the amount of starter N to supply under nursery conditions.

Salinity

Saline conditions can depress N fixation through osmotic withdrawal of moisture from nodules, reductions in bacterial colonization of roots and root hair curling. This depression can be due to a lack of adaptation to saline conditions by either the tree or symbiont. In fact, there is tremendous variation in the response of trees and *Rhizobium* to saline soils. NFS species such as *Prosopis tamarugo* survive and fix nitrogen in extremely saline soils where other NFSs or symbiont strains would not survive.

There are likely to be important interactions between tree and symbiont. For example, Sprent and Zahran (1988) found a host cultivar x rhizobial strain interaction in response to salinity in *Vicia faba.* Balla et al. (1990) demonstrated the suitability of salt-tolerant *Rhizobium* strains in an experiment using 57 strains of *Rhizobium* and several NFS species. Salt-tolerant strains showed better survival, nodulation and nitrogen fixation than salt-sensitive strains under saline conditions. A strain from each of four NFSs, *Acacia nilotica, Dalbergia sissoo, Leucaena leucocephala* and *Prosopis juliflora,* was tested, and all were found to be highly tolerant to saline conditions (9-12 dS/ m).

Lack of Suitable Bacteria

The presence of suitable rhizobia or *Frankia* is essential to nitrogen fixation without them symbiotic biological nitrogen fixation cannot occur. In fields that have not been successfully planted to a given NFS, it is likely that the introduction of suitable bacterial through inoculation will be beneficial. It is important to note that to establish an *effective* symbiosis between NFSs and bacteria, the bacteria must be able to survive in the soil environment of the planting site and be able to fix nitrogen under those conditions. The bacteria must be able to compete with other microbes for nutrients and with other rhizobia or *Frankia* for infection sites. Some species of NFSs are very specific in their requirements for rhizobial strains, for example, *Acacia mangium;* others are nodulated by several strains of rhizobia, for example, *Albizia lebbeck* (Turk and Keyser, 1992).

The genus *Rhizobium* includes fast-growing rhizobia that produce acid on yeast mannitol agar and are most often of temperate origin (Wheeler et al., 1991). The three species in this genus are: *R. leguminosarum, R. meliloti* and *R. loti.* The genus *Bradyrhizobium* contains slow-growing bacteria that do not produce acid on yeast mannitol agar. Bacteria of this genus commonly infect tropical legumes, although numerous exceptions make this a loose generalization.

Competition between strains of rhizobia can result in reduced or delayed nodulation (Lie *et al.*, 1988). The presence of both nodulating and nonnodulating rhizobial strains in the rooting zone can result in infection sites being taken by nonnodulating bacteria, preventing or reducing the establishment of an effective symbiosis.

The persistence of introduced *Rhizobium* strains also can affect the amount of nitrogen fixed. In a study of labeled *Rhizobium* and *L. leucocephala*, South (1982) found that at the time of transplanting 46% to 97% of nodules contained the labeled *Rhizobium*, while 10 weeks later 4% or less contained the labeled *Rhizobium*. In this case, the indigenous *Rhizobium* was a better competitor than the inoculated strain for infection sites on the roots.

Sanginga et al. (1989) found substantial differences in the amount of N fixed by *L. leucocephala* inoculated with two different strains of *Rhizobium.* Strain IRc 1050 fixed 98 kg/ha in 6 months while strain IRc 1045 fixed 134 kg/ha (estimated by the 15N dilution method). Clearly, the type of bacteria present can directly affect the amount of nitrogen fixed by NFSs. There is also some evidence, however, that ineffective symbiont strains can enhance plant growth when they are present with effective strains. Hahn et al. (1990) found that *Alnus glutinosa* seedlings inoculated simultaneously with effective and ineffective *Frankia* strains had greater total dry weight and height than those without the dual inoculation.

RECOVERING FROM STRESS

Symbiotic nitrogen fixation can be affected by a wide range of environmental stressed, but it is clear that many of these effects are both extremely complex and temporary. The direct cause of reduction in N fixation appears to be a reduction in the supply of photosynthate to the nodules. Resumption of higher N fixation levels is dependent upon the ability of the plant to resume production and transport of photosynthate. Gibson (1976) found that the symbiotic system can make a rapid recovery from various stressed such as temperature, light, defoliation and high levels of combined nitrogen. But Wheeler and McLaughlin (1979) found that nitrogen fixation, at least in *Alnus glutinosa,* takes longer to recover from stress than other host plant physiological processes. Numerous examples of N fixation resuming after the removal of stress can be cited for temperature (Rogers, 1969; Pate, 1958), moisture stress (Worrall and Roughley, 1976), defoliation (Moustafa et al., 1969) and salinity (Wilson, 1970)

NITROGEN FIXATION RESEARCH IN INDIA

The word on biochemistry of nitrogen fixation has mainly been done at Haryana Agriculture University, Indian Agriculture Research Institute, Punjab Agriculture University and BARC. At HAU, a pathway for ureide biogenesis in nodules of pigeonpea has been proposed. The further metabolism of ureides in aerial parts particularly leaves and developing pods has also been worked out. Detailed biochemical and physiological studies explaining nodule senescence have also been carried out at IARI. Attempts have been made to understand the regulation of uptake hydrogenase so as to reduce the energy cost of biological nitrogen fixation. In addition, studies are concentrated on increasing nitrogen fixing efficiency by insertion of uptake hydrogenase genes or by increasing nitrogenase activity in nodules. At BARC, early interactions in legume- *Rhizobium* symbiosis have been investigated, according to which, host-induced alterations of capsular polysaccharides and de novo synthesis of specific proteins by both the partners in response to each other, appear to be among the first steps in the series of signal response interactions. At PAU, the role of extracellular invertase of *Rhizobium* in free sugar metabolism in developing root nodules of legumes has been ascertained (Adopted from Dr. Randhir Singh, In: Souvenir, Int. Cong. Plant Physiology, Feb. 15-20, 1988, New Delhi).

The activity of nodule is also controlled by nitrogenase enzyme. This enzyme is about 75% efficient. This is because, when soil aeration is poor and nitrogen availability is limited, electrons instead of being used to reduce nitrogen, now combine with protons forming hydrogen. As a result, nodules evolve hydrogen instead of fixing nitrogen. This is a wasteful process. Sufficient progress is being made to identify strains of *Rhizobia* which evolve less hydrogen and fix more nitrogen.

The most ideal situation would be the one in which the higher plant is able to fix nitrogen without the help of the bacteria. Towards this end, attempts have been made to transfer nitrogen fixing genes (*nif* -genes) to higher plants from *Rhizobium*. However, it appears to be a long way before it is achieved.

In India, intensive forest cultivation of nitrogen fixing trees, with consequent short rotation is needed. This will require replacement of nitrogen reserves at a greater rate than the present secondary succession leading to old-growth forests. With the increased use of trees, genetically selected, it will be necessary to maximize the potential for biological nitrogen fixation in future forest management, so as to maintain the long term productivity of forest lands without significantly accelerating the rate of nutrient depletion. India is in urgent need of additional professional training and better transfer of technology in this context. Further, improved strains and new species of nitrogen-fixing microorganisms should be investigated/developed in order to reduce the use of chemical fertilizers in the fields.

CHAPTER 7

Biological Mobilization of Phosphorus

Phosphorus, the master key element is known to be involved in a plethora of functions in the plant growth and metabolism. The cellular machinery is difficult to be imagined without phosphorus being involved in its metabolic continuity and even perpetuation. Such key functions include cell division and development, photosynthesis, breakdown of sugar, energy transfer, nutrient transfer within the plant and expression. Phosphorus nutrition benefits the plants by producing deeper and abundant roots. So the supply of this element to plant is essential for achieving optimum crop yield. It is supplied through phosphatic fertilizers, animal manures, plant take up their phospho-organic as the primary orthophosphate ion, $H_2PO_4^-$. The secondary orthophosphate ions, HPO_4^{2-} is also believed to be absorbed by plant roots in small quantities.

The P content in average soil is about 0.05% (w/w). The variation being largely due to differences in weathering intensity and parent material composition. Relatively high concentrations are often observed in calcareous soils of arid region whereas relatively low P concentrations are often observed in soil subjected to high weathering intensity. But only 0.1% of the total P is available to plants because of its low solubility and its fixation in soil. Thus the soil is rich in phosphorus which needs to be harnessed. Under these circumstances, microorganisms offers of biological system having evolved the mineral phosphate solubilizing trait(s) to acquire sufficient phosphorus.

FORMS OF P AVAILABLE IN SOIL

In soil P exists chiefly as orthophosphate although phosphine and phosphonates have been detected under some conditions. The diverse soil P forms can generally be categorized as soil solution P, insoluble inorganic P or insoluble organic P. Only a very small fraction of soil P exist in soil solution because of its extreme reactivity. Inorganic P in solution exists mainly as primary or secondary orthophosphate depending on soil pH. Organic P may also make up a large fraction of soluble P as much as 50% in soils within higher organic matter content. The main groups of organic compounds in soil include phospholipid, nucleic acids and phytin. It occurs in humified organic matter, organic residues and microbial biomass that accounts for 2-80% of the P content of surface soil. Inorganic P associated with the solid phase is sorbed to the surfaces of soil constituents or occur in calcium, iron, aluminium and silicate minerals. Calcium phosphate tends to accumulate in neutral to alkaline pH whereas FeP and AIP tend to accumulate in acidic soil.

P AVAILABILITY ISSUE

In agricultural systems, P fertilizers are routinely applied to promote crop yields. The P in these fertilizers is initially available to the plant but it rapidly reacts with soil and becomes progressively less available for plant uptake (as often as much as 90%). This is known as *chemical fixation* of phosphorus. Hence the current trend throughout the world is to explore the possibility of using alternate nutrient sources for increasing the efficiency of chemical fertilizers. Since phosphorus availability from the phosphatic reserves i.e. rock phosphate under neutral and alkaline soil condition is very low or negligible, the phosphate solubilizing microorganisms dissolving interlocked phosphates appear to have an important implication in Indian Agriculture.

Besides fertilization, the availability of P can be achieved by two pathway: (a) the enzymatic decomposition of organic P compound and inorganic P compound, (b) the non-enzymatic solubilization of different rock phosphates and inorganic phosphorus sources.

Many fungi, bacteria, actinomycetes and cyanobacteria are potential solubilizers of bound phosphates in soil. A list of P solubilizers are given in Table 7.1.

Table 7.1: List of some important phosphate solubilizing fungi and cyanobacteria

FUNGI	
Alernaria sp.	*Humicola inoslens*
Aspergillus sp.	*H. lanuginose*
A. awamori	*Mucor meihii*
A. canidus	*Myrothecium roridum*
A. carbonum	*Oidiodendron* sp.
A. flavus	*Paecilomyces fusisporus*
A. foetidus	*Prnivillium* sp.
A. fumigatus	*P. balaji*
A. nidulans	*P. digitatum*
A. niger	*P. funiculosum*
A. terreus	*Rhizoctonia solani*
A. wentii	*Rhizopus* sp.
Cephalosporium sp.	*Sclerotium rolfsii*
Chaetomium sp.	*Trihoderma* sp.
C. globosum	*Trichurus spiralis*
Cladosoprium sp.	*Torula thermophila*
C. sphaerospermum	CYANOBACTERIA
Curvularia sp.	*Anabaena* sp.
C. lunata	*Calothrix braunii*
Cyathusafricanus	*Nostoc* sp.

Cylindrocarpon otusisporum	*Scytonema* sp.
Fusarium sp.	*Tolpothrix ceylonica*
Helminthosporium sp.	

Phosphate solubilizing microorganisms are found in all soils but their number vary with soil climate as well as history. The potentially of different PSMs vary with different P-sources. The solubilization is not restricted to calcium salts, Fe, Al, Mg and Mn and other phosphates are also acted upon. However, calcium phosphate dissolving microorganisms are found in large number as compared to other phosphates. PSMs have been isolated from the rhizosphere of various crops, legume root nodules and compos. Fungi are reported to be the efficient P-solubilizers as compared to bacteria, actinomycetes and cyanobacteria.

Mode of Action of PSMs

Primarily there are two schools following of thought interpreting the mechanism of P-solubilization by microorganisms.

1. Solubilization by production of organic acid.
2. Solubilization by action of phosphatase enzymes.

PSM include various bacterial, fungal and actinomycetes forms which help to convert insoluble/phosphate into simple and soluble forms. Species of *Pseudomonas, Micrococcus, Bacillus, Flavobacterium, Penicillum, Fusarium, Sclerotium* and *Aspergillus* are some of the PSM. Among these, soil bacteria belonging to the genera *Pseudomonas* and *Bacillus* and fungi belonging to the genera *Penicillium* and *Aspergillus* are more common. The solubilization effect of PSM is generally due to the production of organic acids by these organisms. They are also known to produce amino acids, vitamins, growth promoting substances like Indole Acetic Acid (IAA) and Gibberalic Acid (GA) which help in better growth of plants.

ORGANIC ACIDS AND P SOLUBILIZATION

The major microbiological means by which insoluble phosphorus compounds are mobilized is by the production of organic acid which is accompanied by the acidification of the medium. The organic and inorganic acids convert tricalcium phosphates to the di- and monobasic phosphates with the net result of an enhanced availability of the element to the plant. The type of organic acid produced and their amounts differ with different microorganism. Tri- and di-carboxylic acids are more effective as compared to mono basic and aromatic acids. Aliphatic acids are also found to be more effective in P solubilization than phenolic acids and citric acid and fumaric acid has highest P-solubilizing ability. The analysis of culture filtrates of PSMs has shown the presence of a number of organic acids including citric fumaric, malic, lactic, 2-ketogluconic, gluconic, glyoxylic and α-ketobutyric.

The extent of P-solubilization also depends on the accessory minerals present in rock phosphate. The solubilized phosphate may react with Ca and Mg present in rock phosphate as soon as pH of the growth medium increases. Presence of free carbonates in rock phosphates also reduces the extent of solubilization as a large part of organic acids is directed towards neutralization of free carbonates.

Enzymes and P-solubilization

The liberation of P from organic phosphate compounds is mainly due to the action of enzymes of esterase type. PSM along with acid production, produces the phosphatase enzyme which cause the solubilization of P in aquatic environment. Besides these two mechanisms the production of chelating substances, H_2S, CO_2, mineral acids, siderophores and cation exchange mechanisms are also involved in P solubilization by PSMs. In addition, the action of PSMs is not only due to the release of plant available phosphorus but also due to the production of biologically active substances like indole acetic acid, gibberellins, cytokinins production is also correlated with P solubilization.

EFFECT OF PSMS ON GROWTH AND P ECONOMY

Efficient and economic use of P-fertilizers could be achieved by using phosphate solubilizing microorganisms in legumes, cereals and other useful crops. P uptake and P content was augmented by the application of PSM in many leguminous plants In wheat an increase in dry matter production and P-uptake from 10-27% and 15-34% respectively observed as a result of inoculation with *P. bilaji*. Superiority of PSM to uninoculation in respect of nodulation, P-uptake, pod yield, oil yield and net realization in respect of nodulation, P-uptake, pod yield, oil yield and net realization in groundnut production was also observed. Seed inoculation with PSM noted the beneficial effect on the yield of crops as was the saving by the use of 75:25 ratio of MRP:SSP. In addition application of P solubilizers causes a replacement of 25% of phosphate fertilizers. Increased yield was observed in cereals, legumes, potatoes and other field crops on the addition of RP and inoculation with PSM. The yield of soybean in sandy loam alluvial soil was significantly increased by 2.4 q/ha due to rock phosphate and *Ps. Striata* whereas as with 80 Kg P_2O_5 of SSP, increase was only 1.0 q/ha. A synergistic effect of PSB was observed at low P rates whereas at higher P rates the effect was antagonistic. Seed inoculation along with 30 kg P_2O_5 gave similar dry matter yield as 60 kg P_2O_5 with and without seed inoculation. Similarly, increased seed yield was observed with 10 kg P_2O_5 and also with seed inoculation. The release of P from rock phosphate by PSM resulted in higher P uptake and dry matter yield of maize.

DUEL INOCULATION OF P-SOLUBILIZERS AND NITROGEN FIXES

Biological nitrogen fixation depends appreciably on the available form of phosphorus. So the combined inoculation of nitrogen fixers and phosphate solubilizing microorganism (PSMs) may benefit the plant better (by providing both nitrogen as well as phosphorus), than either group of organisms alone. Observed beneficial effect of phosphate solubilizers alone or in combination with *A. chroococcum* on cotton in field experiments. A positive responses of combined inoculation with phosphobacteria and *A chroococcum* on the yield and nutrient uptake of different crops was reported. Dually inoculated plants with *Azotobacter* and PSM showed increased nodulation, growth dry weight yield and P and N uptake. In addition, inoculation with bacterial mixers provided a more balanced nutrition and improvement in root uptake N and P.

Stimulatory effect of combined inoculation of *Rhizobium* and PSB with simultaneous application of RP and also indicated a possibility of saving 10 kg and replacing superphosphate entirely by rock phosphate and PSB inoculation. Combined inoculation of *Rhizobium* and PSM showed increased dry matter content, grain yield, nitrogen and phosphorus uptake significantly over uninoculated control in different legume crops. The inoculation effects were more pronounced in the presence of chemical fertilizers.

Isolation

PSM can be isolated from rhizosphere and non-rhizosphere soil and also from other environments such as rock phosphate deposits and marine environments. PSM vary in their efficiency in solubilizing insoluble phosphates. The initial isolation of phosphate solubilizers is made by using a medium suspended with insoluble phosphates such as tri-calcium phosphate. The production of clearing zones around the colonies of the organism is an indication of the presence of P-solubilising organisms. Such cultures are isolated, identified and the extent of solubilization determined quantitatively. It has been observed that the capacity of bacteria isolated to solubilise phosphate depends on the zone of their origin. Those derived from rhizoplane have the highest, the rhizosphere one has intermediate and non-rhizosphere has the least capacity.

Several rock phosphate dissolving bacteria, fungi, yeast and actinomycetes were isolated from soil samples collected from rock phosphate deposits and rhizosphere soils of different leguminous crops. Many isolates solubilize a very high quantity of rock phosphate. The most efficient bacterial isolates were identified as *Pseudomonsa striata, Pseudomonas rathonis* and *Bacillus polymyxa* and fungal isolates as *Aspergillus awamori, Penicillium digitatum, Aspergillus niger* and *Schwanniomyces occidentalis.* These efficient micro organisms have shown consistently their capability to solubilize chemically-fixed soil phosphorus and rock phosphate from different sources – Mussorie, Udaipur, Matoon, Singhbhum, Morocco, Gafsa and Jordan. In addition, these microorganisms were found to mineralize organic phosphorus to soluble form due to enzymatic activity. By developing improved techniques for isolation of rock phosphate solubilzing micro organisms and by systematic investigations new efficient bacteria such as *Pseudomonas striata, Bacillus polymyxa* and fungi like *Aspergillus awamori* have been selected for preparation of carrier-based inoculant in addition to several basic studies. This preparation is known as 'IARI Microphos Culture' which is now being mass produced and made available to farmers. The first phosphatic biofertilizer under the name of 'Phosphobacterin' incorporating

Bacillus megateriumvar. phosphaticum was prepared at the then USSR.

Crop response

Inoculation of seedlings with phosphate solubilizing organisms resulting in increased uptake of P and crop yields was first shown in 1948 by Gerrtsen. Since then beneficial influence of artificial inoculation with phosphate dissolving organisms has been reported for different crops under diverse agrolcimatic conditions. In India, the microphos culture containing *Pseudomonas striata* and *Bacillus polymyxa* was tested in multilocational field trials with wheat, rice, chick pea, soybean, lentil and potato. Results showed that inoculation of seed and seedlings increased grain yields of crops. The use of cultures increased the efficiency of rock phosphate and superphosphate applied to neutral to alkaline soils.

Table 7.2: Effect of microphos inoculant (phosphobacteria) on grain yield of crops in multi-location trials

Crop	Number of trials	Percent increase due to inoculant	
		P_0	RP_{60}
Paddy	7	10-20	5-15
Wheat	8	7-42	5-50
Gram	3	10-30	15-25
Lentil	2	-	13-15
Potato	2	30-50	13-15

P_0 – Without phosphorus; RP_{60} = 60 kg P_2O_5/ha as rock phosphate
Source: Gaur (1985)

The role of phosphate solubilizing organisms in increasing crop yield is subjected to controversy due to inconsistent results. Of the total inoculation experiments conducted to far, a significant increase in yield of 5 to 10% has been observed only in about 25% experiments. Several factors such as carbon and nitrogen sources, rock phosphate concentration, particle size of rock phosphate, pH, temperature, aeration, incubation period, etc. affect the efficiency of phosphate solubilizers. In general, the response to phosphobacterins was found to be positive only in soils with high organic matter content and low in available phosphorus.

VESICULAR-ARBUSCULAR MYCORRHIZAE (VAM)

The role of fungi particularly those selectively colonizing the root surface of growing plant roots in increasing the availability of phosphorus has received scanty attention. These organisms are known not only to mobilize phosphates but also perform the function of transferring phosphorus from soil into roots as observed in the case of typical ectotrophic mycorrhizae. The term mycorrhizae refers to the association of fungi with rots of higher plants. Mycorrhizae are indeed the chief organisms involved in nutrient uptake of most land plants. VAM occur in most plant species in the two families of major economic importance, the Leguminoseae and Graminae. They are virtually ubiquitous, being present in tropical, temperate and arctic regions.

Their impact in tropical agriculture will be greater because in the tropics phosphate-deficient and phosphate-fixing soils are widespread, phosphatic fertilizers are in short supply and the temperature is more congenial for their activity.

VAM fungi inoculated to crop plants colonize the plant root system and increase the growth and yield of crop plants. The improved plant growth is due to increased nutrient uptake particularly P, Zn and other micronutrients, production of growth promoting substances, tolerance to drought and salinity and resistance to plant pathogens.

Trials conducted with the locally isolated *Glomus macrocarpum* on chilli showed the possibility of saving 50% phosphatic fertilizer without affecting the yield levels. Increase in the nutrient uptake and yield of various crops in India and elsewhere has been common with VAM inoculation. Table 7.3 shows the effect of mycorrlizal inoculation on some crop yields. However, it has not been possible to exploit this organism on commercial scale due to difficulties in culturing the organism on laboratory media.

Table 7.3: Effect of mycorrhizal (VAM) inoculation on crop yields

Crop	P-fertilizer supplementation (kg P_2O_5/ha)	% yield increae over control
Finger millet	19	18
Soybean	25-50	19
Chilli	37.5	55
Chickpea	40	25
Groundnut	40	10-20

Source: Venkataraman and Tilak (1990)

The term *mycorrhizae* comes from the Greek *myco* or fungus and *rhiza* or root and literally means *fungusroot*. Mycorrhizae have been described as perhaps the ultimate in reciprocal parasitism (symbosis) wherein the fungus supplies the higher plant partner with mineral nutrients and plant growth hormones and protects the roots against pathogens, and the higher plant supplies the fungus with energy substrates. Almost all higher plants are dependent upon one of the mycorrhizal associations.

Mycorrhizae are highly evolved, mutualistic symbioses or associations between soil fungi and plant roots. The partners in this association are members of the fungus kingdom (*Basidiomycetes*, *Ascomycetes* and *Zygomycetes*) and most vascular plants. Mycorrhizal plants have been found in every continent and in every major vegetation type. Depending on the plant and fungal species involved as well as distinct morphological patterns, at least seven different types of mycorrhizal associations have been recognised. They are: vesicular-arbuscular mycorrhiza (VAM); ectomycorrhiza (ECM); orchid mycorrhiza; ericoid mycorrhiza; ectendomycorrhiza; arbutoid mycorrhiza and monotropoid mycorrhiza. In addition, dual associations of both ECM and VAM have been found in some trees and shrubs such as *Acacia*, *Casuarina*, *Eucalyptus* etc.

ROOTS AS SINKS AND SOURCES OF NUTRIENTS AND CARBON IN AGRICULTURAL SYSTEMS

Roots are a special category of *soil biota*, being part of organisms that live in two very different types of environment, aboveground and belowground. As they belong to the organisms that form the main target of all agricultural interventions and are directly involved in the efficiency with which nutrients and water are used for crop production, root knowledge should be *at the root* of any discussion on sustainable agriculture. Yet, little direct attention is given to roots, possible because of a lack of:

1. Clear concepts on what to look for
2. Easy (be it quick and dirty) methods to observe properties in actual root systems.

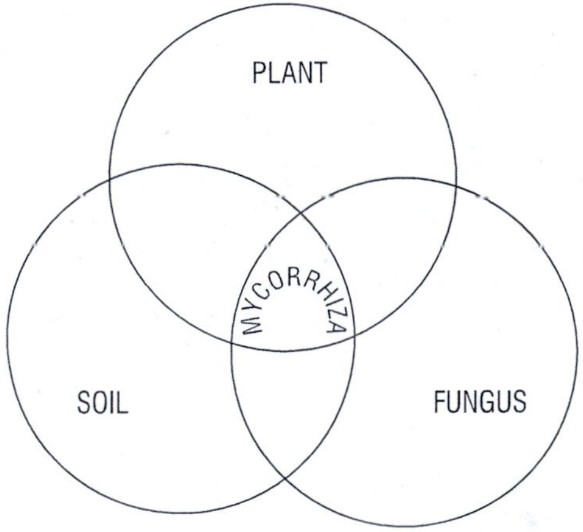

Fig. 7.1: Interrelationship between mycorrhiza, soil and plant.

The need for minimizing resource use efficiency and minimizing environmental pollution has given a new drive for root ecological research. Within the soil, roots are the main sinks for nutrients during the life of the crop, but may also play a role in immobilizing nutrients during the initial stages of their decomposition; subsequently, they will form a source of nutrients for soil organisms and future plants. The efficiency of roots as sinks for nutrients, both during and after their life, partly determines the conflict between environmental and production aims of agriculture.

For a long time the prevailing concept was *the more extensive the root system, the higher crop production*. We know that this is not true across the full range of agricultural production conditions; maximum plant production can be obtained with relatively small root systems if the daily water and nutrient requirements are met by technical means as in intensive horticultural production systems. Roots are, however, directly involved in the efficiency of plants to use available water and nutrient reserves in the soil, and therefore in reducing negative side effects of agricultural production by leaching and losses to the atmosphere. As a first estimate, we may still expect that *the more extensive the root system, the higher may be the nutrient and water use efficiency.* The possibility of obtaining greater resource use efficiency can only be realized if total supply of nutrients and water is regulated according to the crop demands and the attainable resource use efficiency. On a field scale, both resource supply and possible crop production show spatial variability. Inadequate techniques for dealing with this variation may reduce the resource use efficiency well below what is possible in the normally small experimental units considered for research.

If the long-term organic matter balance is secured, system outputs (harvested products, leaching, and gaseous losses) must be balanced by inputs. In the northern temperate zone the main crop growing season normally has a rainfall deficit, and leaching (beyond the crop root zone) is mostly confined to autumn and winter. Denitrification may occur in summer, but in early autumn conditions winter. Denitrification may occur in summer, but in early autumn conditions are especially conducive for this process as relatively high temperatures are combined with wet soil conditions and ample availability of organic substrates.

Rates of net N mineralization are less than required by crop demand during the growing seasons, but continue after the uptake period. Fertilizer requirements for avoidance of any N shortage in the crop have to allow for the weather-induced variability in crop demand and in rates of net N mineralization and also for the spatial variability within a field treated as a single management unit. Possibilities for *response farming* are limited, however, as late fertilizer applications may not reach the roots under dry conditions.

The relevance of various root parameters for predicting uptake efficiency depends not only on the resource studied but also on the complexity of the agricultural system. In intensive horticulture, with nearly complete technical control over nutrient and water suply, fairly small root systems may allow very high crop production in a situation where resource use efficiency ranges from very low to very high. It depends on the technical control over nutrient and water supply. Fairly small root systems may allow very high crop productions in a situation where resource use efficiency ranges from very low to very high, depending on the technical perfection of the often soil less (hydroponic) production system. In field crops grown as a monoculture, the technical possibilities are far lower for ensuring a supply of water and nutrients where and when needed by the crop, and it is difficult to achieve maximum crop production rates at high efficiency. The soil has to act as a buffer, temporarily storing these resources. Root systems are important in obtaining these resources as and when available and needed. Adjustment of supply and demand in both time and space (synchrony and synlocation) become critical factors. Critical values of root length density (root length per unit volume of soil) needed to obtain a specified water or nutrient use efficiency can be estimated from existing models. Normally, the rate-limiting step is formed by the transport from the soil matrix to the root surface and therefore by the geometry of the root-soil system and the required transport distances. In mixed cropping systems (including grasslands), the belowground interactions between the various plant species add a level of complexity to the system. On the one hand, mixed cropping system allow complementary in use of the space and thus of the stored resources, hence, improving overall resource efficiency. On the other hand, it means that root length densities that would be sufficient for efficient resource use in a monoculture may not be sufficient in a competitive situation. The plant mixtures found in agro-forestry system increase the complexity by another dimension, as the perennial and annual components have separate time frames on which to evaluate the interactions.

The soil water balance, as affected by climate, irrigation, and drainage, has a major influence on the required root functions in uptake. As the main crop-growing season normally has a rainfall deficit in the northern temperate zone, dry soil conditions hamper diffusive transport and thus increase the root length density required for uptake. A lack of synchrony between N mineralization and N demand, which would lead to a build up to mineral N in the topsoil, is not a real problem under these conditions, as the main leaching risk occurs after the growing season. The main problem for effective use of organic N pools in the temperate zone is that mineralization is too slow in spring. In the humid tropics, by contrast, with a net rainfall surplus during most of the growing season, any accumulation of mineral N will be leached rapidly from the topsoil to deeper layers. Under such conditions synchrony of N mineralization and N demand is essential for obtaining high N use efficiencies and reducing leaching.

Roots are a poorly quantified source of carbon in the soil ecosystem, and widely different estimates are available in the literature, at least partly due to uncertainties and biases caused by the methods used. Fine root residues of annual crops at harvest time tend to be low in nutrient content, as annual crops generally remobilize nutrients toward the end of their growth cycle, redistributing them to the seeds or vegetative storage tissue. For perennial crops with a less pronounced phenology, root decay may be a continuous process during the growing season. During their decomposition by microorganisms, roots may form a temporary sink of N. Little is known on the timing of net N mineralization from root residues and its role in the synchrony between net N mineralization and demand by the following crop.

Importance of Mycorrhiza

In the sand soil ecosystems of India, pioneer plants are subject to deficiency in major nutrients (N, P, and K) and organic matter, wide fluctuation in soil moisture and temperature. It is widely accepted that arbuscular mycorrhizae are important for the establishment, growth and survival of seedlings, particularly in marginal habitats, where the symbiosis improves stress tolerance as well as conserves soil structures. A lot of work done in the past few decades has enabled these fungi to emerge as a potential biofertilizer; a cheap and environment friendly alternative to expensive, petroleum based chemical fertilizers. This aspect especially gains significance for a developing country like India where judicious and large-scale utilization of the technology can prove very useful for getting maximum and long-term gains in various wasteland reclamation, reforestation and aforestation programmes. In addition it gives a much needed thrust in the production of important agricultural crops, on which the economy of the country is dependent.

Mycorrhizal associations are believed to help plants grow. The potential for manipulating mycorrhizal associations to increase productivity in plantation forestry, or plant establishment during ecosystem recovery after severe disturbance, are the focus of major research initiatives. There is also much interest in their potential utilisation in agriculture and horticulture. However, it could be argued that we do not know enough about the role of mycorrhizal associations in natural, disturbed, or managed ecosystems to evaluate their potential for applied use. Suggested roles of mycorrhizal fungi in natural and managed ecosystems

Benefits to Plants

1. Increased plant nutrient supply by extending the volume of soil accessible to plants.
2. Increased plant nutrient supply by acquiring nutrient forms that would not normally be available to plants.
3. Some ECM and ericoid fungi have the capacity to breakdown phenolic compounds in soils, which can interfere with nutrient uptake.
4. Root colonization by ECM and VAM fungi can provide protection from parasitic fungiand nematodes
5. Non-nutritional benefits to plants due to changes in water relations, phytohormone levels, carbon assimilation, etc. have been reported, but are difficult to interpret.
6. Mycorrhizal benefits can include grater yield, nutrient accumulation, and/or reproductive success
7. Mycorrhizas can cause growth form changes to root architecture, vascular tissue, etc.
8. Suppression of competing non-host plans, by mycorrhizal fungi has been observed.
9. Significant amounts of carbon transfer through ECM fungus mycelia connecting different plant species has been measured This could reduce competition between plants and contribute to the stability and diversity of ecosystems.
10. Networks of hyphae supported by dominant trees may help seedlings become established or contribute to the growth of shaded understorey plants.
11. Nutrient transfer from dead to living plants can occur.

Other Roles in Ecosystems

1. Soil hyphae are likely to have an important role in nutrient cycling by helping to prevent losses from the system, especially at times when roots are inactive.
2. Hyphae are conduits that may transport carbon from plant roots to other soil organisms involved in nutrient cycling processes – cooperating with

other members of the decomposition soil food-web

3. Soil hyphae may have an important role in nutrient cycling by acquiring nutrients from saprophytic fungi
4. Epigeous and hypogeous sporocarps of ECM and VAM fungi are important food sources for placental and marsupial mammals
5. Mycorrhizal roots and fungus fruit bodies are important as food sources and habitats for invertebrates
6. Mycorrhizal fungus hyphae are an important food source for soil invertebrates
7. Mycorrhizas influence soil microbial populations and exudates in the mycorrhizosphere and hyphosphere
8. Hyphae of VAM fungi are considered to contribute to soil structure. Their role in mechanical aggregation has been questioned, but secretions such as glomalin may be more important. Hyphal mats produced by ECM fungi considerably alter sol structure
9. Mycorrhizal fungi contribute to carbon storage in soil by altering the quality and quantity of soil organic matter.

Values of People

1. ECM fungi are economically and nutritionally important s human food resources.
2. These mushrooms have also have been used as medicines and natural dyes.
3. Fungi have aesthetic values and are an important part of the as culture, folklore and appreciation of nature by many people.
4. Fungal diversity is a bio-indicator of environmental quality.
5. Fungi, which have adapted to local soil conditions are required for agriculture, horticulture and forestry.

MYCORRHIZAL INTERACTIONS WITH PLANTS AND SOIL ORGANISMS IN SUSTAINABLE AGROECOSYSTEM

Although farming has been affected by technological development, it remains basically an ecological enterprise. It is an activity in which natural ecosystems, open to the influence of climate, substrate, and wild biota, are modified to increase yields of desired food and fiber products. The greater the changes in the basic patterns of structure and function that prevail in the natural system, the greater is the human effort necessary to maintain the agricultural system. Therefore, at the present time, it has been proved that conventional agriculture products ecological disturbance and lack of sustainability, resulting in a reduction of soil fertility and increased damage by pathogens to cultivated plants. In addition, it has been observed that some traditional agricultural systems have higher sustainability and produce less ecological damage. These systems have been called **low external-input agricultural** (LEIA) systems. Studies related with mycorrhizal research in LEIA agro-ecosystems have been scarce. In sustainable agro-ecosystems, mycorrhizal incidence is low and other kinds of microorganisms are abundant and play important role in the sustainability of the systems. So far, there has been insufficient investigation of interactions between AM and soil organisms in LEIA systems.

As opposed to the conventional systems, LEIA agro-ecosystems have high genetic and cultural diversity, multiple uses of resources, and efficient nutrient and material recycling. The search for strategies to improve yields and to maintain these increases is a great challenge to human population at the present time. The role of biological alternatives, because of the intrinsic nature of farming, is of key importance to the search for reduced use of fertilizers, pesticides, and other chemicals. Among these alternatives, mycorrhiza management is particularly important because it strongly influences the plant nutrition processes and the soil stabilization.

It the physiological and ecological importance of mycorrhiza in natural ecosystems and its beneficial effects on cultivated plants have been widely documented. It is well known that the characteristic dominant plants of each major terrestrial community associate mutualistically with soil fungi to form typical kinds of mycorrhiza. Therefore, cricaceous plants, which form the major component of the heathland biome, form with ascomycetous fungi distinctive *ericoid* mycorhiza. In a similar way, the dominant trees of the boreal and temperate forest biomes associate primarily with basidiomycetous fungi to form ectomycorrhiza, and the natural grasslands and most of the tropical rain forest species of the world form arbuscular mycorhiza in association with fungi of zygomycetous affinitie. As natural ecosystem plants, most of the cultivated plants tend to form mycorrhizal associations. AM is the most widely distributed and colonizes most species of agricultural crops. The objective of this contribution is to discuss the importance of the mycorrhizal fungi and their interactions with plant management and other organisms in in LEIA agroecosystems.

SYMBIOSIS

The term *symbiosis* was first used by Frank in 1877 to describe the regular coexistence of different organisms such as fungi and algaë in lichens. It was used as a neutral term that did not imply parasitism. A decade later De Bary (1887) used symbiosis to include parasite as well but, the meaning of the terms *symbiosis* and *parasite* has been changed later on. Symbiosis was used more and more for mutually beneficial associations between dissimilar organisms, and parasite and parasitism came to be almost synonymous with pathogen and pathogenesis.

In recent years, symbiosis is defined as *the living together of differently named organisms*. This definition includes all associations ranging from *mutualistic*, in which all organisms involved are believed to derive benefit; to *parasitic*, in which one organism benefits to the disadvantage of other member of the association. A more precise definition of mutualism is that associations are *mutualistic* if the fitness (offspring produced) of the associating organisms is greater than they are living apart. In a symbiosis, the organism with the large size is the host and the smaller is the symbiont. The symbiont can either be external to the host (ectosymbiotic) or within it (endosymbiotic). The symbiosis is *obligate* for an organism, which is unable to survive and reproduce in the absence of its living partner, and *facultative* if it is able to do so even if the partner does during the association.

In a dual association, symbiosis is wide-spread especially between higher plants and bacteria in N-fixing symbiosis; between higher plants and fungi in most mycorrhizae; between algae and fungi in lichens; between algae and coelenterate in corals and so on. However, it is not currently known whether symbiosis occurs between higher plants. In addition, the tripartite symbiosis of plants, N-fixing bacteria and mycorrhizae (either vesicular-arbuscular mycorrhiza or ectomycorrhiza) has also been found in some *Acacia, Albizia, Casuarina* and *Leucaena* species. Symbioses are often found in nutrient limiting conditions and nutritional interactions play a key role in most symbioses. There are two important types of symbioses: mycorrhizal symbiosis and N-fixing symbiosis.

Mycorrhizae is the symbiotic association of fungi with roots of vascular plants. The main advantage of mycorrhizae to the host plants lies in the extension of the penetration zone of the root fungus system in the soil, facilitating an increased phosphorus uptake. The interconnected network of external hyphae acts as an additional catchment and absorbing surface in the soil beyond the depletion zone that would otherwise be inaccessible to the plant roots. Endotrophic mycorrhizae have been shown to be present in a wide range of horticultural species including apple, walnut, almonds, citrus, avocado, strawberry and grape. In many cases, the mycorrhizae have been shown to markedly improve the growth of plants.

Some soils are deficit in mineral salts and plants living in them have special means of obtaining nitrogen and other essential elements, for example, the roots of many moorland and woodland plants, particularly coniferous trees, which live in rich humus but deficient in mineral salts possesses a mycorrhizae. Mycorrhizae (**fungus-roots**) are symbiotic associations between fungi and the roots of vascular plants. Two main groups of mycorrhizae are recognized: the **ectomycorrhizae** and **endomycorrhizae,** although the rarer group with intermediate properties, the ectendotrophic mycorrhizae. In the ectomycorhizae, the fungal hyphae form a mantle both outside the root within the root in the intercellular spaces of the epidermis and cortex. No intracellular penetration into epidermal or cortical cells occurs, but an extensive network called the Harting net is formed between these cells. Ectomycorrhizae are common on trees, including members of the families pinaceae (pine, fir, spruce, larch, hemlock), Fagaceae (willow, poplar, chestnut), Betulaceae (birch alder), Salicaceae (willow, poplar, and few other show scaning electron micrographs of two *Pinus contorta* roots, one uninfected and having root hairs and the other infected with an ectomycorrhizae fungus.

Endomycorrhizae consist of three subgroups, but by far the most common are the vesicular arbuscular mycorrhizae (VAM). The fungi present in VAM are members of the Endogonacae, and they produce an internal network of hyphae between cortical cells that extends out into the soil, where the hyphae absorb mineral salts and water. Although VAM fungi seem to penetrate directly into the cytosol of cortical cells (where they form structures called vesicles and arbuscules, giving them their name), the hyphae are surrounded by an invaginated plasma membrane of the cortex cell. The VAM are present on most species of herbaceous angiosperms, whether monocot or dicot, annual or perennial crops, or native or introduced species; they also occur in the gymnosperm genera *Cupressus, Thuja, Taxodium, Juniperus,* and *Sequoia* and in ferns, lycopods, and bryophytes.

ROOT SYSTEM FORM

The main role of mycorrhizal associations is to acquire nutrients by exploring the soil volume with hyphae that are both more responsive and more extensive than the roots themselves. However, roots of some species of plants are also capable of effectively exploring large soil volumes and responding to temporary soil resources, and these species are likely to have faculative mycorrhizal associations, or be nonmycorrhizal. The production of fine roots or associations with mycorrhizal fungi are the two alternative strategies that both require plants to expend metabolic energy on their root systems. It is considered likely that mycorrhizal fungus hyphae should be a more cost effective means of exploring large soil volumes, because of their narrow diameter. Plants generally do not support both high levels of mycorrhizal colonization and root systems with fine/active roots, because of the high metabolic cost that would result.

Considerable variations in root system extensiveness, geometry, depth distribution, and plasticity occur between plant species. In experiments, total biomass is the parameter most often used to quantify roots, but provides much less information than root length or specific root length (length/unit root weight) data. Parameters that could be used to predict nutrient absorption are: total root biomass <fine root biomass <root length <root surface area <rhizosphere volume, arranged in increasing order of predictive ability. When root parameters of plant species are compared, root hairs have been found to be most highly correlated with their ability to absorb P from soils. Plants with extensive (highly branched, fine, long roots with numerous root hairs) have often been observed to derive little benefit from mycorrhizas in experiments.

In addition to have fine roots with long root hairs, there is a tendency for mycorrhizal plans to have roots which are less active (grow slowly, live longer, etc.) than species with little or not mycorrhizal colonization of their roots. The responsiveness, or capacity of plants to exploit small scale or short duration changes in water or nutrient availability by rapidly producing new roots, is thought to be an important determinant of their success during competition for soil resources. Some plants can also change rhizosphere conditions, such as pH, which may influence nutrient availability.

Australian nonmycorrhizal plants in the family Proteaceae have cluster roots that likely promote nutrient uptake by their large surface area and production of leachates, which contain organic acids and other substances that can increase nutrient solubility. The capacity of plants to influence nutrient availability in soils will also depend on the extensiveness and activity of their root system, since young roots are the primary source of exudates.

Host and soil factors which can be used to help predict the potential benefit of mycorrhizas are outlined in the Table 7.4 (below *Assuming that soil nutrient levels are not unusually high and inoculum of appropriate mycorrhizal fungi is available, root strategy differences will primarily determine the magnitude of benefits from mycorrhizal associations.*

Table 7.4: Generalized relationship between features of root systems and the mycorrhizal dependency of plants

Root feature	Mycorrhizal dependency of plant	
	High	Low
ROOT SURFACE AREA	low	high
Root system surface area	low	High
Branching orders of lateral roots	few	many
Branching frequency	sparse	frequent
Root hair abundance and length	few/short	many/long
ROOT ACTIVITY	slow	fast
Root growth rate	slow	fast
Response to soil conditions	slow	fast
Primary root lifespan	long	short
Protective structural features	strong	weak
Root exudation	? less	? more
MYCORRHIZAL FORMATION	efficient	inefficient or inhibited

Notes: This table is based on differences in root system properties between plants of different mycorrhizal status in Canadian forests.

SOIL AND SITE FACTORS INFLUENCING MYCORRHIZAS

Soil factors that can influence the benefits provided by mycorrhizal fungi to plants are considered below.

Mycorrhizal Inoculum

Plants with mycorrhizal associations predominate in most natural ecosystems, so inoculum of mycorrhizal fungi is present in most soils. The quantity inoculum of mycorrhizal fungi which are compatible with a host plant in soils can be measured by bioassay experiments. In these experiments, seedlings are grown in intact soil cores or mixed soil samples for sufficient time to allow mycorrhizas to form, then roots are sampled, processed and assessed to measure mycorrhiza formation.

In Australia, there are fewer species of ECM fungi in sites where eucalypt plantations are established than in undisturbed forests. There are also cases where plantations of trees have been introduced to habitats where they do not occur naturally. For example, Australian eucalypts are now grown in exotic habitats in Africa, Europe, Asia and South America, habitats, which are few compatible ECM fungi, because local fungi are specific to other hosts such as conifers.

Soil Disturbance

Propagules of mycorrhizal fungi maybe absent from soils where severe soil disturbance has resulted in topsoil loss, or where host plants are limited by adverse soil or site factors such as salinity, aridity, water-logging, or climatic extremes (Brundrett 1991). Most studies of mycorrhizal associations in highly disturbed habitats such as mine sites have found reduced levels of mycorrhizal propagules. Less severe forms of soil disturbance, including agricultural tillage, soil animal activities, fire and erosion can also reduce levels of mycorrhizal fungus propagules.

Spores and other propagules of mycorrhizal fungi can introduced to new sites by wind or water erosion or by the activity of animals, which feed on fungi. In a disturbed habitat, the effectiveness of natural vectors will depend on the proximity of undisturbed habitats containing suitable fungi (and their associated animals) as well as the phenology of fruiting of fungi. Effective colonization of disturbed habitats, such as minesites in Australia, by mycorrhizal fungi has been observed, but there is as yet insufficient information about the time required for this process to occur.

Soil Fertility

A key factor, which effects the potential for mycorrhizas to benefit plants in particular sites is the supply of phosphate and nitrogen in soil. Phosphorus is generally considered to be the most important plant-growth limiting factor, which can be supplied by mycorrhizal associations, because of the many abiotic and biotic factors which can restrict its mobility in soils. Reductions in the benefit is provided by mycorrhizal associations to plants are caused by increasing soil phosphorus levels. High rate of P and N fertilizers suppress ectomycorrhiza development in the field. High concentrations of soil N can influence the relative abundance of different ECM types.

Adverse Soil Conditions

Land degradation due to salinity, water-logging, erosion, etc., are serious and growing, problems in Australia and other countries. Excessive NaCl levels in soil inhibit mycorrhizal formation and restrict the activity of most mycorrhizal fungi, but some can tolerant these conditions. Observations in natural ecosystems have shown that plants with mycorrhizal associations are often less common than non-mycorrhizal species in soils which are waterlogged or saline, but that some mycorrhizal plants are normally present in even the worst sols. ECM fungi can be highly sensitive to water logging of sols, while VAM fungi may be less sensitive.

CHARACTERISTICS OF FUNGAL ISOLATES

Populations of mycorrhizal fungi are thought to have occupied the same soil habitats for millions of years, slowly adapting to changes in site conditions. Some mycorrhizal fungi appear to have worldwide distribution patterns and have apparently adapted to a wide range of habitats, but it is known that soil factors such as pH restrict the distribution of many taxa. Unfortunately, information about the soil and climatic conditions where isolates used in experiments were originally obtained is often not provided. These data are urgently needed to allow isolates to be selected for use in inoculation programs and to establish correlations between their taxonomy and physiology.

The fact that most mycorrhizal research has been concerned with plant growth responses with little consideration of the fungi involved has also helped to create the false impression that most of these fungi are functionally equivalent. Properties of individual mycorrhizal fungi result from their adaptations to soil/environmental/host factors. Some Properties of fungi considered to be important are:

1. The amount of soil hyphae produced relative to root colonization.
2. The rates of hyphal growth and root colonization.
3. Physiological characteristics, which regulate nutrient absorption or nutrient translocation by hyphae and exchange with the host.

MANIPULATING MYCORRHIZAL FUNGI

Methodologies have been developed for the inoculation of plants with mycorrhizal fungi. However, there has been less consideration about how we should determine where and when, if ever, it is necessary to do this. The diagram provided below, outlines the steps in a decision making processes that can be used to determine if mycorrhizal fungi that are compatible with host plants are present in soils, or are likely to be introduced by natural mechanisms. This information is required to evaluate the potential benefits of mycorrhizal inoculation in a specific situation. There would only be a need to manipulate mycorrhizal fungi by inoculating plant seedlings of changing site management practices it substantial benefits from these manipulations are expected. Methods for the production of mycorrhizal plants in the glasshouse and nursery to allow experimentation or for use in the field are described in the accompanying manual.

TYPES OF ASSOCIATIONS

At least seven different types of mycorrhizal associations have been recognized, involving different groups of fungi and host plants and distinct morphology patterns.

Table 7.5: Kinds of mycorrhizal associations

Endomycorrhizae	1. Vesicular-Arbuscular (Va)	Higher Plants-Great Majority Of The Vasicular Plants Including Ferns, Monocots, Dicots, Fungi-Endogonaceae (Phycomycetes)
	2. Ericaceous	Higher Plants: Ericales. Fungi-Generally Unknown Although A Basidimycete Has Been Found In Association With Azalea.
	3. Orchidaceous	Higher Plants: Orchids. Fungi: Agaricales and Aphyllophorales (Basidiomycetes).
Ectomycorrhizae	Higher Plants: Great Majority Of The Vasicular Plants Including Ferns, Monocots, Dicots, Fungi-Endogonaceae (Phycomycetes)	

	Fungi: Many thousands Of Species In Basidiomycetes, Ascomycetes, And Endogenaceae (Phycomycetes).	
Ectendomycorrhizae		

1. **Vesicular arbuscular mycorrhizas** (= arbuscular mycorrhizas. VAM or AM) are associations where Zygomycete fungi in the Glomales produce arbuscules, hyphae and vesicles within roots. Spores are formed in soil or roots. These associations are defined by the presence of arbuscules. Fungi within roots spread by linear hyphae or coiled hyphae.
2. **Ectomycorrhizas (ECM)** are associations where Basidiomycetes and other fungi form short swollen lateral roots covered by mantle hyphae. These roots have Harting net hyphae around cells in the epidermis or cortex.
3. **Ectendo-, arbutoid- and monotropoid mycorrhizal associations** are similar to ectomycorrhizas. These associations have not been found in Australia.
4. **Orchid mycorrhizas** consist of coils of hyphae within roots or stems of plants in the family Orchidaceae. Young orchid seedlings and some adult plants which lack chlorophyll are entirely dependant on mycorrhizal fungi for their survival.
5. **Ericoid mycorrhizas** have hyphal coils in outer cells of the narrow "hair roots" of plants in the plant order Ericales. These associations also occur in thicker roots of Australian members of the Epacridaceae.
6. The Australian lily ***Thysanotus*** has a unique mycorrhizal association where fungus hyphae only grow under epidermal cells.

The distinction between ecto-and endomycorrhizae is based upon the morphology of the structure formed by fungus and higher plants. Ectomycorrhizal forming fungi are able to penetrate between, but not into, the cortical cell of the host root, thus forming a network of fungal strands within the cortex commonly called a "Hartig net". The fungus also normally makes a mantle which completely encloses the infected root tip and from which fungal hyphae extend into the soil. This network of hyphae greatly increase the ability of the root to take up nutrients in three ways:

1. The surface area in contact with soil particles is much greater, and the much smaller hyphae can penetrate fissures in soil particles too small for roots;
2. The fungus excretes enzymes which allow it to dissolve soil nutrients otherwise unavailable to the roots; and
3. The fungus extends the life of an absorbing root from a few days to a year or more.

Inoculation of tree seedlings with appropriate fungus species to stimulate development of mycorrhiza has been shown to be critical to reforestation, especially, in a diverse range of sites and climates from timberline in the Alps to subtropical islands like Puerto Rico. The last fiften years have seen great interest in mycorrhiza in the southern states where fungi such as Pisolithus tinctorius are believed to be responsible for increased growth and survival of loblolly pine, especially on harsh sites such as the spoils of mining operations.

Endomycorrhizae differ from the above primarly in that the fungus hyphae penetrate directly into the cortical cells of the root. Neither endo-nor ectomycorrhizal forming fungi are, however, able to penetrate beyond the cortical region – the vascular tissues of the root are not infected.

TYPES OF MYCORRHIZA

Mycorrhizae are classified on the basis of the extent of root penetration, production of external mantle or sheath and the inter- and intracellular structures they form once inside the plant root. Since enormous amount of literature exists on classification and types of mycorrhizas only salient features of different types are mentioned here. So far seven types of mycorrhizae are recognized.

Ectomycorrhiza

Here the fungal members belong to 25 families of Basidiomycotina 7 families of Ascomycotina or 1 genus of Zygomycotina i.e. *Endogone.* The plant symbionts include both Gymnosperms and Angiosperms. A fungal sheath or mantle is present. There is no hyphal penetration of cells. Fungal hypha is generally septate. A distinct Hartig's net is present between the cells. Hartig's net is plexus of fungal hyphae between epidermal and cortical cells. It provides a large surface area for the interchange of nutrients between host and the fungi.

Ectendomycorrhiza

The fungi belong to Basidiomycotina, which covers both Gymnosperms and Angiosperms plants. The fungal sheath if present is very thin. Septate fungal hyphae penetrate the root cells and coil inside it. The fine network of fungal hyphae (Hartig's net) is present in the intercellular spaces and thereis intracellular penetration of living hyphae in cells.

Endomycorrhiza/Arbuscular Mycorrhizal Fungi (AMF)

Fungi are confined to the order Glomales of Zygomycotina. The plant symbionts range from Bryophytes to Angiosperms. There is no fungal sheath. Aseptate hyphae enter the root cortical cells and form characteristic arbuscules and/or vesicles hence the name. The plasmalemma of the host cell invaginates and encloses the arbuscules.

Arbutoid mycorrhiza

Here the funal partner is Basidiomycetous and colonises members of order Ericales (Arbutoidae). Fungal sheath present. Septate hyphae penetrate only the epidermal cells and coil inside it. Hartig's net is present in the intercullular spaces.

Monotropoid mycorrhiza

The fungi belong to Basidiomycotina colonising achlorophyllous member of Angiosperms belonging to family Monotropaceae. Fungal sheath present. Branches from septate hyphae penetrate the the epidermal cells forming haustorial pegs inside. Hartig's net not seen in mature roots.

Ericoid mycorrhiza

Fungal members are usually Basidiomycetous and one, *Hymenoschyphus*, belongs to Ascomycotina. This is found in roots of plants belonging to order Ericales. Rootlets are covered by a loosely woven mesh of dark brown septate hyphae from which branches penetrate the cortical cells, forming intracellular coils invaginating and enclosed by the host's plasmalemma. Hartig's net formed in the intercellular spaces.

Orchidoid mycorrhiza

Fungi belong to Basidiomycotina and mycelia sterlia. These colonise only members of family Orchidaceae (Angiosperm). Septate hyphae enter the cells and develop into coils often called "peltrons". This association is probably pseudomycorrhizal but plays an important role in the establishment of orchid seedlings.

Host Plants

The relative importance of different mycorrhizal associations for plant species is graphically summarized below using data for Angiosperms from the mycorrhizal literature. Many Pteridophytes are also known to have VAM associations, while Gymnosperms may have ECM or VAM associations. Most information about the mycorrhizal status of plants if from the northern hemisphere in temperate regions, especially for crop plants and weeds.

Place of Availability of Mycorrhizal Plants

Field surveys have found that plans with mycorrhizal associations predomniate in most natural ecosystem sin Australia and throughout the world.

Table 7.6: Briefly summarises the importance of mycorrhizal associations throughout the world

Association	Occurrence
Vesicular Arbuscular Mycorrhizal (VAM) Plants	Plants with VAM are common in most habitats It is easier to say where they are not found
Ectomycorrhizal (ECM) Plants	Trees with Ecm are dominent in coniferous forests, especially in cold boreal or alpine regions ECM trees and shrubs common in many broad-leaved forests in temperate or Mediterranean regions ECM trees also occur in some tropical or subtropical savanna or rain forests habitats
Nonmycorrhizal (NM) Plants	NM plants are most common in disturbed habitats, or sites with extreme environmental or soil conditions NM plants appear to be more common in Australia than in other continents.

Mycorrhizal Fungi

Members of the fungus kingdom obtain nutrition from many sources, including decomposition of organic substrates, predation and parasitism, and involvement in mutualistic associations (Christensen 1989, Kendrick

1992). Many soil fungi are saprobes with the enzymatic ability to digest organic substrates of varying degrees of complexity, but some subsist on very low levels of organic or inorganic substrates (Wainwright 1988). Mycorrhizal fungi are a major component of the soil microflora in many ecosystems, but usually have limited saprophytic abilities (Harley & Smith 1983). Mycorrhizal fungi are considered to have many important roles in natural and managed ecosystems.

Very little is known about the taxonomic or functional diversity of mycorrhizal fungi. Indeed, there is a paucity of knowledge about all Australian fungi, with only about 10% of these organisms having been named, while an estimated 80% of them remain undiscovered. Laboratory and glasshouse experiments have shown that different fungal taxa vary in their capacity to utilize resources, withstand adverse conditions etc. For this reason, mycorrhizal fungus diversity is thought to contribute to the resilience of ecosystems, but there is very little comparative data that can be used to test this hypothesis. More information about fungi forming ECM and VAM associations is provided in the relevant sections.

Arbuscular Mycorrhizal Fungi (AMF)

AMF are the most widespread in their distribution both among plant species and geographically. They are believed to be disseminated inter-continentally prior to continental drift as supported by fossil record of earlier plants. They occur more frequently in cultivated soils usually in top 15-30 cms. Because most economically important plants form this type of mycorrhiza, this area of research has attracted much attention. Commercial use of AMF may be an alternative to the rising agricultural and fertiliser costs, because of their potential in increasing the crop yields.

Establishment of Symbiosis

These associations do not usually cause observable external changes in the morphology of plant roots, and can not be detected without appropriate staining. They colonise a wide range of host plants that include most agricultural and horticultural plants both annual and perennial, fruit trees and a large number of forest tree species. AMF do generally not colonise plants that enter into other types of mycorrhiza. They lack host specificity from the host point of view as shown by their wide host range. However from fungal view point the specificity does exist. Of the 150 species that have been described in *Glomales* only a small proportions are presumed to be mycorrhizal. Since these fungi lack sexual reproduction recognition, characterization and classification is based upon their chlamydospores which show variation in size, shape, cooler and wall characteristics depending upon the species.

These fungi are able to colonise the epidermal and cortical cells of the root, but do not enter the vascular system or the root meristems. The morphological stages of development are variable, depending mainly on the plant species involved.

In general, fungal hyphae growing from spores in the soil or from adjacent plants roots contact the root surface, where they differentiate to form an appressorium. This is the first indication of recognition between the fungus and the plant, and appressoria are not formed in the roots of non-host plants or on synthetic media or membranes. Penetration of the root occurs via the appressoria and the fungus may produce intracellular coils in the subepidermal cell layers, followed by intercellular growth into the inner cortex in the subepidermal cell layers, followed byintercellular growth into the inner cortex of the root (although sometimes only intercellular hyphae are observed). On reaching the inner cortex, hyphal branches penetrate the cortical cell walls and differentiate within the cells to form highly branched, terminal structures known as arbuscules. Here the symbionts are in intimate contact, separated only by the their membranes. This interfare is assumed to be the site of carbon and phosphate exchange between the symbionts, and the arbuscule generally regarded as a key structure in the symbiosis. Following the formation of arbuscules, some species of fungi also form vesicles in the root. Vesicles are small darkly staining globular or spherical structures formed later. They store oil and Polyphosphate granules, which are utilized by the plant under phosphorus deficiency conditions.

Classification of AMF

There are only six genera of fungi that contain species, which are known to produce AMF with plants. Two of these genera, *Glomus* and *Sclerocystis*, produce chlamydospores only. Four genera form spores that are similar to azygospores: *Gigaspora, Scutellospora, Acaulospora* and *Entrophospora*. Their spore characteristics and the relationship of the spore to the associated hyphal attachments distinguish the genera. The existent genera were establishment to reflect the manner in which the spores were produced, except for the genus *Sutellospora*. Species in this genus are distinguished from species in the closely allied genus *Gigaspora* by wall characteristics, method of germ tube formation, and auxiliary cell characteristics.

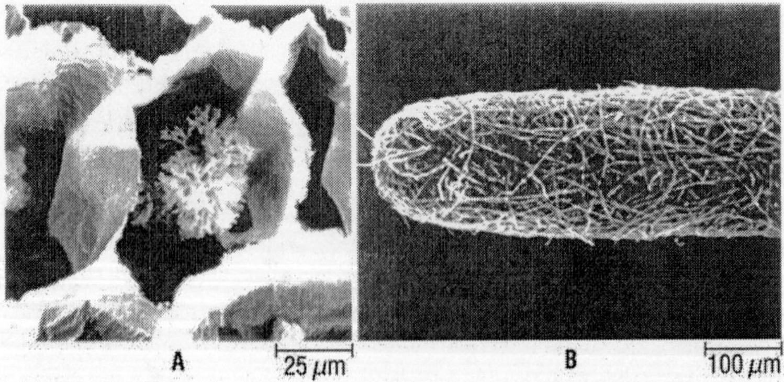

Fig. 7.2: Scanning electron micrographs of mycorrhize. (A) A fully developed arbuscule of an endomycorrhiza in a plant cell. (The term arbuscule means little bush). As the arbuscule decomposes, it releases accumulated nutrients from plant. (B) The mycelial mantel of a typical ectomycorrhizal fungus surrounding a eucalyptus tree root.

The fungi forming AM are are placed in the class Zygomycetes, order Endogonales, family Endogonaceae. The type genus for Endogonaceae is *Endogone,* a genus in which species form zygospores and azygospores but do not produce arbuscular mycorrhizae.

Over the years, numerous attempts have been made to monoculture AMF on artificial media. These attempts, however, have met with little success. Thus, AMF must still be collected from field soils, increased in the greenhouse on living host plants (nurse crops) and collected once again from these "pot culture" soils, where the TERI is specialized. Clearly, the production and collection of propagules from soil remains an important, time-consuming and fundamental part of AMF research, both for inoculum preparation and population estimation. On the other hand, all AMF are not morphologically and physiologically the same, so it is important to know the identity of the fungus with which one is working. Although these fungi have extremely wide host ranges, they are very specific in a single plant species and they can differ also in their effect on plants in different ecosystems, soil types, and in similar soils with differeing biological, chemical, and physical components.

ECTOMYCORRHIZAE (ECM)

They are characterised by dense mycelial sheaths around the roots and a Hartig net between root cells, are limited to mostly temperate forest trees and *Basidiomycetes* and some other fungi. Mycorrhizae involve an intimate association of the host plant's root tissue and the fungus, but there is also fungal tissue that extends into the soil as individual hyphae and in some species as more complex strands. The external hyphae can take up mineral nutrients from the soil and transport them into the host root. Mycorrhizal associations can therefore be of great potential benefit to the host plant in nutrient limited systems.

VESICULAR-ARBUSCULAR MYCORRHIZA (VAM)

Zygomycete fungi form arbuscules and/or vesicles and external hyphal networks in the soil and grow extensively within the cells of the cortex, are formed by nearly all vascular plants. The endomycorrhizae, a fungus, (arbuscular mycorhizae) found in close association among angiosperms. The VA mycorrhizae are formed by the non-septate phycomycetes fungi belonging to the genera *Glomus, Gigaspora, Acaulospora* and *Sclerocystis* in the family Endogonaceae of the order Mucorales. They produce vesicles and arbuscules inside the root system. Arbuscules are highly branched fungal hyphae while vesicles are the bullous swellings of these hypae. These VAM fungi are beneficial to the plant which they colonise. They make more nutrients available to the plant, improve soil texture, disease resistance and help in better plant growth. Besides, mycorrhizae also help in the biological control of root pathogen.

An extensive research on VA mycorrhizal association in horticultural crops have been done at the Unviersity of Agricultural Sciences, Bangalore and Indian Institute of Horticultural Research (IIHR), Bangalore. Tomato, brinjal, okra, bell pepper, chilli and onions have been studied for their response to VAM inoculation (Sukhada, 1993). He also reported beneficial response of VAM fungi in cowpea has been reported (Sukhada, 1993).

In India the beneficial effects of vasicular-arbuscular mycorrhizae (VAM) have been observed in fruit crops like citrus, papaya and *litchi*. Inoculation with *Glomus fasciculatum* fungus was found to increase the dry matter content and the uptake of Fe, Zn, Mn by different rootstocks of citrus (Manjunath et al., 1983).

In litchi, positive correlations of VAM population in the rhizosphere of soil and mycorrhizal infection in the roots with leaf area and fruit have been noticed (Rana and Srivastava, 1984). In papaya, inoculation with *Glomus fasciculatum* and *Glomus mossae* was found to increase the plant height, dry matter content and nutrient uptake with lower doses of applied phosphorus.

Recent studies have shown the possibility of domesticating *Mycorrhizae* in agricultural systems is not in doubt. Lack of a suitable inoculum production technology is the major limitation for commercial exploitation of this system.

Table 7.7: Effect of mycorrhizal (VAM) inoculation on crop yields

Crop	P-fertiliser supplementation (kg P_2O_5/ha)	% yield increase over control
Finger millet	19	18
Soybean	25-50	19
Chillies	37.5	55
Chickpea	40	25
Groundnut	-	10-20

Source: Venkatraman and Tilak (1990).

HOW MYCORRHIZAS WORK

Mycorrhizal associations are defined in the introduction section. Mineral nutrient acquisition from soil is considered to be the primary function of mycorrhizas, but other roles for these fungi have been suggested. These roles are listed below, but many have not been quantified in natural ecosystems. Only the uptake of nutrients is discussed further here.

Nutrient Depletion Zones

The demand for a particular mineral nutrient depends on plant internal requirements, while the supply of that nutrient primarily depends on its availability and mobility in soils. Mineral nutrients such as phosphorus have very limited mobility in soils so that depletion zones – where all the available nutrient has been utilized, quickly from around roots. Thus to obtain more phosphorus, plants must bypass these depletion zones by further root activity elsewhere in the soil. The outcome of this quest for phosphorus (and other relatively immobile soil resources) should largely be determined by the surface area of a plant's root system. The most important role of mycorrhizal fungus hyphae is to extend the surface area of roots.

Hyphal Activity

Mycorrhizal fungus hyphae are considered to function primarily by increasing the soil volume from which available forms of phosphorus are absorbed and provided to roots. Hyphae of VAM fungi can respond to localized sources of soil nutrients mor erapidly than roots and to produce tine highly-branched *absorptive* hyphae in decomposting organic substrates. ECM fungus hyphae also proliferation in decomposing soil organic matter.

It was considered that hyphae of VAM fungi utilized the same forms of nutrients as roots. However, there now is evidence that these associations have a greater benefit when phosphorus is present in less-soluble forms. Hyphae of ECM fungi can utilize both inorganic and simple organic sources of nitrogen and phosphorus. There are some ECM fungi which specialize in growth is soils enrichment with ammonium. Soil mineralisation by organic acids produced by hyphae has been measured within Ecm fungus mats. ECM fungus hyphae can also grow within rocks where they contribute to mineral weathering. Thus, mycorrhizal fungi often have access to forms of nutrients which are not directly available to plants, as well as acquiring nutrients which are spatially or chronologically separatd from roots.

Mycorrhizal Dependency

Mycorrhizal dependency is A measure of the benefit provided by mycorrhizas and will depend no relative contribution of root and mycorrhizal mediated nutrient uptake to plants (Janos 1980). Mycorrhizal dependency has often been quantified by calculating the yield ratio between mycorrhizal plants and uninoculated control plants grown in a particular soil at a single soil P level. However, it is better to analyse mycorrhizal benefits across a range of soil P levels, by producing nutrient response curves.

The concept that plants have varying degrees of dependence on mycorrhizal associations is gaining acceptance. Detailed examinations of plants in natural ecosystems often show consistent differences between host plant species in the intensity and consistency of mycorrhiza formation (proportion of root system involved). These observations have shown that species generally either have:

1. Consistently high levels of mycorrhizas
2. Intermediate, or variable levels of mycorrhizas.
3. Or are not mycorrhizal.

Plants belonging to these categories can be called obligatorily mycorrhizal, facultatively mycorrhizal, or nonmycorrhizal, as are defined below.

In natural ecosystems, plant with facultatively mycorrhizal associations or nonmycorrhizal roots are more common in very dry, wet or cold habitats where plant productivity is limited by soil/environmental conditions, or in disturbed habitats where mycorrhizal fungus inoculum islimited. Nonmycorrhizal trees are rare (members of the Australian family Proteaceae are one exception). There are a number of non.ycorrhizal genera which are important in agriculture and horticulture including members of the families Chenopodiaceae, Amaranthaceae, caryophyllaceae, Polygonaceae, Brassicaceae, Scrophulariaceae, Commelinaceae, Juncaceae and Cyperaceae.

Obligatorily Mycorrhizal Plants

Obligatorily mycorrhizal plants are those plants which will not survive to reproductive maturity without being associated with mycorrhizal fungi in the soils (or at the fertility levels) of their natural habitats. These species consistently support mycorrhizal colonization throughout most of their young roots.

Facultatively Mycorrhizal Plants

Plants that benefit from mycorrhizal associations only in some of the least fertile sols in which they naturally occur. In ecosystem surveys, inconsistent mycorrhizas or lo w levels of mycorrhizal colonization have been used to designate facultatively mycorrhizal species when soil fertility levels could not be manipulated.

Nonmycorrhizal plants

Plants with roots that consistently resist colonization by mycorrhizal fungi, at least when they are young and healthy. Nutrient levels and other soil properties and mycorrhizal propagules dynamics can also reduce mycorrhizal formation, but usually do not prevent it completely.

MYCORRHIZAL FUNGI AND PATHOGEN INTERACTION

Tomato plants inoculated with VAM fungi were highly resistant to *Phythium aphanidermatum* causing damping off. Control of nematodes was also reported due to VAM inoculations. Accumulation of phytoalexin in cowpea root during the infection of *G. fasciculatum* and *G. aggregatum* which could be correlated with the degree of infection. Nematode infection had reduced significantly in mycorrhizal plants. The root extract of mycorrhizal plants caused mortality of nematode larval compared to non-mycorrhizal root extracts. Tomato seedlings inoculated with VAM fungi were more resistant to *Meloidogynae javanica.* A similar control of nematode infestation in tomato and brinjal by inoculation with *G. mosseae* and neem cake.

Table 7.8: Crop typs and responses of inoculation

Crop	Inoculation and Response	Reference
Banana	Responds excellently to VAM inoculation. The *G. mosseae* and *G. fasciculatum* inoculationin combination with phosphate solubilizers and *Azospirillum* improved bunch weight in bale Elakki	Sukhada (1997)
Banana Dwarf Cavendish	Treated with *G. fasciculatum* 7 days prior to nematode treatment was able to offset the ill effects of the latter and had more vigorous root nematode population reduced.	Umesh *et al* (1988)
papaya, Citrus, Mango and Pomegranate	plans could be inoculated effectively with VAM fungi in the nursery and transplanted to the field The inoculation improves the plant establishment, improves of phosphorus in soil and their uptake (Sukhada, 1992). Due to the increased availability of phosphate in soil a saving in the addition of inorganic phosphate in the soil can	(Sukhada, 1997). work conducted at IIHR,

Crop	Inoculation and Response	Reference
	be brought about.	Bangalore
papaya, Citrus, Mango and Pomegranate	plans could be inoculated effectively with VAM fungi in the nursery and transplanted to the field The inoculation improves the plant establishment, improves of phosphorus in soil and their uptake (Sukhada, 1992). Due to the increased availability of phosphate in soil a saving in the addition of inorganic phosphate in the soil can be brought about.	(Sukhada, 1997). work conducted at IIHR, Bangalore
Litchi *(Litchi chinensis* Son)	Distribution of endomycorrhizal spores in the rhizosphere	Rana and Srivastava (1984)
Papaya (Coorg Honey Dew)	Studied the uptake of ^{32}P labeled super phosphate by endomycorrhizal and derived that mycorrhizal fungi besides acting as scavengers of phosphate also increased the availability of phosphates in the soil	Sukhada *et al* (1995)
	Standardized PCR amplification technique to identify different genera of VAM fungi	Sukhada *et al* (1997)

NITROGEN TRANSFER IN MYCORRHIZAL PLANTS

The idea of nitrogen fixed by a legume component may be available to its associated non-legume plant originated from earlier studies in the late of. In several of their experiments they found that as much as 10 to 30% of the total N fixed in field pea was deposited in the growth media of sand, predominantly as amino acids, and a major part was taken up by an associated cereal. In general, the process of N-deposition and uptake in a neighbouring plant is often termed **N-transfer**. Since then substantial N-transfer have been observed in several legume/crop intercropping systems such as soybean or cowpea/maize or sorghum, cowpea/rice, gram clover/wheat, barley or oat/vetch or lupin etc., although some experiments showed little or no N-transfer between them

Compared to pure stand cultivation, possible benefits obtained by non-legumes or trees may partly be due to transfer of the symbiotically fixed N by legumes, either through directly release from nodulated roots or through decomposition of dead nodule and root tissue in the soil or through a common mycorrhizal network or even enhanced by mycorrhizal hyphae. However, the mechanism of N transfer and the role of mycorrhizal hyphae in the direct transfer of nitrogen are not well established. Therefore, studies in identifying the pathways of N transfer and whether mycorrhizal hyphae play a direct role in facilitating N transfer are needed.

NITROGEN NUTRITION IN MYCORRHIZAL PLANTS

Forms of Nitrogen Used by Mycorrhizal Associations

The two most available forms of inorganic N for plants and both VA mycorrhizal and ectomycorrhizal fungi, are NO_3^- and NH_4^+. Originally, NH_4^+ is mainly derived from the reduction of atmospheric dinitrogen symbiotically fixed by nitrogen fixing plants, either by *Rhizobia* in legumes or by *Frankia* in actinorhizal plants. In general, NO_3^- is the dominant form of nitrogen available to plants and fungi in almost all agricultural soils due to the rapid nitrification of NH_4^+. In contrast, NH_4^+, which is released from soil humus and other organic N sources by ammonifying organisms, predominates and NO_3^- may be almost entirely absent in many undisturbed or very acidic soils. NO_3^- is highly mobile and is readily transported towards the plant roots by mass flow and diffusion. NH_4^+ is absorbed to negatively charged soil particles and transported towards the plant roots mainly by diffusion.

Mycorrhizal Effects on Nodulation and Nitrogen Fixation

The possibility of a direct involvement of ectomycorrhizal fungi in N acquisition by plants was first suggested by Frank in 1894. Then after 50 years of Frank's work, the probably first observation of growth, nodulation and mycorrhizal status of a large number of legumes was made by Asai in 1944. In most cases, improved nodulation and N fixation in mycorrhizal plants appears to be the result of relief from P stress and possibly uptake of some other essential micronutrients, which result in both a general improvement in growth and yield and indirect effects upon the N-fixing system. The differences between mycorrhizal and non-mycorrhizal plants usually disappear if the latter are supplied with a readily available P source.

It was known that cereals intercropped with grain legumes generally benefit from the association in terms of increased grain and nitrogen yields per unit area compared with monocropped cereals. This indicates that N-transfer from the nitrogen fixing plants (donor) to an associated non-nitrogen fixing plant (receiver) is most likely to happen in nature either via the interception and uptake of released fixed nitrogen from donor plant by the roots of receiver plant or through the root mycorrhizal between donor and receiver plants, or even enhanced by mycorrhizal hyphae. Therefore, in agricultural and/or natural ecological communities, mycorrhizal associations may be important factors influencing the performance of both donor and receiver plants through the acquisition and translocation of nitrogen by the mycorrhizal fungus, particularly when the relatively immobile NH_4^+-N rather than the mobile NO_3^--N is the major source of plant available N.

It is therefore, concluded that Nitrogen fixing and non-nitrogen fixing plants can provide a good example for investigating the N-transfer between the plants where neighbouring plants may have differing nitrogen status. In general, with or without a split root system, combined with the fine nylon mesh to allow the direct mycorrhizal link but not root contact, together with ^{15}N-isotope labelling technique to enrich the N in the donor plant, some experiments have demonstrated a below-ground N-transfer whereas others found no evidence for transfer of nitrogen between donor and receiver plants via mycorrhizal hyphae (see Smith and Read, 1997). Furthermore, there is controversy about the extent to which direct N-transfer is actually being facilitated by mycorrhizal hyphae or by some other indirect means, and whether it is of agricultural or ecological significance if N-transfer do occur between these plants through mycorrhizal hypha.

The following questions still remain to be answered:

1. Does N transfer occurs between mycorrhizal plants at all? If so, how

much N is transferred and how much N-transferred is enhanced by mycorrhizal hyphae?.

2. How much net N transfer is from nitrogen fixing to non-nitrogen fixing mycorrhizal plants and *vice versa*?

P Response Curves

Using the most efficient fungus for inoculating a particular crop, the amount the P fertilizer that can be saved ma be worked out. Found very large grain yield responses to VA mycorrhizal inoculation in unfertilized soil, but no responses in plots give P fertilizer. In this and several other trials, mycorrhizal responses have been recorded in unfertilized plots, but not in highly fertilized plots. These results highlight the need to describe the mycorrhizal growth response against a growth response curve obtained from a range of P fertilizer rates. It should then be possible to select the fertilizer level at which the responses to inoculation can be optimized. These kinds of experiments suggest the amount of P fertilizer that could be saved with mycorrhizal inoculation to produce the same amount of growth or yield. Mycorrhizal inoculation will only be useful in agriculture if it enables the farmer to achieve optimum yield level with a significantly lower level of P fertilizer. The savings from the reduced fertilizer use will be dependent on the shape of the P response curves, and the same could be compared with the costs of supply and introduction of the mycorrhizal inoculant.

INOCULATION METHODS

Various methods have been tried for introducing VA mycorrhizal inoculum into a field crop

A. Trasplanted crops

Seedlings are raised in sterilized or unsterilized soil supplied with selected VA mycorrhizal fungi in small nursery beds or containers and planted out when mycorrhizal colonization is well established. This method has been successfully used to produce worthwhile and economic growth response in agronomically important crops like chilli, finger millet, tomato and tobacco horticultural crops like citrus, mango, asters and marigold, and forest tree species like leucaena, *Tamarindus indica, Acacia nilotica* and *Calliandra calothyrsus*.

It can thus be used in many other transplanted crops of importance in agriculture, horticulture and forestry. The possibility of introducing efficieng VA mycorrhizal fungi to cereals through forest tree species in alley cropping needs investigation.

B. Plant directly sown in the field

Coating seeds with VA mycorrhizal inoculum

Coating seeds with mycorrhizal inoculum similar to that which is produced with *Rhizobium*, would be the easiest method of inoculating plants with VA mycorrhizal fungi if it would provide consistently good infection. Seeds are coated with an adhesive, ex. Methyl cellulose, to which inoculum is expected to stick and subsequently infect the emerging radicals. Unfortunately, because of their large size it is much more difficult to attach VA mycorrhizal propagules than bacteria in this way. However, it has been proved satisfactory for large-seeded crops such as citrus in field nurseries.

Mycorrhizal pellets

Rather than coating seeds with VA mycorrhizal inoculum, it seems technically more feasible to incorporate seeds into the inoculum to form multiseeded pellets. These pellets, about 1 cm in diameter, consist of soil inoculum from pot cultures stabilized with clay, or peat inoculum. Good infection has been achieved by this means. Hall and Kelson have described a system for producing some 5000 VA mycorhizal infected soil pellets per man-day with seeds attached by gum arabic.

Fluid drilling

Incorporation of seeds and inoculum into a uniform suspension in a various fluid, ex 4% (w/v) methyle cellulose, and applycing in furrows as a slurry is another means of placing seeds and inoculum in proximity. This has proved successful under field conditions, but germination of the seeds must be rigorously controlled to prevent damage to the emerging radicle. The advantage of this technique is that the inoculum is less bulky and does not increase appreciably the volume of material drilled and can be combined with rhizobia for legumes.

Inoculation in furrows

Placing inoculum under or beside seeds sown in a furrow is probably an effective method of inoculating field sown crops. For field experiments, research workers have placed mycorrhizal inoculum below the seedling point. The rates of inoculum they used varied from 0.8 to 167 tonnes per ha, with rates of 20 to 30 tonnes per ha commonly used. Most of these rates would be unacceptable in practical agriculture.

Used a tractor drawn drill (Model Nodet) to handle fresh or granulated VAM inoculum for onion and maize crops. Mycorrhizal inoculum placed 3 cm below the seed at the rate of 1.9 tonnes per ha increased onion bulb yield by 17 present. With modification in the drill the amount of inoculum used could be reduced and it was found that inoculum at the rate of 400 kg per ha and placed along with the seed, increased the yield of maize by 15 present. These field trials show that the existing farm machinery can be modified to accept mycorrhizal inoculum and give biologically and statistically significant increase in crop yield in the field. The least amount of inoculum (400 kg/ha) used in this field trial is also too high to be used in practical agriculture.

Precropping

Populations of VA mycorrhizal propagules can be raised *in situ* by growing strongly mycorrhizal host plants and leaving the infected roots and associated spores in the soil to infect the next crop. Hence, where mycorrhial benefits look possible, judicious crop rotations might be considered as a means of manipulating the native mycorrhizal fungi. These can be increased in bulk if they are effective, otherwise efficient introduced endophytes can be built up on the pre-crop. The top soil could be used as crude inoculum elsewhere. Suggested to partially sterilize microplots by burning straw in furrows (2 kg straw for a plot of 1m x 1m with 5 furrows, each 15 cm deep) and then raise sorghum or Rhodes grass inoculated with a desired VA mycorrhizal fungus. After 75 days, the top 15 cm soil along with roots (chopped) can be used as mycorrhizal inoculum. In this way farmers can raise their own inoculum in their field from a starter culture obtained from an authentic source

Appropriate VA mycorrhizal Technology

A. Transplanted crops

As pointed out earlier, there is an immediate potential for farmer use of VA mycorrhizal inoculation for transplanted crops. VA mycorrhizal inoculum of suitably selected strains can be used for inoculation in the nursery bed. Growers only need to incorporate inoculum in the nursery bed or seedlings trays or polythene bags at the appropriate rate by hand. Seedlings thus raised will be colonized by the introduced fungus and can then be planted out in the field. In developed countries seedlings are usually raised in fumigated soil or potting mix.

Inoculation of such substrate, with selected VA mycorrhizal fungi for the crop, is an easy and appropriate technology. Commercial use of VA mycorrhizal inoculation, by this method, in the nursery production of citrus is done in USA. Inoculation of the nursery with efficient VA mycorrhizal fungi has been successful in the large scale production of white ash seedlings for reforestation in and many horticultural plants in France.

In many developing countries and sometimes in developed countries, seedlings are raised in unsterile soil or substrate. Inoculation with appropriate fungi in seedling trays containing unsterilised potting mix was found to give vigorously growing seedlings of asparagus which performed well when planted out in the field. In trails in India it was found that inoculation of unsterile nursery beds with efficient VA mycorrhizal fungi enhanced the growth rate of slow growing tropical tree species and rootstocks of citrus and mango. This technology which can easily be adopted by farmers is now being tried on a large scale.

B. Field sown crops

Development of appropriate technology for mycorrhizal inoculations is more difficult for crops sown from seed in the field. Work of showed that farm machinery can be modified to accept mycorrhizal inoculum and to give significantly increased yield to directly field sown crops like maize. So technology for applying VA mycorrhizal inoculum using farm machinery for directly field sown crops is already available should be developed further. The main limiting factor would be production of large quantities of inoculum. VA mycorrhizal inoculation by hand, in developing countries, may be feasible for crops with low field densities (because of larger spacing between plants) such as fruit trees and root crops like yam, cassava etc.

CONCLUSION

The efficiency of phosphatic fertilizers is very low (15-20%) due to its fixation in soil. Besides native soil phosphorus is mostly unavailable to crops because of its low solubility. The introduction of efficient P solubilizers in the rhizosphere has been found to increase the availability of phosphorus from both applied and native soil P. The phosphorus which gets fixed in soil is made available to crops by the action of microorganisms which solubilize insoluble/fixed forms of phosphates into forms which are readily taken up by the plants. Another factor which affects the effective utilization of P is its low mobility in soil. Fortunately some filamentous organisms like *vesicular arbuscular mycorrhizae* (VAM) which form symbiotic association with plant roots have the ability to mobilize phosphorus from soil and thereby helping in absorption of P by plant roots. Commercial exploitation of phosphatic biofertilizers can play an important role particularly in making the direct use of abundantly available low grade phosphate possible.

CHAPTER 8

The Cyclic System of Nutrient Management

Plants require 13 essential nutrient elements (6 major, 7 micronutrients) and can profit from some beneficial nutrients. The amount of nutrients required by crops depends largely on the target yield level. Only under rare circumstances the nutrients required for high yields are "automatically" supplied from the soil nutrient reserves.

The response of crop growth and yield to the nutrient supply can be expressed in growth or yield curves and described by "yield laws". In practical cropping, nutritive minimum factors are often responsible for unsatisfactory yields and thus present growth problems. Therefore, a proper nutrient management (of the nutrients added to the soil as well as in order to keep many nutrients, especially microelements, on a sufficient natural supply level) is of fundamental importance for effective, successful and sustainable agriculture.

The non-compensated use of soil nutrients is usually regarded as exploitation or "mining" of nutrients leading to degraded soil fertility. However, two different concepts should be distinguished: exploitation vs. utilization. Their difference can be compared with pumping water from wells. If a water level is slightly or considerably lowered, water is exploited and the system is not sustainable. If the water level is not noticeably lowered even after long period, water is "just" utilized, the system could be called sustainable.

Exploitation of soil nutrients is frequently applied which decreases the level of available nutrients quickly or slowly and is therefore non-sustainable.

Utilization of soil nutrients consists in partly removing nutrients without (significant) improverishment of the nutrient supply or reduction of soil fertility. This can continue on fertile soils for a long time (centuries for some microelements) and creates the impression of permanence of natural nutrient supply.

GEOGRAPHICAL NUTRIENT TRANSFER

In smaller or greater region there are natural nutrient flows and nutrient transfers by trade, e.g., into cities or even export into other countries. These mainly unilateral transfers are of environmental concern since they are not only the cause of nutrient losses from a farming area (and partly completely lost for production), but also unwanted from water eutrophication point of view.

Natural Transfers (Nutrient Flows in Land-Scapes)

A steady flow occurs naturally with surface or ground water movement in hilly or mountainous areas as part of the slow natural erosion process even under natural vegetation cover. The annual losses from the hilly pat of the landscape are relatively small, and so re the gains for the low-lying land. In geological periods, however, this transfer has produced many fertile alluvial sols in river basins that now belong to the best agricultural lands. If this process is accelerated by recent man-made strong soil erosion, it can lead to serious soil fertility decrease in the hilly area and to an excessive input on nutrient (especially N and P) into water.

The extent of nutrient transfers in small watershed areas or whole river basins is difficult to determine because of many factors involved (sources not only from agriculture, but also from towns, industry, etc.). Anyhow, several studies of this kind are conducted in Europe in order to estimate those amount of nutrients transferred into rives or lakes for which the agricultural production of this area is held responsible (e.g., as a basis for restrictive regulations on fertilization).

Nutrient Transfer into Cities

With food and other farm products plant nutrients are transferred from farm areas to villages, towns, big cities and megacities. The rate of recycling usually decreases with increasing size of the settlement and depends on the disposal system for sewage and garbage. Many cities are proud of their perfect sewage disposal system (leading biologically treated sewage water into rivers or the sea and filling garbage into deposits). This system of just getting rid of the waste materials, however, means an enormous loss of plant nutrients with the unwanted secondary effect of eutrophication (by N and P).

Very few cities have a direct recycling of treated sewage onto fields, because recycling of sewage sludge s compost seems to be more adequate. But even so, farmers are increasingly reluctant to apply composted sewage or garbage as nutrient source (even if offered free of charge) because of the well-known problem of toxic heavy metals (and possibly even toxic organic substances), not only in view of decreasing food quality, but as a potential danger for soil fertility. With the past experience of environmental laws getting ever stricter, farmers suspect, that the critical limits for soil contamination might be decreased for political reasons and thus putting otherwise fertile land out of production by law. In addition, farmers dislike being accused by city people of "poisoning" the soil and thus decreasing food quality while helping them to dispose their waste products.

Theoretically, the measures required for an adequate recycling, are well known and feasible, namely to provide (almost) uncontaminated organic materials via separating the different waste materials and excluding toxic ones. Although this poses many problems in detail, a more or less quantitative recycling will be essential from the big cities in the future.

Export into other Countries

The nutrient export with feedstuff and food into other countries can reach substantial amounts that are lost from the national nutrient budget (although not from the global one). Many developed countries with high animal production, even those which could produce sufficient feedstuff by their own agriculture, import substantial amounts of feedstuff and corresponding plant nutrients because of the cheaper price. For the Netherlands this import is outstandingly high, namely about 2/3 of P and K of the total fertilizer import (Cooke, 1989).

On the other hand, as long as this unilateral transfer with agricultural products can be compensated by re-import of mineral fertilizer, the nutrient loss of the developing countries will not really be a serious problem since there seems to be no urgent need for self-sufficiency of countries in plant nutrients.

THE CONCERN

First, there are the scientists concerned with agriculture chemistry who produce research data and present theories. The validity of their information is usually taken for granted, however, some conclusions from field experiments and regional studies are not as reliable as claimed by the authors. Some yield increases attributed to better N-supply correspond in fact to secondary effects (problem of wrong x-axis), the result being an over-estimation of the required N-amounts and higher avoidable losses. As another example, there seems to be a tendency to over-estimate the part of river pollution that must be attributed to the agriculture of the region (whereas other sources tend to be underestimated). Consequently, the scientific information on fertilization, nutrient flows and losses should be critically examined and evaluated.

Second, advisors are supposed to "translate" the theoretical suggestions into a practical "know-how". Unfortunately, the information obtained from experts is not always very precise (especially as for the conditions under which the results are valid) or not really prepared for this purpose. Anyhow, advisors have to combine the information available with their own experience, express it in terms of farmer's knowledge and formulate it for practical use.

Third, there are the farmers, eager to produce as high crop yields as possible and who, if they care for their land as owner or tenant, want to maintain its production potential for crops and animals for a long time. In addition to their own experience, they want intelligible and definite guidelines rather than sophisticated models with plenty of vague factors. They know about the difficulties, uncertainties and disappointments of primary production on the field and do their best to get along with the information available for sustainable crop production. On the other hand, farmers are also entrepreneurs, living in a world where economical considerations have priority over correct farming practice. It would be a long list describing the deviations from what farmers might want to do and what they are forced to do for economical reasons (in order to maintain their income and their farm).

Least, but not last, there is the general public that is more and more represented by city people. Most of them live rather far away from agriculture and are hardly aware of biological and chemical crop production principles. These consumers demand a lot of good and cheap food, pure drinking water, etc. They try to impose on the farmers their ideas of nature, landscape and farming, as well as their demands for town waste recycling. There should be limits to these demands, however, because many deplorable developments in agriculture are the result of the consumer's pressure. Consumers should realize, that farming has its own rules and that high soil productivity is a man-made valuable property that must be sustained, especially by proper plant nutrient management for generations to come. The world population can be fed in much higher numbers than today, but the pollution problems will set stricter limits.

Future Nutrient Shortages

The customary approach of agronomists is to adapt the essentials of food production (especially nutrient use) to the requirements of the population being determined by prevailing socio-economic conditions. In many areas, an increasing number of people must be fed according to their nutritional habits, their location, mainly in cities and their desire for cheap food.

If this finally happens on a global scale, there will be shortages of at least some plant nutrients in the not so far future. However, long time before nutrient shortages become really acute (which may first be the case of P, the bottle-neck of world hunger) within a few hundred years, the remaining reserves should rather be used more to replace unavoidable losses than to comply with wasteful nutrient use and to serve an over-extending food demand.

In the long run, instead of adapting the plant nutrient consumption to food demand, the food demand may have to be adapted to the amount of nutrients that can be annually used for the remaining reserves. Sustainable agriculture is not a matter of decades or centuries, but should go on indefinitely. The future agricultural production must not only be sustainable, but its extent and form should also be adapted to the limited resources. With increasing awareness of limited global plant nutrient resources, some conflicts of interest are unavoidable in different respects.

First, the necessity for farmers to produce as much and cheap in order to be competitive on the market may lead to a certain disregard of many well-proven agronomic principles, and even more of good animal husbandry (a list of society-induced "sins" of agriculture would be rather long).

Second, due to the high demand of animal protein, much more grain and other crops must be produced for vegetative food (e.g., the production of 1 kg of meat requires 3 to 4 kg of grain). This is a well-known fact, of course, but rarely regarded from the nutrient supply point of view.

In this context it may be quoted, that promoting long-term sustainable nutrient management will ultimately require radical changes in the way agriculture and society are organized. These include reintegration of livestock and crop farms and the encouragement of closer physical association of people with farmland. Establishing and implementing performance expectations for contemporary patterns of nutrient flow is especially difficult because the scale of transactions is broad and control of most aspects is far removed from the farm level (Magdoff *et al.*, 1997).

Finally, in a distant future, the old "chinese village" principle may have to be applied to the whole world. If food production (even on all available land and by using all possible means) is limited and there is no population outlet neither to east nor to west nor to other planets, to eating habits may have to shift to predominantly vegetarian food – and even the population number may have to be adapted to the remaining resources.

FROM NUTRIENT FLOWS TO CYCLES

A main objective is converting unwanted flowers into useful cycles. Recycling of nutrients within the farm area tends to be not uniform. For example, slurry is preferably deposited on fields near a farm because of transport costs. On grassland with cows, the manure excreted in distributed very unevenly. Another example is the spotty distribution of ashes after the burning of straw or of the natural vegetation.

None of the cycles is closed, most of them are leaking. Some losses are unavoidable (and may be considered as natural load, a minimum tribute for primary production), whereas, others are avoidable by adequate nutrient management. There is a widespread opinion that decreasing N-fertilization is the best preventive measure for reducing leaching losses of nitrate. This holds only true, however, for fertilization systems where N is over-emphasized (unfortunately not a rare practice in intensive fertilization cropping), whereas, even high application of N together with a sufficient supply of other nutrients (i.e., with no yield limiting minimum factor) may result only in still tolerable losses.

Cycling is a very complex system. Although it is not so difficult to design a complex model of nutrient flows, their quantitative assessment will partly have to be based on some more or less uncertain estimates. Therefore, the result can only be approximate. For practical use, cycling models should be restricted to the important nutrient fluxes.

In the case of big cities (especially megalopolis) the recycling of imported nutrients to the farm area is rather limited. Many waste products are rather deposited into landfills or led into rivers or the sea (after "biological" treatment, but mostly without extracting phosphate)

Organic Farming and Waste Recycling

Organic farming is the backbone of sustainable agriculture. Organic farming mainly depends on organic recycling. Industrial agricultural chemicals like fertilizers, pesticides, herbicides, etc. are not used or used to the minimum extent necessary in this kind of farming. According to the definition of FAO, organic farming should involve successful management of resources for agriculture to satisfy changing human needs while maintaining or enhancing the quality of the environment and consuming natural resources. Organic farming may result in comparable performance as conventional agriculture and crops grown with high organic manure application could tolerate the pest and disease attack better. It is an ecologically sound and sustainable way of growing more food.

Harnessing the nutrient energy from biological and industrial wastes is of prime importance for maximizing production. When these wastes are recycled as manure for crop production and are subjected to the degradation and assimilative capacity of soil, pollution of streams and/or rivers receiving these wastes is reduced to a large extent as compared to their direct disposal in water bodies. Organic recyclable wastes include crop residues, animal wastes, farm/industrial wastes, municipal and sewage wastes. They are valuable sources of plant nutrients and humus. In tropical and sub-tropical soils found in India, there is a general deficiency of organic carbon and plant nutrients due to rapid loss of these components by biodegradation. To make up for these losses, extensive utilization of organic residues in agriculture is essential. In addition, they also protect the soil from erosion. The manurial value and quality of these wastes could be improved by composting and enriching these organic sources along with inexpensive materials such as rock phosphate.

In India, there is a great potential for utilization of crop residues/straw of some of the major cereals and pulses. Approximate availability of straw is to the tune of 141.2 mt, which contribute about 0.7, 0.84 and 2.1 mt. of N, P_2O_5 and K_2O respectively. Considering that about 50 % of these residues are utilized as animal feed, the rest could be very well utilized for recycling of nutrients. Crop residues have wide C:N ratio and when applied directly, nutrients may get immobilized leading to initial adverse effect on crop growth. Hence, adequate care is required to use the residues after proper composting with efficient microbial inoculants. Inoculation of these wastes with efficient cellulotic microorganisms produces enough simple sugars for growth and multiplication of beneficial microflora like free living N fixers and phosphate solubilizers in the soil which ultimately increases crop yields.

While the incorporation of crop residues e.g. wheat and rice straw, as such or inoculated with fungal species had beneficial effects on crop yields and improvement in physico-chemical properties of soils. Incorporation of legume residues viz., green gram or soybean stalks in rice, cowpea stalks in pearl millet or wheat saves energy input besides improving the soil physical and chemical properties.

THE CYCLIC SYSTEM FOR NUTRIENT MANAGEMENT

With a system approach i.e. synergistic integration of more than one component, the management of nutrient is to be done in such away so that there is minimum loss along with addition (in organic form) of the nutrient in the soil, taken out in terms of yield. This makes a cyclic system of nutrient management, which has major role in long term sustainable production.

MINIMIZING NUTRIENT LOSSES

Important loses of nutrients occur through removal of the harvested products, soil erosion, gaseous losses and leaching. In the following chapter we will especially deal with losses through soil erosion, because this is the major form of undesired nutrient loss in Sub-Sahara Africa (Stoorvogel and Smaling 1990).

Factors affecting soil erosion are rainfall intensity and quantity of rainfall, erodability of the soil, length and steepness of slopes and soil cover (Young 1991). Traditional soil conservation projects have been emphasizing the use of physical structures (terraces, etc.) which mainly reduce soil erosion by changing the steepness and length of slopes. Current approaches to soil conservation, however, give more attention to reducing soil erosion by improving soil cover rather than the construction of physical structures (Young 1989, Hudson 1992).

Ongoing research explore technologies to modify slopes and improve soil cover. The following section gives an overview of the current approaches.

Changes in Soil Slope

Reducing soil slope by terracing has been the traditional approach to control soil erosion in many soil conservation projects. Terracing is a very effective way of reducing soil erosion if properly constructed. The major disadvantage is the high labour requirement (Douglas 1994). The current approach is to give less emphasis to complete terracing and instead promote the use of progressive terracing. Progressive terracing implies that terraces are formed gradually as sediments are deposited in front of a contour barrier. This will often also tend to increase the fertility of the soil in front of the barriers because the deposited sediments are often rich in nutrients. The barriers can be made out of stones, earth, crop residues, vegetation strips, trash etc. In the case where the barriers are made out of stones or vegetation, the barriers are semi-permeable allowing water to filter through. They will, therefore, to some extent take care of excess water and thereby avoid complete inundation of the field.

The best choice of barriers depends on the circumstances. It stones are easily available they can be used as building material. Success stories with the use of stone barriers have been reported from Niger (Pretty 1995). The height of the stone barriers will have to be increased as sediments are deposited and the structures will also need to be repaired.

Vegetation strips can also be used as an effective barrier especially on low to moderately sloping land. The most simple way to establish vegetation strips is to leave some strips of the land unploughed (Osunade and Reij 1996). Vegetation strips can also be establishing by planting. However, vegetation strips are difficult to establish in areas where there is a high grazing pressure. Under such circumsntances non-palatable grasses or herbs should be used as strips. In order arise an interest among the farmers to plant vegetation belts, the plans should have some economic value by, for example, providing thatching material. Farmers should be encouraged to identify local plants, which can be used in barriers. Species used as vegetation strips should not be too aggressive as they can then easily turn into weeds. If grazing can be controlled, use of palatable species can be used in the vegetation strips. This will offer a much greater range of species to choose from. The barriers should contain species which have a high economic value.

Another simple way to form barriers is to crop residues, weeds or any other material, which is readily available (trashline). Such barriers are not very suitable on steep slopes as they easily break. The major advantage is, however, that they are easy to make and repair.

It is also possible to combine different materials, either using them in alternative barriers or combining them within the barrier. Trashlines can for example be strengthened by using stones at weak points.

No universal answer exists with regards to which material should be used in physical structures. Farmers and project staff need to continuously test and screen technologies.

Soil Cover

Improving soil cover can be just as effective in controlling soil erosion as changing the slope (Young 1989). The initial process in water erosion is detachment of soil particles by the force of raindrops. Improving soil cover will reduce the velocity of the raindrops, thereby halting soil erosion in its initial phase. Soil cover can be altered by changing the crop, increasing the yield by use of soil fertility enhancing methods and use of surface mulch.

Controlling soil erosion by changing species composition from annual to perennial crops is an important way of controlling soil erosion. Annual crops are especially inefficient in controlling soil erosion early in the growing season before they become well established. Perennial plants will cover the ground for a longer period of time and have a more dense cover. It has been reported from Rwanda that coffee is in the order of 18 times as effective as maize in controlling soil erosion (Lewis 1987). Farmers should therefore be stimulated to use perennials such as coffee, tea and fruit trees. It has, however, not been very easy to convince farmers to plant trees in their farm land. Alley cropping, a technology to which great expectations were attached in 1980s has not been successful among farmers.

The main reason for using soil fertility enhancing methods is to increase yield. However, IPNM technology will reduce soil erosion because plans in vigorous grown give a good soil cover. Erosion can be reduced to less than half under a well managed maize crop as compared to a maize crop grown under poor management (Roose 1977).

Another way of improving soil cover by increasing yield is to undertake appropriate and timely tillage. In Ethiopia, oxen are used for ploughing. However, there is often a shortage of oxen and the individual farmer does not always have access to the two oxen considered necessary for tillage (Aune and Aragay 1998). This causes delayed sowing leaving the land unprotected early in the growing season. Research by the International Livestock Research Institute (ILRI) has shown, however, that one ox in good shape can perform the task just as efficiently as two oxen in bad shape. Reduced tillage can be a method for reducing soil erosion because soil left undisturbed is more resistant to the erosive forces of the rain

HARNESSING UNTAPPED NUTRIENT SOURCES

If used appropriately, the recycling of organic waste from urban to rural areas is a potential, largely untapped, source of nutrients for farm and crop needs, especially on agricultural lands near urban centers. For example, environmentally undersirable wastewater has been used to irrigate fields and return nutrients and organic matter to the soil (Tandon 1992). Like organic manure, urban waste sludge is a source of primary nutrients, albeit a relatively poor source in comparison with commercial fertilizers. Stabilized municipal waste sludge typically contains about 3.3 percent nitrogen, 2.3 percent phosphorus, and 0.3 percent potassium, although some concentrations can reach as high as 10 percent nitrogen and 8 percent phosphorus on a dry weight basis (EPA 1984). Actual nutrient content, however, varies widely and depends on the source of the waste. Urban waste also has a number of other benefits. Like other organic manures, it helps improve soil structure by adding organic matter to the soil. It is also a source of the secondary nutrients and micronutrients that are necessary in small quantities for proper plant growth. In addition, urban waste transforms material that would otherwise be slated for costly disposal into a useful farm product.

Urban waste also contains organic compounds such as dyes, inks, pesticides, and solvents that are often found in commercial and industrial sludge. These pathogens have been shown to cause genetic damage, while others, such as bacteria, protozoa, and viruses can cause salmonellosis, amoebic dysentery, and infectious hepatitis (Conway and Pretty 1991). Untreated urban waste can put these pathogens in contact with fruits and vegetables. One option is to compost the sludge. Composting concentrates nutrients and helps to kill disease-causing organisms, slow the release of nitrogen that might otherwise percolate into groundwater, and eliminate aesthetically objectionable odors (Kurihara 1984). Another option is to use ionizing radiation to kill pathogens in and on food without affecting taste. Despite some public concern about the safety of food irradiation, the technique is likely to be adopted more fully in the future in order to protect public health, improve the shelf-life of food, and make it more feasible to apply treated, nutrient-rich urban waste to farmland.

Currently, effective use of urban waste is hampered by its high water content, bulkiness, distance from rural areas, contamination with non-decomposable household items and high handling, storage, transport, and application costs. However, given the cost and the lack of availability of inorganic fertilizers in some areas, the relative abundance and benefit of urban waste as a soil amendment, and the rising cost of environmentally safe waste disposal, economies may make urban waste an appropriate fertilizer choice in areas where agricultural lands are near urban centers.

Alternative sources of inorganic fertilizers will also be required in the future, particularly in those parts of Africa where the fertility of the soil needs to be rebuilt and high costs and supply constraints limit the application of fertilizer. Soil infertility (particularly phosphorus deficiency) in parts of semi-arid West Africa, for example, limits crop productions more than the lack of moisture. Phosphorus application of 15-20 kilograms per hectare can substantially improve crop yields. Medium-reactive and partially acidulated, less-reactive phosphate rock found in Mali, Niger, and Senegal are as agronomically effective as commercial superphosphate fertilizers (Bationo and Mikwunye 1991). Low-cost technology needs to be developed so that phosphate fertilizer can be locally produced from these and other untouched or currently uneconomical phosphate rock reserves. Where phosphate deficiency is sever in Sub-Saharan Africa, government assistance in developing low-cost technology and in applying phosphate fertilizer should be evaluated. Because phosphorus and phosphorus rock bind to the soil and thereby reduce the opportunity for leaching, and because these fertilizers release nutrients slowly over time, their use to preserve the long-term productive capacity of the land should be considered more of a capital investment than a subsidy or an environmentally undesirable government intervention (Mokwunye 1995; Gerner and Baanante 1995; Teboh 1995).

EXPLORING INTERNAL NUTRIENT SOURCES

Although new sources of nutrients can be developed, genetic engineering offers the potential for plants themselves to generate some of the nutrients they require through nitrogen fixation. In this process, rhizobium bacteria infect, invade, and draw energy from leguminous plants, and in return the bacteria convert and store atmospheric nitrogen in a form that the plant can use for growth (Rao 1993). Besides helping the plants themselves, cereals grown in rotation with leguminous plants can absorb the nitrates released from the decaying roots and nodules of the leguminous plants. Experiments have shown that rice-legume rotations can result in a 30 percent reduction in chemical fertilizer use (Pingali and Rosegrant 1994).

Genetic research has begun to identify the genes responsible for such nitrogen fixation and assimilation. Further research offers the opportunity of altering or developing microorganisms that can fix nitrogen in nonleguminous plants, such as cereals. As with leguminous plants, plant nitrogen needs could be partially met by the plant itself, such that farmers would then simply need to top-up crops with inorganic nitrogen fertilizers (Rao 1993). The task is considerable. Some 17 genes code the enzymes involved in nitrogen fixation. Since these genes, as well as the genes necessary for nodule formation, need to be transferred, the process is complex and its realization will be costly. Furthermore, because the amount of energy required to fix 150 kilograms of nitrogen per hectare could reduce wheat yields by 20-30 percent, an appropriate balance needs to be found between the nutrient-supply-enhancing benefits of nitrogen fixation and the potential reduction in yields (Greenley and Farrington 1989; Lipton and Longhurst 1989).

BETTER USE OF NUTRIENT SOURCES

Soil Nutrient and Soil Fertility Aspects

Soil nutrients are the basic source of the farm nutrient supply. Part of them is utilized by crops, i.e., the easily available portion (water-soluble, exchangeable) as well as the easily mobilizable fraction. The mobilization of nutrients from (mineral and organic) slowly available sources can be enhanced to a certain extent by activating soil life (in general by organic matter of by special biofertilizers), by crop varieties with strongly mobilizing capacity, by better accessibility of nutrients after structure improvement, by deepening of the plow layer or by fallow periods.

The best use of nutrients for crop growth can be obtained on the basis of a high soil fertility level. Soil fertility is a complex term (not very precise, but very useful even in its vague form) which includes many components: soil depth, texture and structure (pore space for supply of oxygen and water), soil reaction, organic matter content and composition, activity of soil organisms, nutrient content, storage capacity for nutrients, content, respectively, absence of detrimental or toxic substances. The result of an optimum combination of these factors is a high soil fertility that mean as high crop production potential.

For the practical use of soil nutrients, the basic problem is the estimation of the amount that will be available during the growing season (in order to fill the remaining gap with additional nutrients). The main diagnostic method is soil testing for available nutrients, and most countries have their soil testing system calibrated for special areas. Whereas, the analytical methods are rather precise, due to the heterogeneity of most soils, the diagnosis can only provide an approximate estimate. Nevertheless, it is possible to draw useful conclusions on the necessity of additional nutrients. For more precise information, systems of plant analysis have been developed.

Very fertile (and thus productive) soils are rarely found in nature. Unfortunately, most soils are far from being ideally fertility and should therefore be improved, not only by adding nutrients, but also by other soil amendments, e.g., lime for correcting strong soil acidity, organic matter for maintaining the activity of "soil life", etc.

Soil fertility may be rather uniform in a field, but it sometimes varies within a short range (in some fields visible by considerable differences in plant height). In such cases, soil productivity studies (by soil and plant analysis) are suitable to determine the cause of such differentiated growth and offer the possibility of eliminating the limiting factor.

Optimum Soil Reaction

Many soils, due to a continuous acidification caused by several factors, are a little or much too acidic for high productivity. Therefore, liming is required and the soil should have its optimum soil reaction which varies with soil texture, cropping intensity, etc. The usual (local) recommendations for optimum pH are generally higher with a higher clay content and yield level.

Optimum pH values or ranges are often a compromise between two different goal: first a good soil (crumb) structure which requires rather natural pH range on heavy soils and second a better rate of mobilization of most nutrients, especially micronutrients like Fe, Mn and Zn, which requires a somewhat lower pH range.

In rare cases of over-limed soil in intensive cropping areas, the pH can be temporarily lowered by application of acidifying N-fertilizers (mobilizing nutrients like Mn of Zn on tiny spots).

Soil Organic Matter

The qualitative importance of organic matter for soil fertility is rather well known, but many quantitative aspects are still to be solved. The main functions of soil organic matter are physical soil improvement (increase in water and air capacity, protection against erosion, etc.) and several chemical functions (increase of sorption capacity, mobilization of nutrients by mineralization of organic matter and from minerals, short term immobilization into soil organisms (which thus compete with crop for N) and long term fixation into stable humid substances (e.g., improvement of quality via C/N ratio), supply of beneficial organic substances (growth promoters, antibiotics, etc.)

The importance of soil organic matter differs in climatic regions, the role of nutrient source often being a dominant one. It should not be decreased, but maintained at an acceptable level, above critical values corresponding to structural stability and biological activity. In rainfed tropical agriculture, soil organic matter is the corner stone for the building of any sustainable annual cropping system, but there is not yet a definite answer to the critical level that should be maintained (Pieri, 1995).

In areas of intensive cropping, an increased level and better quality of soil organic matter has significantly contributed to the high productivity level.

ORGANIC MATERIALS

There is a large variety of materials produced on the farm as well as "imported" from outside, especially communal and food processing waste materials.

Crop Residues and Farm Manures

Crop residues vary considerably in their nutrient contents (e.g., grain straw vs. green leaves). Residues and manures should be returned to the fields with only little nutrient losses. As for animal manures, the change from the old-type animal manure/straw mixture to slurry (induced by the labour-savings but otherwise somewhat doubtful progress due to straw-less stubbles) has substantially contributed to N-losses. Main losses of animal sludge distribution occur from ammonia volatilization (sometimes more than half), but losses can be reduced substantially by application of the sludge into the soil instead of spreading it on the surface, the latter procedure being unfortunately still widely used in practice.

A general question of nutrient management is the alternative to using waste materials directly or improving them by special treatments, especially by effective composting.

In many regions, there is a competitive use of organic matter that could be used as feedstuff, fertilizer, fuel for direct burning or for biogas. As a rule, organic matter fit for soil application should not be burned, although in some situations, there may be little choice for a farmer.

Organic materials should be applied to crops which make the best use of them, e.g., the immediately available ammonia of slurry given to crops before the period of rapid growth, or slowly acting N-sources to crops with a long growing season and with extra profit from the structural improvement for root growth.

Under humid conditions, crop residues are generally mixed into the topsoil as a nutrient source and also for soil improvement. Input of organic material should not only increase, but also improve (higher C/N ratio, etc.) soil organic matter (Burgos, 1995). In arid regions (although an increase in soil humus would even be more important), crop residues are often left on or near the surface for soil erosion control and for water conservation (Prasad and Power, 1991).

Under high yield conditions it may be difficult to get the straw of the previous crop decomposed in a short time and there can be negative effects like temporary fixing of available N (due to the large requirements of the decomposting bacteria). The practice of burning a surplus of straw and just using the ash as a mineral fertilizer is now forbidden in some countries for environmental reasons (smog).

Commercial and Industrial Waste Products

The waste products that can be used as nutrient sources are materials produced by composting from sewage sludge or (organic) garbage.

It is in the interest of towns and food processing industries to return their waste products to farmers' fields. These waste materials may even be offered as cheap nutrient sources which can save mineral fertilizer input. The advantage for a farmer can be estimated by taking into account the nutrient content and some additional organic matter effect.

There is a growing concern, however, about some toxic elements and even toxic organic substances that might decrease soil fertility and quality of food and fodder products. Especially in the case of some heavy metals farmers should be very conscious of not enriching their soils above the toxic limit in an irreversible way since amelioration will prove to be very difficult. Some protection against this danger is given by keeping the contents of toxic substances below tolerable limits and by using these materials only on soils which still are rather uncontaminated.

Other Organic Fertilizers

It is well known, that with some biofertilizers the fixing of atmosphere N can be increased by organisms such as *Rhizobium. Azotobacter,* blue green algae, *Azolla/Anabaena* (Gaur and Singh, 1995).They are described in detail in the previous chapters.

BETTER NUTRIENT MANAGEMENT FOR CROPS AND CROP ROTATIONS

Adaptation of crops and crop rotations to nutrient supply

For cropping conditions of low natural nutrient supply and restricted input special emphasis must be placed on the nutrient efficiency of crops, mainly uptake efficiency, but possibly also internal use efficiency.

Crop species and even varieties differ to a certain extent in their ability for mobilization and uptake of soil nutrients, especially under conditions of low natural supply. Since under low input conditions the contribution of soil nutrient reserves (especially P and K) is of eminent importance (Syers, 1995), varieties should be chosen that provide special mechanisms like root excretions for their mobilization and for improved uptake. The same holds true for cases of rather immobile single micronutrients (e.g., Fe, Mn, Zn, Cu deficiency).

In intensive agriculture more emphasis is placed on high yield performance and disease resistance than on nutrient efficiency since it is more economical to supply extra nutrients (for micronutrients even by leaf spray).

The uptake from only slowly available soil reserves can be further improved by choice of crops with an extra potential of nutrient mobilization via mycorrhiza or rhizosphere associated organisms. Symbiotic mycorrhiza fungi (VAM = vesicular-arbuscular mycorrhiza) are widely present in fertile soils (or may be inoculated for special crops). The effect of this symbiosis is an improved nutrient uptake in general, mainly a better P-uptake of the roots due to better mobilization of soil P reserves and also to more efficient use of added P fertilizers, furthermore an improved uptake to other nutrients with low mobility like some micronutrient metals. VAM improves the productivity by reducing the input, whereas relying more on soil nutrient reserves, but also by enhancing plant survival under nutritional stress conditions (Barea, 1991).

The most striking example of crop adaptation to nutrient deficiency is the cropping of legumes with rhizobia for fixing nitrogen. An example typical for paddy rice is the *Azolla-Anabaena* symbiosis.

Not only a crop has to be adapted to nutrient supply, but also general agronomic measures like proper weed control (saving nutrients for crops) and efficient crop protection (for obtaining and protecting high yields) are required to make the best use of nutrient sources.

Adaptation of Crop Rotations to Nutrient Supply

Although monocultures (like tree plantations or sugar cane) offer the easiest way of production on arable land, many crops are not self-compatible and a farmer usually has to consider the nutrient supply of crop rotations (cropping systems).

Rotations are planned for several (an partly competing) reasons: for special products wanted due to economical reasons, for maintenance of soil fertility, for plant disease and pest control, for nutrient supply, etc.

From the crop nutrition point of view, different aspects have to be considered. Rotations can consists of crops with high or low nutrient requirements or of crops known either to improve or decrease soil structure and activity of soil life.

The well-known example of legumes alternating with non N-fixing crops should also be mentioned. Rotations may be chosen which do not leave the sol without green cover (intermediate crops, green manure or even weeds) in order to prevent nutrient losses form bare fallow in the period with pronounced leaching.

The soil nutrient supply may also be improved by not using one crop, but crop combinations like intercropping or mixed cropping (Francis, 1989). A widely applied practice is the simultaneous growth of grasses with legumes in fodder production.

NUTRIENT MANAGEMENT FOR PRODUCT QUALITY AND STRESS RESISTANCE

Higher Quality of Food and Fodder Products

Nutrient management, beyond the yield aspect, includes also aspects of quality since food or fodder should contain sufficient minerals elements for healthy growth of men and animals, especially since some critical values for quality are higher than those for high yields. As a general rule, better nutrient supply of crops improves food quality, i.e., increase the mineral and organic components required by men. On the other hand, an unwanted surplus of some nutrients (especially N) may decrease quality parameters. Food quality also includes aspects of undesirable toxic elements like cadmium, the uptake of which should be kept as low as possible.

Nutrient Management of Grassland for Fodder Quality

The nutrient cycling in grazed pasture ecosystems has been reviewed by Haynes and Williams (1993). A prominent feature is the recycling with a very uneven distribution because of the spotted deposition of excretions by grazing animals.

As for the quality aspect, an adequate nutrient supply of grassland must take into account not only the yield (of milk, meat, etc.), but also the health and fertility, e.g., of cows. The nutrient requirements of grassland plants and animals are more or less similar, but they can also differ in part. For example, for highly productive milking cows a higher content of P, Mg, Mn, Zn, Cu in the fodder is required than for a high grass yield. In addition, the special requirement of cobalt, selenium or even chromium for animals should be considered, especially in deficient are as (unless mineral supplements are given).

If grassland is included in crop rotations, there should not be too much carry-over of such nutrients, which may disturb the nutrient balance of fodder, e.g., surplus of K may induce Mg deficiency.

Higher Stress Resistance of Crops Due to Better Nutrition

The usual concept of nutrient supply is based on "normal" growth conditions. In fact, however, there are frequently constraints for crop growth under stress conditions, which require special nutrient management.

The stress situation is often aggravated by combination of several factors, e.g., low temperature at the time of crop protection spraying.

Resistance of crops against cold stress can be increased by optimizing the supply of some of the nutrients (K, P, Mn, Cu).

For arid conditions, it is important to realize that with better crop nutrition less water is required and that especially with good K supply the transpiration of crops is reduced. Saline soils need a special nutrient management for reducing the uptake of unwanted Na, etc. The water-use efficiency depending on plant nutrient supply is reviewed by Davis (1994).

Soil acidity, problems are wide-spread in humid areas, the most striking example being the acid sulfate soils. Soil below about pH 4.5 mostly produce low yields (if at all) and the plants are more susceptible to diseases due to the soil acidity syndrome (lack of some nutrients, especially P, and surplus of toxic substances like Al).

Soil-borne stress can be eliminated or at least decreased by soil amelioration. On soils polluted with toxic heavy metals their negative effect on crops can be decreased by special amendments (stronger fixation of the mobile portion by liming and metal fixing agents).

A balanced nutrient supply (instead of a relative N-surplus) generally increased the resistance against certain bacteria and fungi (resulting in a lower requirement of pesticides).

NUTRIENT MANAGEMENT IN FARMING SYSTEMS

All farming systems have their qualifications under special natural or economical conditions. Of the determining factors nutrient management may not have priority, but it nevertheless plays an important (an increasing) role, namely to make the best use of soil and fertilizer nutrients within the framework of the dominating factors for a special farming system.

For the present purpose, various types of farming systems and agricultural land utilization are shortly listed in alternative groups with some corresponding problems of nutrient management added.

Efficient Nutrient-Recycling with Agroforestry

Agroforestry systems can lead to more closed nutrient cycling than agriculture, and hence to more efficient use of nutrients. Under natural woodland and forest ecosystems, a large quantity of nutrients circulate annually between plants and the soil; in a mature (climax) community the plant-soil system as a whole is in equilibrium and the inputs and outputs of nutrients are small. A high-input agricultural system, in contrast, has inputs processes (leaching, gaseous loser and erosion) and removal in harvest. The element of recycling, through decomposition of root and, perhaps, crop residues is relatively small. By the inclusion of trees, agroforestry systems can achieve a condition intermediate between these extremes, by reducing losses and increasing plant uptake.

This can be expressed quantitatively as the nutrient-recycling ratio, treating the plant-soil complex as a system. Gains and losses are external to the system, uptake and return are internal transfers, respectively soil to plant and plant to soil. Treating the organic and mineral parts of the soil together and indicating where processes affect certain nutrients only, flows in the plant-soil nutrient cycle.

Gain	Atmospheric: rain and dust Fertilizer Organic additions fixations (nitrogen) Symbiotic fixation (nitrogen) Deep capture (uptake by roots of trees from below the rooting depth of crops) Rock weathering (excluding nitrogen)
Losses	Erosion Leaching Harvest (including for fodder) Removals not treated as harvest Gaseous losses: denitrification and volatilization (nitrogen) Burning (nitrogen, sulphur) Net fixation on to clay minerals (phosphorus)
Internal flows	
Uptake	Uptake by plants from soil
Return	Mineralization of liter Humification of liter Throughfall and stenflow Ash from burning

For any specified nutrient

Nutrient balance (kg ha^{-1} $year^{-1}$) = Gains – Losses

$$\text{Nutrient recycling ratio (\%)} = \frac{\text{Uptake} + \text{Return}}{\text{Gains} + \text{Losses} + \text{Uptake} + \text{Return}} \times 100$$

The nutrient balance of the system is the gains minus losses, as kg ha-1 year-1 of the nutrient, and may be positive or negative. The recycling ratio is the average of the internal flows (uptake and return) as a proportion of the external plus internal flows, expressed as a percentage. The recycling ratio would be 100% in a closed system, with neither gains nor losses, and 0% if there were no internal recycling. In rotational systems, both these indices can be averaged over a cycle of land use.

Mature natural forest and woodland have nutrient balances close to zero and can attain recycling rates as high as 90%. For example, for the phosphorus cycle of a rain forest in Panama, the flows (kg phosphorus ha^{-1} $year^{-1}$) were measured as gains – 1.0, losses = 0.9, uptake = 11.0, return – 11.9, a phosphorus recycling ratio of 92%. In savannahs, the trees act as a stabilizing influence, having a slower nutrient cycle than the grasses. In natural ecosystems, the capacity of trees roots to scavenge nutrients and concentrate these beneath their canopies results, over time, in a mosaic of islands of higher fertility, which influence community structure and ecosystem junction.

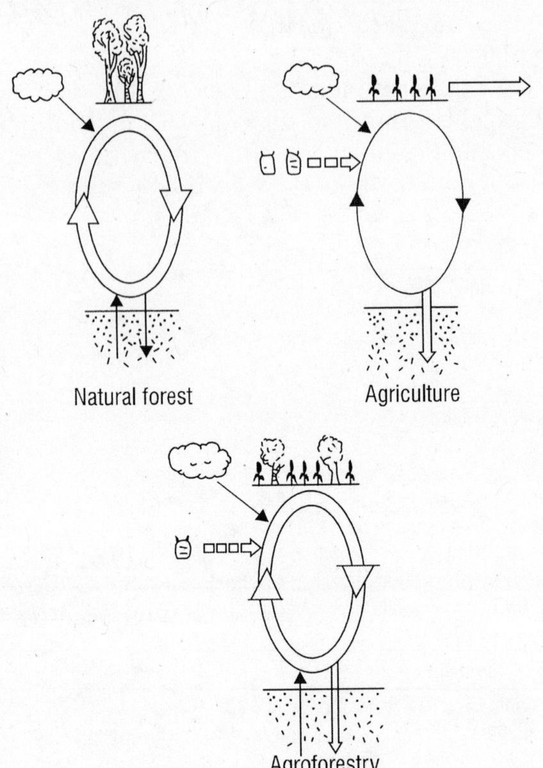

Fig. 8.1: The agroforestry nutrient-cycling hypothesis.

In contrast, the nitrogen budget for a fertilizer cereal crop, ignoring smaller flows, might be of the order of (kg N ha^{-1} $year^{-1}$).

Gains	Fertilizer	100
Losses	Leaching, etc. Harvest	50 100
Uptake		150
Return	Root residues	50

The nitrogen balance is 100 – 150 = -50 kg N ha^{-1} $year^{-1}$, and the recycling ratio (200/450) x 100 = 44%.

To reduce harvest conflicts with the objective of managed land-use systems. Increases in nutrient-use efficiency through agroforestry can therefore be achieved primarily by

- Increasing the cycling of nutrients from tree liter and pruning, via the soil, into crops
- Reducing losses by leaching;
- Reducing losses by erosion.

On sloping land without erosion control, the nutrient budget is dominated by losses in eroded soil. Where erosion is moderate to serve, these may amount to an order of 10-100 kg ha-1 of nutrients. The large savings in nutrients that can be made by controlling erosion have been noted above.

Since leaching is a loss in the course of the tree-to-crop pathway, recycling and leaching can be considered together.

Recycling Paths: Short-Term and Longer-Term

There are two paths for nutrient recycling from plant to soil and back into plants: the shor-term path, via mineralization of plant residues, and the longer-term path, via humification and storage as soil organic matter. The short-term or mineralization path takes place within the time span of the growth of an annual crop, in the tropics largely within the first 3 months. In the longer-term or humification path, nutrients stored as organic molecules are released slowly, with a half-life of the order of 10 years in the tropics and up to 50 years in the temperate zone.

In agroforestry, the short-term path concerns nutrient transfer from tree residues to crop grown in the same year. This involves questions of plant-residues decomposition, leaching loss and synchrony between nutrient release and uptake. This aspect has tended to dominate recent agroforestry research. The longer-term path, in which nutrients recycled by trees may reach crops in subsequent years, should not be ignored. It is one of the two major roles played by soil organic matter, along with that of maintaining soil physical properties. Work is in progress to compare the short-term and longer-terms paths with respect to their effects on soil fertility.

AGROFORESTRY SYSTEMS

Numerous agroforestry systems have been developed in different areas. One system differs others in respect of structure, composition, age, intensity, technology etc. Classified agroforestry systems on the basis of structure, function, socio-economic and ecological factors. Suggested classification based on structure, function, physiognomy, floristic characters, ecology, socio-economic scales and history. Each of these criteria has merits and demerits and no single classification method can be accepted as universally, applicable. Therefore, classification schemes will have to be purpose oriented. Since there are three basic sets of components viz. trees, agricultural crops and animals therefore, agroforestry classification based on nature of components is a logical one. Suggested following common agroforestry systems in India on the basis of nature of components.

1. Agrisilviculture (Tree + crops)
2. Boundary plantation (Trees on boundary + crops)
3. Block plantation (Trees + crops)
4. Energy plantation (Trees + crops during initial years)
5. Alley cropping (Shrubs + crops)
6. Agrihorticulture (fruit trees + crops)
7. Agrihortisilviculture (Trees + fruits trees + crops)
8. Agrisilvipasture (Trees + crops + pasture/animals)

9. Hortipasture (Fruit trees + pasture/animals)
10. Silvipasture (Trees + pasture/animals)
11. Forage forestry (Forage trees + pasture)
12. Shelter belts (Trees +/- crops)
13. Wind breaks (Trees +/- crops)
14. Live fence (Shrubs/ under trees on boundary)
15. Homestead (Multiple combination of trees, fruit trees, vegetable etc.)
16. Entomoforestry (Trees + sericulture)
17. Aquaforestry (Trees + fishes)

Besides the aforesaid important common systems of agroforestry, there are many more component combinations.

Role of Agroforestry

Agroforestry fulfil both production and service roles. Agroforestry can provide a more diverse farm economy and stimulate the whole rural economy, leading to more stable farms and communities. It increases the income of farm family by providing various products like food, fodder, fruit, nuts, fuel wood, fibres, timber, moles, mulch, medicines, cosmetics, oils reigns, lac, honey, silk etc.

Agroforestry can play a very efficient role in the utilization of all the natural resources in a most effective manner for sustainable crop production. It provides services such as food security, conservation of soils, improvement of soil fertility, moderate microclimate, provide, living fences for crops and fruit trees, demarcation of boundaries, reclamation of degraded lands, stabilization of watersheds and protection of biodiversity.

Agroforestry for Soil Enrichment

Agroforestry system has the potential to improve the fertility of soil. Agroforestry trees enrich the soil by adding organic matter through litter fall in the form of dead and falling leaves, twigs, branches, and dead roots. The tree helps in recycling of nutrients in the soil. The nitrogen fixing trees species enrich the soil through biological nitrogen fixation. Some trees species have found to fix 100-500 kg N per ha. *Leucaena leucocephala* has been reported to fix as much as 400-500 kg N per ha. The important contribution of trees under agroforestry can be to furnish sufficient quantity of fuel wood thereby diverting 300-400 million tonnes of cowdung for use as manure in the agricultural fields rather than burning it as fuel.

Ecological Balance

Our ecological balance is under threat as a result of reducing area under forest. Once forests covered more than half of the Earth's land. Now less than one-third of our planet is forested. Every year more than 15 million ha (37 million ac) of World's forest area are converted to other uses, mostly for in urbanization and large-scale agricultural enterprises but some times by poor, small-scale farmers searching for agricultural lands. In India, forests are confined mostly to hilly soil and poor soils, which are not suitable for agriculture. Marginal lands, environmentally fragile areas, more suitable to forests than agricultural has been brought under cultivation, which has further aggravated the problems of ecological imbalance. The forests now cover only 19.47 per cent of the geographical area of country whereas National Policy provides that the forest in hilly region should be 66 per cent and for the country as a whole 33 per cent. Agroforestry offers considerable scope of harnessing advantages of forest by growing trees on marginal/sub-marginal lands with crops on agricultural lands. Agroforestry therefore, can restore ecological balance without reducing the area under agriculture as such. Non-arable marginal and degraded lands, which are about 43 million ha in India, can be brought under agroforestry systems.

Organic Matter and Nutrient Addition to Soil

In agroforestry system, major recognized avenue for addition of organic matter and nutrients to the soil is through aboveground biomass like literfall/pruning. Twigs, branches, fruits etc. and belowground biomass such as roots.

Agroforestry systems have the potential to maintain organic matter of soil. Estimated the amounts of plant residues (both above and below-ground) that are needed to maintain the soil organic matter in the three major climatic zones of the tropics. This general of estimates gives a required annual addition of above ground plant biomass, in round figures of 8000 kg, 4000 kg and 2000 kg DM/ha/yr for humid (forest), sub-humid (savanna) and semi-arid environment, respectively. These figures may be compared with the potential of agroforestry trees to produce such amounts of biomass. The biomass production potential of multipurpose trees indicates that agroforestry systems can provide required amount of plant biomass for maintaining soil organic matter.

The literfall/purning add nutrition to the soil under agroforestry systems. Reviews indicate that leguminous trees in alley cropping system produced up to 20 t/ha/yr dry matter or pruning, can supply as much as 358 kg N, 28 kg, P, 232 kg K, 144 kg Ca and 60 kg Mg. The nutrient contents of plant biomass vary widely. It depends mainly on tree species, relative proportions of leaves and stems in the pruning and their respective nutrient concentrations. In leguminous trees the N concentration of leaves varies from 1.5 to 3.4 per cent or more. The nutrient content in biomass of some multipurpose tree species (MPTs) in agroforestry systems.

Biological Nitrogen fixation

The nitrogen fixation by the trees represents a clear gain to the nutrient economy in agroforestry systems. The nitrogen fixing trees are introduced in agroforestry system to provide nitrogen to the companion crops. In a green house experiment under N-limiting nutrient medium. Estimated 40 per cent increase in N content over a period of 185 days in sorghum intercropped with nodulated *Leucaena leucocephala.* The increase in N content in sorghum suggested that fixed N was transferred from *Leucaena* to sorghum roots via soil. Two possible sources of N from legume are root/nodule turnover and root/nodule exudes.

About 650 trees and large shrub species belonging to 9 plant families are known to fix nitrogen. The legume (*Luguminosae*) dominates this list with about 320 mimosoids, 170 papilionoids and 25 eaesalpinoids. Non-leguminous families include Betulaceae, Casuarinaceae, Coriariaceae, Elaeagnaceae, Myricaceae, Rhamnaceae, Rosaceae and ulmaceae. The N-fixing potential of trees varies widely. The reported amount of N-fixed by most trees suitable for agroforestry is ranging from 20 to 300 kg N/ha/yr. Some of tree could fix as high as 600 kg N/ha/yr. The estimate of N-fixed per hectare per year is 13 kg for *Gliricidia spium* 20 kg for *Acacia albida*, 35-40 kg for *Inga jinicwil*, 60 kg for *Erythrina poeppigiana*, 60-100 kg for *Casuarina equiselofolia*, 83-109 kg for *Sesbania rostata*, 200 kg for *Acacia mearnsii*, 350 kg for *Sesbania sesban*, 100-500 kg for *Leucaena leucocephala*. In general, the fixed nitrogen is 20-65 per cent of the total plant nitrogen.

Nutrient Cycling

The combined production of agricultural and tree crops in agroforestry is more efficient in nutrient cycling than monoculture cropping systems. The trees have more deeper and spreading root systems than herbaceous plants of monoculture cropping systems and thus have more potential to capture and recycle nutrients from deeper soil layers. The deep roots of trees may serve as safety net for capturing nutrients that leaches below the rooting zone.

The primary inputs of nutrients for soil plant systems in agroforestry are pruning/literfall, root biomass and biologically fixed nitrogen. The nutrient additions through these inputs are discussed in previous section. The trees are also involved with inputs of nutrients from the atmosphere and from weathering of primary rock as wet deposition (dissolved in rainfall) and dry deposition (in dust). The trees augment dry deposition of nutrients by reducing near-surface wind velocity. The nutrients also enter into the agroforestry system from the weathering of primary rock minerals. The deep-root systems of trees uptake newly released nutrients from weathering of rocks.

Reviewed the nutrient-cycling potential of agroforestry system. Studies on *Erythrina poeppigiana* as shade trees over *Coffea arabica* in *Coasta Rice* have proved the nutrient cycling potential of agroforestry system.

Some trees and shrubs can accumulate certain nutrients even in soil that contain very low levels of these nutrients. Palms accumulate potassium, tree ferns accumulate nitrogen, *Gmelina arborea* accumulates calcium, and *Cecropia* species appear to accumulate calcium and phosphorus in acid soils. Introduction of these nutrient conserving tree/shrub species in agroforestry may help to enrich nutrient status of the system.

Conclusion

Keeping in view the potential of agroforestry systems to meet the increasing demand for food, fodder, fuel and other useful products from shrinking per capita availability of land resources, more emphasis should be given on efficient nutrient management for higher farm productivity on sustainable basis. The nutrient management is more difficult in agroforestry systems than in annual agricultural crops or forest systems as these are structurally and functionally more complex. Some efforts have already been made toward nutrient management in agroforestry by various means e.g introduction of nitrogen fixing tree species, nutrient inputs in organic forms through litterfall/prunings, application of mineral fertilizers, improved designed systems which maintain and improve soil productivity by protecting soil from erosion. However, these efforts are not sufficient for efficient nutrient management in various agroforestry systems. Therefore, more research works should be carried out for better management of nutrients.

Studies have shown that nutrients, which are recycled through tree roots and their mycorrhiza in forest stands can be higher than above-ground biomass, but similar information is lacking in agroforestry system. There is little information regarding nutrient contribution through tree roots in agroforestry. Thus, further research is needed on the role of roots in carbon and nutrient cycling in agroforestry.

In agroforestry, woody legumes, are used as nitrogen source for the systems. In comparison to nitrogen, phosphorus role is limited in agroforestry as recycle of phosphorus through litterfall/pruning is low and removal by the produce is relatively high. The responses of phosphorus application in agroforestry systems have infrequently studies. Therefore, further study should be undertaken in the direction. The emphasis should also be given to study on response of other nutrients including micronutrients its information on these are lacking in agroforestry systems.

The nutrient release patterns of organic inputs are mainly dependent on quality of plant biomass. The synchronization of nutrient release patter with crop demands is most important to achieve higher productivity from agroforestry systems. Some management strategies have already been developed on nutrient synchrony. However, more information is needed on nutrient synchrony for better utilization of plant biomass and recycling of nutrients to increase productivity.

EXPLOITATION CROPPING VS. SUSTAINABLE AGRICULTURE AT DIFFERENT YIELD LEVELS

In this section some examples will be given of farming systems with nutrient exploitation and of sustainable systems at low, medium or high yield levels.

Exploitation Cropping

Exploitation of soil nutrient basically means cultivating crops until available soil nutrients are exhausted and the yields are so strongly declining after some years that the field must be abandoned and left again to natural vegetation.

A typical example of exploitation and thus non-sustainable cropping is the (standard-type) shifting cultivation used by subsistence farming in tropical forest areas (e.g., described in Africa by Nye and Green-land, 1960, in South America by Sanchez, 1976). After clearing of the land (usually by burning the forest trees) a short rotation of annual crops is grown which uses the available nutrients from soil organic matter. The cropping period is generally short due to nutrient removal and especially to losses by leaching below the root zone of annual crops. This nutrient depletion (together with other causes) results in substantial successive yield decrease. If after 3 to 5 years cropping yield levels of only half or one-third of the first year are obtained, the field is left to natural vegetation again. The "forest fallow" usually restores the soil nutrient status (also by recovering nutrients from the subsoil) and the humus content within a period of 15 to 20 years.

This system of alternating short cropping and long (forest) fallow period is not very efficient, producers only low yields (except for the first few years) and is exploiting because losses are not compensated. Nevertheless, to a certain extent, it is stable as long as there is sufficient land available for long regenerative phases under natural vegetation provided that no serious soil deterioration takes place during the cropping period (the population needs at lest seven times more land than actually used).

The poor reputation of shifting cultivation as a misuse of soil resources is mainly due to deviation from the original concept by shortening the forest fallow period and thus soil regeneration, mostly as a result of an increased population pressure.

Exploitation cropping is still widely used due to economical and other constraints, but at least the systems with insufficient regeneration should be replaced as far as possible by a sustainable type of agriculture.

Sustainable Agriculture at Low to Medium Yield Level

The concept of sustainable agriculture is getting a high priority. It involves the successful management of resources for agriculture to satisfy human needs while maintaining or enhancing the quality of the environment and conserving natural resources (Plucknett, 1990). Obviously, systems of this kind involve complex interactions and require integration (Edwards 1990).

A prominent concept is called by the acronym LISA, i.e., Low-Input Sustainable Agriculture, LISA is supposed to optimize the management and use of internal production inputs (mainly on-farm nutrient resources) in order to obtain a satisfactory level of sustainable crop yields and profitable returns (Parr, 1993). It is in fact a production at a somewhat lower end of the crop response curve.

Low input agriculture may be required for compelling natural and economical reasons (extensive sustainable agriculture in vast areas of developing countries) or deliberately used because of ideological reasons (bio-or ecofarming in developed countries).

Extensive Sustainable Farming

In areas with strong yield limiting factors, e.g., in dryland farming, extensive farming with low input and low to medium yields holds still its lace.

The main emphasis lies on the use of mobilized soil nutrients and internal nutrient cycling via organic substances, although a complete cycling is difficult to achieve because of some unavoidable losses. Some exceptions may occur on small subsistence farms. The input of mineral fertilizer is kept low since their efficiency is know to be low under stress conditions induced by dryness, etc. Some information on yield limiting factors can be drawn from soil productivity studies on farmer's fields.

It has been pointed out, that many marginal ecosystems are not precisely sustainable because of low input (Miller and Larsen, 1990). Whereas at least in some years, they could profit from an improvement of the yield limiting factor by fertilization, in general, a fertilizer input is risky because the V/C (value/cost-ratio) of 2 required for economic reasons may not be reached.

Bio-or Ecofarming (Deliberate Extensive Farming)

In developed countries the term biofarming (organic farming, ecofarming indicates special farming systems without application of chemicals such as factory-made fertilizers or pesticides, but with animal manures, compost, legumes ad nutrient sources and with great care for nutrient cycling in order to produce better, i.e., allegedly non-contaminated food and to protect the environment against unwanted pollution from agricultural chemicals. Somewhat lower yield (30 to 50%) are compensated by corresponding higher product prices.

Although the claims for superior quality food by avoiding chemical fertilizers are not substantiated (Finck, 1992) nor those by avoiding chemical crop protection (Goring, 1990), a limited number of consumers maintain this production of so-called "natural" food.

The further claim that these types of bio-or ecofarming are less (water polluting because of no chemical fertilizer input also needs correction. Although a lower amount of N-leaching often is achieved per unit of land, it rarely holds true per unit of crop product (because almost twice an area of land is required for cropping than with conventional farming).

Sustainable Agriculture at High Yield Level

Sustainability is not confined to low-input conditions. The input, however, should be adequate to the goal that may be even a very high yield level. Such systems could be called AISA (Adequate-Input Sustainable Agriculture) as demonstrated in Western Europe and elsewhere, a high but adequate use of fertilizers results in sustainable production with high yields and yet without significant adverse effects on soil fertility or environment (Finck, 1992). Farming systems of this kind are rather diverse (e.g., rainfed or irrigated), but have many similarities of nutrient management.

The goal is the same as of maximum yield experiments, i.e., obtaining as far as possible the yield potential of high-yielding crops by eliminating all nutritional minimum factors (via fertilization and maintaining a high soil fertility) while simultaneously protecting the crop against all kinds of diseases and insect damage.

The details need not be described here since they are presented in modern text books, but a certain negative aspect of practical nutrient management should be mentioned, namely the tendency for neglecting the soil nutrient reserves and those available in farm waste materials (e.g., slurry), especially if cheap chemical fertilizers are available. This has led to the misleading pubic opinion in the past that intensive cropping essentially is a "nutrient-easting" system.

Sustaining crop productivity at a high yield level has proved to be possible in many progressive agricultural areas, e.g., in India by integration of chemical, organic and biological nutrient sources and their effective management, partly while even saving mineral fertilizer input (Hedge and Dwivedi, 1993).

Food production can certainly be increased by better nutrient cycling and prevention of losses; but for the food supply of an increasing population, demands cannot be met with the farm nutrient sources nor with the present use of mineral fertilizers, but require an increasing nutrient input (Ange, 1995).

In fact, the principles of modern comprehensive fertilization concepts and the basic principles of sustainability will have to be combined. Integrated plant nutrition systems, as part of integrated crop production, will be a decisive factor for the final goal: a sustainable high yielding and profitable crop production without avoidable negative effects on environment.

NUTRIENT MANAGEMENT ON ARID LANDS

In rainfed dryland farming, the yield level is limited by shortage of water (rainfall being not only scarce but also variable and thus unreliable). Since better nutrition of crops reduces their water requirement, a proper management of farm nutrients is of farm greater importance than in humid areas.

The main nutritional problem is the ubiquitous shortage of total and available N due to the rather low soil humus content. In order to make the best use of the scarce soil N-source, the crop N-requirement should be adapted (via sowing time) as far as possible to the nitrate flash occurring from rapid mineralization at the onset of the rainy season (although it must be admitted, that this suggestion is not easily put into action because of the uncertain progress of the rainy season). There is also a certain upward movement of nitrate from the subsoil by evaporation (formerly mistaken for a chemical N-fixation at the soil surface by a hypothesis of Dhar in India).

The natural N-supply may be sufficient for low yields, e.g., of 1 to 3 t ha^{-1} of grain, but for medium yields additional N-sources like farm waste materials or even mineral N should be added (if there is sufficient moisture).

For adding organic material, depending on priority, there is the choice of mulching for soil protection or mixing it into the top soil as nutrient source. an increase of the very low soil organic matter level would be very desirable, but the possibilities are rather limited because of the high mineralization rate. The application of organic substances, however, is often limited by competitive use of crop residues, etc., for manuring, for animal feed or for fuel.

Another possibility to improve the natural nutrient supply (as well as of water) is the inclusion of a bare fallow period for storage of water and mobilized nutrients.

Besides N, also the phosphate supply is often insufficient, especially on sandy soils. Since P is specially required for (early) root growth and since deep rooting may be decisive for survival during dry spells, a good P-supply (from fertilization) is therefore important beyond the acute crop requirement. A good K-supply is also essential for reducing the transpiration of crops, but for dryland farming on many arid soils (practically without leaching), there is often sufficient available K for at least low to medium yield levels. The same holds true for magnesium and sulfate.

As for micronutrients, due to immobilization at neutral soil reaction of arid soils, there is a frequent lack of available Fe or Zn. A certain improvement can be obtained by using strongly acidifying N-fertilizers such as ammonium sulfate or (to a lower extent) urea. In addition, special fertilization may be required, more effective by leaf spray than via soil. In some cases the problem of micronutrient deficiency can be solved by using crop varieties with high uptake efficiency.

Whereas many arid sols are rather coarse-textured, the vertisols represent the clay-type. Their natural supply of all nutrients is abundant even under intensive cropping with irrigation. The only exception is nitrogen because of the low soil organic matter content that therefore needs special nutrient management (Finck and Venkateswarlu, 1982; Dudal, 1989).

Although nutrient management on arid soils with many different crops includes a vast complex of problems in detail, only few can be mentioned in this overview.

Agrosystems in semi-arid regions with common agricultural practice may not be always sustainable, but can be made sustainable with the existing research knowledge and with producer's experience (Steward and Robinson, 1997; Ryan, 1997).

Saline Soils

Since on parts of the arid lands the soil salinization adds even more problems, some comments maybe added on the management of deleterious soil substances. If the excessive surplus of soluble salts (Na, Cal, SO_4, etc.) cannot be removed by amelioration, cropping must take into account a certain degree of soil salinization by adapting fertilization to this stress environment.

Whereas, the (salt-induced) water stress can hardly be reduced, at least the excessive uptake of Na may be lowered to a certain extent by (antagonistic) K-fertilization with the additional advantage of an improved K-status which reduces transpiration losses. K-fertilizers, however, should be chosen which cause the least increase of salinity stress, i.e., rather K-sulfate than K-chloride or even better, K-nitrate (if N is required in corresponding amounts).

As for the micronutrient situation, in addition to deficiencies often occurring on neutral soils, at some sites boron toxicity may prove to be a special problem because it is difficult to ameliorate. Detailed information on nutrient management under conditions of salinity is given by Oertli (1992).

NUTRIENT MANAGEMENT IN HUMID TROPICS

Of this wide topic only some problems may be selected. Since strategies of nutrient management must be based on soil fertility properties, these will be discussed briefly. The widespread ferralsols (oxisols of tropical forest), in spite of their diversity in many respects, have some common properties due to the hot and moist enviroment.

As for soil acidity, in order to reach a normal pH-range of at least 5.5 to 6, considerable amounts of Ca-carbonate are required, which may, however, not always be available. The ash from burning the forest trees produces at lest some transitory effect of liming. Another source is fertilizers with an additional liming effect like rock phosphate.

Liming experiments, even if the pH is increased above the toxicity level of Al, sometimes give rather disappointing results due to some (not accounted for) yield limiting factor (whether major or micronutrient), which has to be added in order to get satisfactory yields.

A typical example of very acid soils are the acid sulfate soils with often very low yields, sometimes less than 1 t ha^{-1} of grain. However, even small amounts of lime in combination with rock phosphate may increase the yield level to 2 to 3 t ha^{-1}.

For compensation of the humus mineralized, crop residues should be left on the field. In farming systems with cash crops, there are often considerable amounts of residues from the processing factories that can be recycled. The nutrients of the added organic materials are rapidly mineralized (N- and P-source).

In order to reduce the losses by leaching, the period with bare soil between annual crops should therefore be kept as short as possible.

P-fixation results in a strong P-deficiency and an inefficiency of water-soluble P-fertilizers because the phosphate is quickly fixed and thus rendered unavailable. To overcome the P-fixation, large amounts must be applied, or rock phosphate maybe more effective.

Sustainability of highly productive farming systems can be obtained by using a nutrient management that takes care of these problems caused by production limiting soil properties and the humid hot climate. (For incoming information on sustainable crop production in Asia see Johnston and Syers, 1998).

The nutrient management of special annual or perennial crops on arable land or of tree plantations cannot be discussed here, except for a short comment on two topics.

Rice Cropping on Paddy Soils

Irrigated rice presents special and large problems of nutrient management. A central and not yet really solved problem is the low efficiency of N-fertiizers (urea being the main source used). A recent review (de Datta and Buresh, 1989) pointed out that usually only less than half of the N added is taken up by the crop (compared with about 70% in intensive wheat cropping). Therefore, the losses by ammonia volatilization, denitrification as well as runoff and leaching should be reduced. However, in the paddy environment it is difficult to suggest a single superior method, but some progress seems to be possible by placement of urea, the use of nitrification inhibitors and top dressing at later stage.

The possibility of supplying N via the Azolla/Anabaena symbiosis appears to be very promising and could probably replace a large portion of mineral N-fertilizer.

Agroforestry

For the nutrient management, the advantages of agroforestry are protection against nutrient leaching and soil erosion, nutrient mobilization and transfer from trees to annual crops and pumping nutrients from deeper soil layers (Buresh, 1995).

The potential of alley-cropping systems has been critically reviewed by Lal (1991). The advantage of alleys of shrubs or trees for erosion control is obvious, but obtained at the price of slightly lower yields from the annual crops (e.g., 10% for maize under special conditions). The tree cuttings (as an organic nutrient source) are less effective than mineral N. The general failure of alley-cropping to give higher yields than conventional cropping results from shading effects, root competition for nutrient and water, N-immobilization by the mulch, etc. Nutrients from deeper layers are only transferred to the surface soil if there are substantial nutrients in the subsoil.

In general, alley-cropping does not sustain production without substantial input in chemical fertilizer. Agroforestry has a potential to be a sustainable alternative to shifting cultivation, but its main advantage correlates with fertile soils.

INTEGRATED NUTRIENT MANAGEMENT

Integrated nutrient management (INM) aims at efficient and judicious use of all the major sources of plant nutrients in an integrated manner so as to get maximum economic yield without any deleterious effect on physico-chemical and biological properties of the soil. The basic concept underlying INM is maintenance and possible improvement in soil fertility for sustained productivity on a long term basis. Major sources of plant nutrients are fertilizers, FYM/composts, green manures, biofertilizers and crop residues /recyclable waste. Proper integration of one or more sources will ensure optimum nutrient supply.

Fertilizer is the most important component in integrated nutrient supply system under intensive cropping. Use of balanced fertilization is required to boost the crop productivity. Chemical fertilizers are to be properly integrated with other sources.

Integration of fertilizers with bulky organics

Farmyard manure and compost are the traditional organic sources that have been used in crop production. About 50 % of the manurial potential of the livestock population is only utilized for crop production. Sizable amount of plant nutrients are lost because of improper storage, handling and incorporation techniques. Results of various field experiments have clearly indicated the importance of integrated use of FYM and fertilizers for overall improvement in soil fertility. Results also indicate that a part of fertilizer requirement of rainy season (kharif) crop can be met with the use of FYM in different cropping systems without any adverse effect on the total productivity. Based on a number of experiments on rice, it is concluded that 1 tonne of FYM supplying 12 kg nutrients ($N+P_2O_5 + K_2O$) produces as much grain as 3.6 kg fertilizer nutrients (Singh and Dwivedi, 1996). In addition to supplying plant nutrients to the current crop, FYM leaves substantial residual effect on succeeding crops in the system. Meelu and Morris (1984) reported that 12 t FYM and 80 kg N/ha gave residual effect on the succeeding wheat equivalent to 30 kg/ha each of N and P_2O_5. There was also a build up in soil fertility with FYM application. Kulkarni et al. (1983) on the basis of 240 trials conducted in cultivators' fields in rice growing regions of the country concluded that application of 12 tonnes of FYM along with 90 kg N/ha to the kharif crop followed by 60 kg N/ha to rabi crop could produce an yield comparable to 120 kg N, 60 kg P_2O_5 and 60 kg K_2O/ha in kharif and 60 kg N/ha to rabi crop in rice-rice cropping system.

Besides supplying plant nutrients, FYM application also improves the physico- chemical and biological properties of the soil. Application of FYM along with recommended level of fertilizers enables to maintain the organic matter in the soil compared to fertilizer application alone.

Green manuring with inorganic fertilizers

Before the introduction of inorganic fertilizers, green manuring was invariably practiced for crops like rice, wheat, sugarcane, etc. With the easy availability of fertilizers and adoption of highly intensive cropping, the practice was almost given up. In the recent years, there is a renewed interest on green manuring. Sunnhemp and dhaincha are the most common green manure crops in India. Green manures produce large amount of green biomass of narrow C:N ratio. When incorporated into the soil, green manures contribute sizable amount of plant nutrients, particularly N and also improve the soil physical conditions. Experiments carried out at several locations have distinctly established the advantage of introducing green manure crop as a supplement to fertilizers in cereal crop rotations. Green manuring along with the recommended level of fertilizers benefits the crops in the rotation and increase the total productivity of the system to the tune of 10-20 %. Incorporation of 10 tonnes / ha of green manure through *Sesbania* or *Crotolaria* supplement fertilizer N equivalent to 40-45 kg/ha. Integrating P with green manure in the first crop of rice produces higher yields of second season rice.

Green manuring in drylands is rare, but is practiced in small pockets in India. On vertisols receiving more than 1000 mm of rainfall in Khammam district of Andhra Pradesh, green manure is grown during the South-West monsoon season in June followed by its incorporation before sowing sorghum in September. Depending on sowing of green gram, it is either incorporated as green fodder or harvested as pods. This practice has been observed to supplement 50 % N (Venkateswaralu, 1987). Inclusion of a grain legume has been found to be advantageous for sustained yields under rainfed situations. e.g. cowpea–fingermillet (Bangalore), green gram-finger millet (Bhubaneswar), sunnhemp-wheat (Hoshiarpur) and black gram-wheat (Rewa). The benefit of green manuring is felt only after continued applications for 3-5 years (Singh and Venkateswaralu, 1985).

Wherever it is not possible to include green manure in the multiple cropping system, there is a possibility of inclusion of legume in the system. Because of their shorter duration, small stature and erect growth habit, legumes could be easily accommodated as intercrops. Legumes of short (Vigna group) or long (pigeonpea) duration group offer greater advantage as intercrops in cereals by providing additional profit to the farmers. The Vigna group is harvested earlier and provides some nitrogen to the longer duration companion crop. Pigeonpea is of longer duration. It is least competitive during the early growth phase and starts using the space effectively by putting up additional growth after harvest of the early duration companion crop. These pulses provide N upto 50 kg/ha to the associated crops in vertisols. Higher benefits have been reported for crop combinations of cowpea (fodder) + sorghum, green gram/black gram + sunflower and pigeonpea + groundnut.

Green manure crops can be very well accommodated as intercrops in wide-spaced crops like sugarcane and maize. Since initial crop growth upto 100 days after planting of sugarcane is very slow, this period can be utilized for growing legumes in inter-row spaces as intercrops for grain or fodder purposes. Summer legume crops like black gram and green gram intercropped with spring planted sugarcane increased the cane yield markedly besides producing 0.4 – 0-5 tonne of pulse grain/ha (Yadav, 1980).

In addition to their capacity to fix atmospheric N, the legumes are capable of extracting less soluble form of P from the soil. They also extract the higher valency nutrient cations better.

Biofertilizers with inorganics

Various biofertilizers can be used according to needs of the crops. Biofertilizers act as complimentary and supplementary sources of plant nutrients. Biofertilizers help in increasing the biologically fixed atmospheric N and enhancing native P availability to the crops. Among the biofertilizers for increasing N supply, N- fixing bacteria (Rhizobium, Azotobacter and Azospirillum), blue green algae (BGA) and Azolla are important. The availability of P is improved by P solubilizing microorganisms (PSM) and mycorrhizae.

Rhizobium is of vital importance to various pulses and some of the oilseeds and fodder legumes as 80 to 90 % of their N requirement is met through biological nitrogen fixation (BNF). Inoculation of legumes with specific species of Rhizobium is widely used and is a successful practice throughout the world. Responses to Rhizobium under field conditions have been quite variable and depend on soil conditions, quality of inoculum and effectiveness of native Rhizobium. It fixes atmospheric N ranging from 50 to 60 kg/ha in groundnut and 100-300 kg/ha in alfalfa. Pulse crops fix 50-110 kg N/ha under optimum conditions. It is generally accepted that the fixation is about 50 kg/ha in a season. Field studies have indicated an increase in pulse yields by 14-53% due to Rhizobium inoculation.

Azospirillum and *Azotobacter* are the other important groups of bacteria. *Azospirillum* fixes N in associative symbiosis with the plant roots and *Azotobacter* is a free-living N fixer. On an average, they contribute 15-20 kg N/ha to different crops when inoculated with efficient strains. Application of organic manure along with Azospirillum enhances N fixation to the tune of 20-40 kg/ha. Promising results have been obtained with the use of Azospirillum in crops like wheat, rice, maize, barley, sorghum and millets. Azotobacter has given promising results in rice, sugarcane, wheat and pearl millet. Response to the bacterial inoculant appears to be enhanced when applied along with N and P in rice and cotton and with P in sugarcane (Gaur, 1982)

Blue green algae and Azolla are very well suited for rice ecosystems. In lowland rice, the average N contribution by these N fixers is equivalent to 25-40 kg/ha. Various field studies have shown positive influence of BGA on rice plants under different agro-climatic conditions. If this technology is popularized, nitrogen equivalent of 1.5-1.7 m.t. urea could be supplied to rice through BGA (Venkataraman and Tilak, 1990). Besides supplying N, BGA also synthesizes and liberates growth promoting substances such as auxin and amino compounds which stimulate the growth of rice plant. It also supplies additional humus to the soil. Results of the studies conducted by the International Network on Soil Fertility and Fertilizer Evaluation for Rice (INSFFER) have shown that incorporation of 10 tonnes of fresh azolla before or after transplanting is equivalent to split application of 30 kg fertilizer N/ha. It can be applied along with fertilizer N to step up rice yields. Based on field trials, an increase in rice yield by 13-54 % has been reported, besides building up organic carbon, total N and availability of P in the soil (Satapathy and Singh, 1980).

In recent years several strains of P solubilizing bacteria and fungi have been isolated. Phosphobacteria solubilize the residual and insoluble phosphates available in the soil-water solution very near to root surfaces. The phosphate solubilizing power is further enhanced by addition of organic matter. Studies reveal that integrated use of P solubilising cultures with low grade rock phosphate can add 30-35 kg P_2O_5/ha in neutral to mildly alkaline soils. Vesicular-arbuscular mycorrhizae (VAM) exploit the soil phosphate better, leading to more efficient use of added fertilizers, particularly of slowly mobile ions like NH_4^+, Cu^{2+} and Zn^{2+}. VAM infected sorghum had significantly high P in roots and shoots and increased the dry matter production. Because of problems in isolation and multiplication of pure strains of VAM fungi, large-scale use on commercial basis is yet to be established.

Integrated Approach in Biofertilizer Use

It may be possible to integrate the use of more than one biofertilizer for increasing crop yields. *Rhizobium Azotobacter,* mycorrhizae, phosphate solubilizers, etc. could be used together or separately. In India, where mixed cropping and double cropping is followed in many places, the combined application of more than one biofertilzer holds great promise. Integrated use of biofertilzers, organic manures and chemical fertilizers would not only give higher yields but help to maintain soil health by narrowing down the gap between nutrient removal and supply. It may be mentioned here that no single nutrient source, be it chemical fertilizer, organic manure or biofertilizer is in position to meet the increasing nutrient demand. Some possible fertilizer equivalents of diverse plant nutrient sources are given in the following Table which may be helpful in adjusting fertilizer dose.

LOW-COST TECHNOLOGY FOR ENRICHMENT OF COMPOST

Recent experiments conducted at I. A. R. I. have shown that inoculation with *Azotobacter chroococcum* and phosphate solubilising microorganisms *(Aspergillus awamori* or *Bacillus polymyxa)* improved the manorial value of compost (Gaur et al., 1978, 1980). Inoculation of compost with nitrogen fixer and P solubiliser was done when the temperature had stabilized around 30-35°C after one month of composting. Low-grade rock phosphate (10 kg/tonne of raw material) was applied at the initial steps of filling up of compost pits.

The results indicated that 31.5 per cent carbon was decomposed in control during the first month whereas the loss in different treatments ranged between 33.3 to 41.5 per cent. This was further increased during second month showing maximum loss of 52.4 per cent of organic carbon in rock phosphate. *Azotobacter* and P solubiliser inoculated series. In this treatment C/N ratio was reduced to 14.1 and 12.0 during I[st] month and IInd month after inoculation respectively.

There was appreciable gain in nitrogen content per tonne of finished compost in control as well as inoculated series ranging from 60.5 to 111.6 per cent over the initial raw material. Addition of powdered rock phosphate alone resulted in an increase of 28.3 per cent over control. Inoculation with *A. chroococcum* and *A. awamori* amended with rock phosphate showed an increase of 32.0 per cent in nitrogen content over the control resulting in gain of 22 kg N/5 tonnes of finished compost. It was also noted that humus content increased in all the treatments including control. The best effect in humus content 9.3% was observed in case of *A. chroococcum* + *A. awamori* after 1 month of inoculation.

In another study the number of *Azotobacter chroococcum* in compost was increased due to its inoculation and use of rock phosphate. The mean numbers of nitrogen fixers of nien weekly intervals per gram of dry compost in different treatments were as follows: Control 1,380, rock phosphate 3,590; *Azotobacter* 6,070; rock phosphate + *Azotobacter* 21,070. This also resulted in an increase in the nitrogen content of compost by 25 to 44 per cent over control. These findings clearly show that quality of compost can be improved by inoculation with *Azotobacter* and P solubilisers suitably amended with rock phosphate.

Sugarcane trash compost

Sugarcane trash left out after harvest is about 19-38 m. t. It is either burnt in situ or used as domestic fuel. It can be recycled after composting suitably. Technology has been developed for production of enriched compost from sugarcane trash at Pune (Bhavanase et al., 1995). In the compost pit of 1× 1× 1 m^3, chopped sugarcane trash is filled in layers. In the bottom of the pit, trash is filled to 30 cm thickness. A mixture of cellolytic fungi @ 500g / tonne of sugarcane trash and Azotobacter are added. These biofertilizers are mixed with cow dung. Sufficient quantity of water is applied on the surface of trash. A mixture of cellulolytic fungi are inoculated: *Trichoderma viride, Trichurus spiralis, Aspergillus awamori, Pencillium* spp and *Paecilonmyces fusiparus*. These inoculants improve the rate of decomposition and mineralization of nutrients. Another layer of sugarcane trash is then added. Above this layer, urea (to supply 8.33 kg N /ha) and rock phosphate (5% of trash) are applied dissolving in sufficient quantity of water. Cow dung with biofertilizer mixture should not be applied simultaneously with the fertilizers. Thus the pit is filled with trash in layers. Water is regularly added on alternate days to maintain 60% moisture in the compost pit. Adequate moisture is important for proper activities of the microbes. Turning of the material is to be done at regular intervals (at least once in a month) to improve aeration and maintain temperature at 50-60°C which is optimum for composting.

Table 8.1: Changes in Carbon (%), total Nitrogen (%), C/N ratio during enrichment of compost by*Azotobacter* and P solubilizing cultures

Treatment	Ist month				IInd month			
	Carbon	Nitrogen	C/N ratio	Humus %	Carbon	Nitrogen	C/N ratio	Humus %
Control	31.4	1.36	23.0	7.8	28.5	1.38	20.6	4.85
Control + Rock Phosphate @ 10 kg P_2O_5/tonne	30.7	1.61	19.0	9.15	27.5	1.77	15.5	6.90
Azotobacter + P solubilizer	30.5	1.51	20.1	9.30	26.6	1.73	15.4	6.00
Rock phosphate + *Azotobacter* + P solubilizer	26.9	1.89	14.1	9.0	21.9	1.82	12.0	6.95

Initial sample: Organic carbon 46.0% total nitrogen 0.86%, C/N ratio 53.3.

Fresh sugarcane trash contains 0.36% N with a wide C:N ratio of 122:1 Small bits of trash decompose earlier than big size trash. Addition of rock phosphate, urea, biofertilizer and moisture play significant role in early composting. The composted trash contains higher content of N (1.09%) with reduced C:N ratio (20:1).

Biogas slurry

Organic manures from animal wastes are very important nutrient sources in building up the soil fertility. In India, the estimated production of dung and urine from bovine population works out to 1002 and 658 million tonnes respectively. They contribute about 5.71 m. tonnes of N, P and K. Proper method of collection and preservation is required to fully utilize the manurial value of these materials. Cow dung is an input in biogas plants. The biogas route has the dual advantage of providing both fuel (gas) and fertilizer (slurry).

The residual slurry that comes out of the digestion tank has a manurial value superior to that of FYM. The wet slurry should not be used as such. it should be dried for usage as manure. While drying, there is some loss of N and the dry manure contains about 1.8 %N, 1.10 % P_2O_5 and 1.50 % K_2O.

Vermicompost

Vermicompost is an organic manure produced by the activity of earthworms. Vermicompost is a mixture of worm casts which are rich in macro and micronutrients, vitamins, growth hormones and immobilized microflora. The average nutrient content of vermicompost is much higher than that of FYM. Vermicompost contains 1.60% N, 5.04% P_2O_5 and 0.80% K_2O with small quantities of micronutrients. The C:N ratio of vermicompost is much lower (1:16) than that of FYM (1:30). Application of vermicompost facilitates easy availability of essential plant nutrients to crops. Numerous species of worms have been identified for vermicomposting.Detals are given in chapter 3.

Industrial wastes

Among the industrial by products, spent wash from distilleries and molasses and pressmud from sugar industry have good manurial value. It is important to use only well- decomposed pressmud at 10 tonnes per ha. Addition of pressmud improves the soil fertility and enhances the activity of microbes. It is also used as an ameliorating agent in saline and sodic soils.

Coir waste is the byproduct from coir industry. Its production is about 1 million tonne in India. Coir waste can be used as manure after proper decomposition. Composted coir wastes (inoculated with *Pleurotusspp.* and 5 kg urea / tonne) is recommended for irrigated and dry crops. It has been reported that addition of 12.5 t/ha of composted coir waste with recommended fertilizers increased the yield of groundnut and maize.

Municipal and sewage wastes

Municipal and sewage wastes form an important component of organic wastes. In India, the total municipal refuse is about 12.m.t. per annum containing about 0.5 %N, 0.3% P and 0.3 %K. Sewage sludge is available to an extent of 4 million tonne per annum containing 3% N, 2% P and 0.3% K (Bharadwaj and Gaur, 1985). Sewage sludge, particularly from industrialized cities, is contaminated with heavy metals and thus pose hazard to plants, animals and human beings. Segregation of toxic wastes at the source will minimize the concentration of such elements in the sludge. High N input through excessive hydraulic loading of sewage can lead to ground water contamination. Further, survival of pathogens in soil and on crops causes health risk. This can be minimized by proper treatment, design of hydraulic loading, selection of suitable crops and adoption of modified irrigation system.

MANAGEMENT OF MAJOR NUTRIENTS (N, P, K) WITH ORGANIC SOURCES

Organic farming system is based on the dynamic interaction between the soil, the plants, animals, humans, the ecosystem and the environment. The system is directed towards enhancing natural life cycles rather than suppressing nature. It relies largely on locally available resources and able to provide agricultural products of good quality while minimizing negative side-effects. Organic farming is gaining popularity not only among the farmers but also fast expanding market and it very useful for the readers to have knowledge of nutrient management in organic farming.

It is a difficult task to do well in conventional agriculture it is difficult to do well in sustainable agriculture, which is characterized by profitable farming while preserving the environment as much as possible. In fact, switching from conventional to sustainable agriculture is not a simple substitution, such as replacing insecticide with a predator insect or replacing commercial KCl fertilizer with green sand. It is reasonable to say that although more in tune with nature than conventional agriculture, sustainable agriculture is information intensive and requires strong management skills. For example, cultural and biological pest control requires detailed knowledge about a pest life cycle as well as economic threshold levels of the crop. Similarly, using chicken manure to replace urea as a N source for crop must take into account the release pattern of organic nitrogen from the manure and the nitrogen demand pattern of crop, so that the two can be synchronized. Soil fertility and plant nutrients are important to farming, whether it is conventional or sustainable. Although sustainable farming is much broader than organic farming, the two usually use the same standards for soil amendments. In the organic farming the use of non synthetic sources of nutrients is essential for certification point of view. It is also useful in terms of ecological and economical returns to use natural sources of nutrient. Nitrogen, Phosphorus and Potash are the nutrients required comparatively in large quantity. There management with organic sources is the gateway of organic or sustainable farming

Nitrogen

Nitrogen is needed by all plants and usually in large quantities. In fact, nitrogen is so important to plant growth, and thus to food and fiber productions of the world, that the two German scientists who first invented the process of making NH_3 from atmospheric nitrogen gas (N_2) and natural gas-derived hydrogen (H_2) won Nobel prizes for their work in the early of this century. For example, average N concentrations needed for normal growth are about 3% for corn and coffee, 4% for tomato, and 2% for macadamia. If plants do not have enough nitrogen, they are stunted. Their leaves are small and pale-colored, sometimes even yellow or reddish-tinted. The reason is that nitrogen is a component of chlorophyll, less nitrogen results in less chlorophyll, thus less green. Normal corn leaf is on the right side, N deficient leaf is on the left. Since proteins are continually being synthesized and degraded in the plant, meaning that nitrogen is quite mobile in the plant. Thus, N will move from older leaves to younger ones, trying to meet the N need of the most active growth regions. The consequence is that nitrogen deficiency will begin on older leaves then progress upward toward the growing points

In soils, nitrogen must be present as either NH_4+ or NO_3- before plants can absorb and use it. That explains why urea [$(NH_2)_2CO + 2H_2O \rightarrow 2NH_4OH + CO_2$] and ammonium nitrate ($NH_4NO_3$) have been so popular as N fertilizers in conventional farming systems.

In organic farming systems, urea cannot be used nor NH_4NO_3, nor any synthetic chemicals for that matter. Alternatives must be sought. Table 8.2 lists the total N content of various organic sources.

Table 8.2: Total N concentration in common organic sources as compared with urea (46% N).

Organic source	Total N (%)
Poultry manure	1.5-3.0
Pig, horse, cow manure	0.3-0.6
Green manure	1.5-5.0
Compost	0.5-2.0
Seaweed meal	2.0-3.0
Sewage sludge	1.0-5.0
Fish waste	4.0-10.0
Blood (slaughter house)	10.0-12.0
Human urine/night soil	1.0-1.5

(Adapted from Caplan, 1992; Hue, 1995; Tandon, 1992)

So on the basis of total N, organic sources would have somewhere between ten and hundred times less N than urea. This means that in order to meet your crop N requirement, say 100kg N/acre for a crop of sweet corn, need:

Urea (46% N) $\rightarrow$ 100/0.46 = 217 kg/acre as urea

or compost (1% N) $\rightarrow$ 100/0.01 = 10,000 kg/acre as compost

This simple calculation shows that there is need of large amounts of compost (or most organic N sources) relative to urea in order to supply an equal amount of N.

Transportation cost might become a factor here. A close inspection, however, reveals that the assumption for the above calculation is incorrect. N from the compost is as readily available for plant uptake as N from urea, which is not true. Unlike urea which can be hydrolized into NH_4 nearly instantaneously, nitrogen from compost, first must be released from its organic substrates, and this process may take from several days to several months to complete. In general, depending on types of organic materials and

Environmental conditions, such as temperature and moisture, only 10 to 50% of N from an organic source would be converted to NH_4 in a six-month period (Hue, 1995). Thus, in order to provide 100 kg N/acre that a crop of sweet corn needs over a growing season of approximately 3 months, it need to apply from 20 to 50 ton/acre of compost instead of 10 tons as the first calculation had indicated. If it is practical or economical to apply 50 tons of compost to each acre of land for a crop of sweet corn. Fortunately, this is more of a requirement only during the transition period (one or two years) from the conventional farming to sustainable farming systems. After about 2 years of using composts or green manures as a N source, N release from previous years applications would reduce the current application rates significantly. That explains why many sustainable farmers have seen their crop yields increase with time

Synchronization

The need for N fertilizers would be reduced significantly if the crop's N requirement can be timed closely to the nitrogen release from organic sources. Using ourselves as an example, it is commonly know that babies and older folks require much less food intake than teenagers or young adults. The same principle applies to plants. Plants require different amounts of N at different stages of growth: seedlings and senescing plants definitely require much less N than flowering or fruiting plants. We also have known that different organic manures have different N releasing patterns. N release patterns of 10 organic sources; three of them are shown here as an illustration.

First, don[1] t count on wood-based composts (blue line) as a source of N. In fact, adding such composts to soils may even cause N deficiency in fast growing crops such as tomato or lettuce as shown below.

Second, N from chicken manure and chicken manure-based composts (red line) was readily available right after application, then declined gradually with time. The implication is that chicken manure can provide quite large quantities of N the first 4 to 6 weeks after application. Thus, its application should be timed closely to the fastest growing period of the crops.

Third, in the green manure it took about two weeks for the N releasing reaction to start (green line). The green manure released most N between 3 and 7 weeks after application. The release remained low and nearly constant for another month (between 7 and 11 weeks) then started to increase again. This second phase came from the decomposition of microbially more resistant substrates in the green manure. The main point is that the N release patterns of organic manures depend on many factors and require experience and knowledge to predict. This translates into the following conclusion:

Successful use of organic manures (whether it is compost, green manure or animal waste) as a N source requires more experience and better knowledge about the manure, the crop, and environmental conditions than the use of chemical fertilizers (e.g., urea).

GENERAL PROPERTIES OF NITROGEN FROM ORGANIC SOURCES

1. The availability of N declines as manures age or are composted.
2. Nitrogen from poultry manures (non-composted) is more quickly available than N in manure from cow, horses, and sheep. This is because poultry manure contains significant amounts of uric acid, which is readily decomposable. Also, roughage in cattle and horse diets results in manure that is less readily decomposed due to lignin and cellulose.
3. Presence of bedding or litter lowers N content by dilution, but its effect on N availability cannot be easily predicted.
4. Moist manure, when exposed to the air, undergoes significant loss of N as volatile NH_3. Nitrogen losses after spreading can be significantly reduced by incorporating the manure even shallow incorporation is adequate.
5. Nitrogen in composts is in a more stable form than N in manures. Thus, there is a decreased likelihood of losing compost N from the system. On the other hand, composts containing less than 1.5% total N supply little or no N to crops during the first few weeks after application.
6. Green manures are fast growing catch crops like clovers, mustard, rye, and vetches that are incorporated into the soil while they are still green in order to improve the soil and to supply N to the associated/subsequent cash crop. A properly managed legume cover crop can fix up to 75 to 100 kg N/acre in 3 to 6 months.
7. Both manure and compost can improve soil tilth, but the timing and mechanisms may differ.

COVER CROPS AND RIDGE TILLAGE IN SUSTAINABLE FARMING

In practice, cover crops such as legumes (e.g., clover, vetch) and grasses are often grown either to supply N to subsequent cash crops or to reduce nitrate leaching and soil erosion. Crop rotation and intercropping have been used to achieve this goal. Work in tropical humid regions of India has shown that corn yields and N uptake increased by 15 to 20% by intercropping with soybean or cowpea. Legume intercropping was equivalent to adding 80 kg N/ha of fertilizer N. A marked synchrony between plant growth stage and microbial population in the rhizosphere 60 days after seeding was observed in this study.

The N release from cover crops, however, depends on many factors, including the C to N ratio of the residue, its degree of incorporation into the soil and soil temperature and moisture regimes. It may take one to three weeks after cover crop incorporation before N release exceeds N immobilization. The farmer may have to decide if additional fertilizer N is needed and, if so how much. On the other hand, when the soil is not cropped, cover crops, especially grass, can reduce the potential loss of N through leaching or denitrification. In the US mainland, rye is often used as a winter cover crop to control weeds through competition and allelopathy.

Along with crop rotation and intercropping, ridge tillage is often practiced in sustainable farming. This technique only tills a zone of soil 15 to 20 cm wide and 5 to 10 cm deep at planting time. The practice creates a desirable seedbed environment. It also covers crop residue and weeds in the area between rows with soil.

Subsequent between-row cultivation a couple of weeks after planting promotes mineralization of soil organic residues, controls weeds and reforms the ridge with a raised bed for future crops.

Phosphorus

Along with nitrogen, phosphorus is a nutrient that plants need in relatively large quantities for normal growth the two genetic entities that are essential for growth and reproduction of living organisms. Living organisms, whether plants or humans, also derive their internal energy from P-containing compounds, mainly adenosine diphosphate (ADP) and adenosine triphosphate (ATP). This means that inadequate P supply will result in a decreased synthesis of RNA, the protein maker, leading to depressed growth.

Phosphorus-deficient plants, therefore, are stunted with a limited root system and thin stems. In many plants such as corn, guava, and tomato, seedlings look stunted, and older leaves may turn purple because of the accumulation of anthocyanins or purple pigments. Fruit trees deficient in P have few and short new shoots and malformed fruits and seeds. Thus not only low yields but also poor quality are the results of P deficiency.

Internally, most crops need from 0.2% to 0.5% P in the dry matter for normal growth. Macadamia is an exception, iron chlorosis may occur if P content is greater than 0.15% in leaves. Table 8.3 lists the sufficiency range pf P in leaves of selected crops (Tamimi et al., 1994).

Table 8.3. Sufficiency P levels in leaves of selected crops.

Crop	Plant P %
Cabbage	0.4-0.6
Coffee	0.2-0.3
Corn	0.3-0.5
Head lettuce	0.4-0.6
Macadamia	0.08-0.10
Soybean	0.3-0.5
Sugarcane	0.2-0.3
Tomato	0.3-0.8

Unlike animals and humans who can obtain P from various food sources, plants must extract P exclusively from the soil where they grow. Also unlike humans whose gastrointestinal enzymes can break up organic P in food, plant roots can absorb P only when P is in the soil solution and in either H_2PO_4 - or HPO_4 form. This fact is emphasized because not only do plants like H_2PO_4 - and HPO_4- but so do soils. There is a strong ompetition between plants and soils for P in the soil solution. And the winner usually is soils, especially highly weathered soils like those of tropics. Most of our soils here contain large amounts of iron-and aluminum-oxides or amorphous alumino-silicate clays, which tie up P firmly, making P virtually unavailable for plant uptake. For example, in order to provide enough P for many of our favorite vegetables, such as lettuce and tomato, first time growers must apply approximately: Of course, amounts of P required would vary, depending on how much P the soil has, to begin with. That s true, and that is why you need a soil test before trying to amend or fertilize your soils. In fact, by adding P to the soil year after year, you might build up soil P to a point that it becomes detrimental to your crops because of excess. The upside of this phenomenon is that once you have built your soil P to a good level, that level would remain for many years without any additional P input. The reason is that unlike nitrogen, P does not move easily with water. It virtually stays put. Leaching is minimal.

In conventional farming you can use such fertilizers as superphosphate ($Ca(H_2PO_4)_2$) which contains about 20% P or 46% as P_2O_5 or diammonium phosphate (DAP) which has coincidently the same amount of P as in treble superphosphate. These fertilizers are highly soluble, dissolve quickly in water and can provide enough P to your crops before being tied up by the soil. These P fertilizers, however, are not acceptable in organic farming.

Table 8.4: Total P concentration in organically acceptable sources as compared with superphosphate (20% P)

Source	Total P, %
Rock phosphate	17-26
Bone meal	20-30
Fish meal	5-10
Wood ash	2-5
Poultry manure	0.5-1.5
Green manure	0.2-0.5
Compost	0.2-0.5
Sewage sludge	0.4-2.5

Table 8.4 shows P content of selected sources that are acceptable to organic farming community in most states. Among these, the first two, rock phosphate and bone meal, have reasonably high total P content: between 20 and 30%. However, P in these two sources is very insoluble, thus much less plant available than P in superphosphate. More specifically, P in rock phosphate and bone meal has the formula: $Ca_5(OH)(PO4)_3$ which is hydroxy apatite or apatite for short. A couple of things should be mentioned about rock phosphate. First, it is a source of P for long-term soil improvement, don't expect any noticeable effect from it within weeks or months, unless huge amounts of it used. Second, its solubility, and thus plant availability, depends strongly on soil pH and particle size. It is more effective in acid soils than in calcareous or alkaline soils; more effective when it is fine than coarse, and more effective in the presence of mycorrhiza than without

Mycorrhiza

By contrast, P contents of chicken manure, compost, and sludge are relatively low, usually below 3%. Thus, large amounts would be needed to meet P requirement of the crop. P from these organic sources is quite available to plants; sometimes even more effective than superphosphate. Organic matter from the manure interacts with clay minerals and reduces P sorption by the soil, thereby enhancing P availability to plants

Potassium

Compared favorably with nitrogen, potassium is needed in large quantities by many crops. Table 8.5 lists the sufficiency range of K in selected crops.

Table 8.5: Sufficiency range of K in selected crops.

Crop	Tissue K, %
Banana	3.0-5.0
Cabbage	4.5-7.5
Coffee	2.0-3.0
Corn	1.5-3.0
Lettuce	4.0-7.5
Papaya	3.0-5.0
Tomato	3.0-5.0

(Adapted from Tamimi et al., 1994)

In plants, potassium is required for maintaining osmotic potential of cell. That is, K makes plants look turgid. Since K regulates the osmotic potential of cells, and the close or open conditions of stomata, it plays an important role in water relations in the plant. Potassium is involved in water uptake from the soil, water retention in the plant tissue, and long distance transport of water in the xylem and of photosynthates in the phloem. Potassium affects cell extension. With adequate K, cell walls are thicker, thereby improving plant resistance to lodging, pests and disease. Fruits and vegetables grown with adequate K seem to have a longer shelf life in the grocery store.

Consequently, K-deficient plants show low resistance to disease, their seeds and fruits are small and shriveled. In tomato, K deficiency results in smaller fruits whose flesh development is incomplete. In corn, the maturity is delayed and ears are smaller when K is deficient.

The stalks are weak and lodging is common. The most visual K deficiency symptom is the scorching or firing along leaf tips and margins as shown here for soybeans.

In alfalfa and clovers, the first sign of K deficiency shows up as small white or yellowish dots around the outer edges of leaves. As the deficiency intensifies, these edges turn yellow, then brown, and finally die.

In soils, potassium is quite mobile as compared to phosphate. It exists as K+ in soil solution and is absorbed by roots in that form. Although K+ can be retained to some extent by negative charges on clay surface, it can be displaced into the soil solution by Ca_{2+} or Mg_{2+}, when gypsum or dolomite is added. Thus if K is not taken up by plants, it might be lost by leaching. One way to reduce K leaching is to add organic matter such as compost or green manure to your soil. These organic materials usually have large cation exchange capacity which can retain K effectively. Among soils, K deficiency occurs more often in sandy than in clayey soils, more often in highly weathered soils than in young soils, and more often in volcanic ash-derived soils If soil is low in K, in that case commercial KCl is an easy choice in conventional farming, but not acceptable in organic farming.

Table 8.6: Total K concentration in selected organically acceptable sources.

Source	Total K %
Sul-Po-Mag [Mg, K, SO_4]	22.0
Polyhalite [Ca, K, SO_4]	10-15
Wood ash	5-10
Green sand	5-7
Green manure	2-5
Seaweed meal	2-3
Compost	5-2.0

The first two sources are natural deposits of potassium minerals. Sul-Po-Mag as the name implies is a double salt of magnesium and potassium sulfates. It has about 22% K. Polyhalite is a double salt of Ca and K sulfates. It has about 10 to 15% K. Like Sul-Po-Mag, polyhalite is very soluble, not much different from KCl. In terms of plant response, this material is as good as or even better than KCl or gypsum Wood ash also contains considerable amount of potassium, and this source is also readily available (and alkaline).

BIOLOGICAL APPROACH FOR SECONDARY AND MICRONUTRIENTS ACQUISITION

Secondary and micronutrients can be supplied through biological systems. Use of organic manures as a complete plant nutrient supply system is a time tested practice. Vasicular Arbuscular Mycorrhizae, plant growth promoting rhizobacteria are some examples by which availability of secondary and micronutrients to the crop plants can be augmented.

Long term field expression show that no soil can support medium, high yields indefinitely without external nutrients input. The available nutrients gradually decline from optimum level to deficiency levels. A series of Long Term Field Experiments conducted during 1970s showed that application of a single nutrient did not support similar crop response over a period of time as deficiencies of other macro and micronutrients develop. Secondary (Mg, Ca, S) and micronutrients although needed in smaller quantities than N, P_2O_5 or K_2O are removed by every crop. There is ample data available to show that the nutrient reserves of Indian soils are being depleted. In India at present deficiencies of five nutrients (N, P, K, S, Zn) need the attention of agricultural scientists. The depletion of micronutrients particularly zinc in the soils is predicted due to the introduction of high yielding varieties (HYV), intensive cultivation practices and multiple cropping systems.

Deficiency of secondary nutrient like S and micronutrient like Zn is widespread in rice growing areas. Acid tropical soil are generally deficient in molybdenum. The status of micronutrients in alkaline soils is poor because calcium carbonate renders many micronutrients, even if in good supply, insoluble. Table 8.7 gives the status of micronutrients in India soils.

Increased cropping intensity and decreased availability of organic matter leads to micronutrients deficiencies. A series of Long Term Field Experiments initiated during 1970's clearly showed that application of single nutrient die not support similar crop response over a period of time due to deficiency of macro and micronutrients, therefore, application of micronutrients in addition to the recommended NPK dose ensures significant improvement in crop yields and over all fertility status of the soil. It is also estimated that mineral fertilizer input accounts for 60 per cent and the rest is contributed by biological resources.

VAM

Most species of crop plants are normally colonized by mycorrhizal fungi. Numerous studies have indicated that VAM improves plant growth through assimilation of immobile soil nutrients (P, Cu, Zn) especially in soils which are poor in these micronutrients. Enhanced accumulation of plant nutrients like phosphorus, zinc, copper, manganese, iron through greatest exploration by mycorrhizal hyphae imported. Uptake all micro and secondary nutrients increase due to VAM inoculation when their efficiency exists in the soil (Tables 2). Improved plant uptake of potassium, present low concentration in the soil solution having low coefficient of diffusions increase in copper concentration mycorrhizal Hevea braziliensis root status was observed. However, a reduced in root colonization by arbuscion mycorrhizal fungi (AMF) under condition of heavy fertilization results in a decrease in the concentration of Ca and Zn in tissue. Thus AM fungi acts as a sensitive uptake system for various ions.

Table 8.7: Micronutrient uptake by crops following arbuscular mycorrhizal inoculation

Crop	Microorganisms (Arbuscular mycorrhizae)	Experimental conditions	Response
Sorghum	VAM fungi	Different temp. tested.	Root colonization by VAM fungi was temp. dependent. Higher phosphorus, potassium, zinc uptake by VAM plants than non-VAM plants observed.
Maize	*Glomus mosseae*		Phosphorus and zinc cotent of plants improved
Maize	*Glomus acrocarpum*	Soil having different concentration of zinc tested	Absorption of zinc by mycorrhizal maize was maximum in soils deficient in zinc
Wheat	*G. mosseae*& Phosphotase producing bacteria		Higher content of copper, manganese and zinc showing synergistic effect of both in inoculants on micronutrient translocation
Wheat	VAM + Phosphatase		Enhanced shoot concentration of copper

Crop	Microorganisms (Arbuscular mycorrhizae)	Experimental conditions	Response
	producing fungi		and zinc
Wheat Maize Potato	*Glomus macrocarpum*	Zinc deficient in soil	VAM infected plants significantly assimilated more zinc from soil than non-mycorrhizal plants
Barley	*Glomus Caledonium*		Increased dry matter production phosphorus, zinc, copper and bromium uptake
Groundnut			Significant gains observed in uptake of zinc, copper, iron and manganese
Field bean spring wheat	VAM fungi	Green house grown	Better uptake of micronutrients like Zn, Cu, Fe and Mn
Sorghum	VAM fungi		Higher amount of P, K, Mn, Mg, S, Ca, Fe Cu, Zn in mycorrhizal than non mycorrhizal plants

Different mechanisms suggested for hanced micronutrient uptake mycorrhizal plants are greater physical ploration of soil, faster ion movement into mycorrhizal hyphae and modification of root environment. Increased transpiration rates of plants and hence greater transpiration pull for nutrients may also contribute to better micronutrient uptake.

Table 8.8: Microbial species known to produce siderphores

Bacteria Enteric sp.	Enterobactin, Aerobactin
Agrobacterium tumefaciens	Agrobactin
Pseudomonas sp.	Pyoverdine, Pseudobactin
Bacllus megaterium	Schizokinen
Arthobacter	Arthrobactin
Azotobacter vinelandii	2-3 dihydroxybenzayl Lycine
Acinomyces sp.	Ferrioxamines
Mycobacterium	Mycobactins
Penicillum, Aspergillus, Neurospora	Ferrichromes, Coprogen
Rhodotorula sps.	Rhodotorullic Acid

Plant Growth Promoting Rhizobacteria (PGPR)

Plant growth promoting rhizobacteria (PGPR) a heterogenous group of microorganisms are known to improve plant growth by their ability to colonize the rhizosphere. Besides their effect as biocontrol agents and producers of plant hormones. PGPR also alter the plant physiological processes resulting in enhanced nutrient uptake. These organisms possess the ability to produce siderphores, a class of high affinity iron transport molecules which also act as a growth promoting factor. The ability of PGPR to produce siderophore or to affect the activity of siderophores produced by other organisms influences the uptake of iron by the plants when grown under Fe-deficient condition siderophore producing pseudomonads form a yellow, green fluorescent siderophore iron complex. Some microorganisms known to chelate iron through production of siderophores are given in Table 8.8:

Organics

Large potential for the exploitation of manurial value of organic wastes exists in India. Efficient use of organic like biogas shurry, farm yard manure, vermicomposts, pressmud and poultry manure can provide a greater level of essential plant nutrients. Presented from sugar factories, exalic acid industry wastes, phosphogypsum and sulphur sludge from myon industry are an efficient source of nutrient sulphur.

Biogas slurry, poultry manure, compost, pressmud are reported to be better source of zinc compared to zinc sulphate under calcareous soil condition. Poultry manure is also known to improve iron availability to crop plants. Micronutrient availability in some organics is given in Table 8.9.

In a longterm rice-wheat cropping system based experiments conducted on a sodic soil (pH 8.8) application of organic manure along with mineral fertilizers increased the pH, electrical conductivity and improve micronutrient cation (Cu, Mn, Zn, Fe) status of the soil.

A significant increase in the uptake of 'S' and Fe was found over control following application of farm yard manure along with iron pyrites. Application of farm yard manure @ 20 t/ha increased the availability of hot water soluble boron in different the textured soils (clay, clay loam, sandy loam at sandy). The effect was found in the ordinary clay > clay loam > sandy loam > sand. Increase in the boron content over the control was of the tune of 21.05 per cent.

However, use of organics like sewa sludge, industrial wastes for crop show be done after proper treatment to avoid to entry of pathogens, toxins and heavy manuals into the food chain. In this direction optimization of organic manures in co-junction with mineral fertilizer without having negative environmental impact will the right approach.

Table 8.9: Micronutrient availability in some organics

Source	Fe	Zn	Mn	Cu
	Concentration (ppm)			
Farm yard manure	24.66	40.00	12.0	2.8
Sewage sludge	-	216.2	15.0	49.5
Biogas slurry	-	103	231.0	50.6
	-	237	-	112.4
Vermicomposts (mean	21.6	12.65	19.17	5.83
Pheritima elongata	27.28	18.00	16.40	7.64
Eisenia foetida	17.82	19.20	24.60	7.56
entration	19.80	0.87	16.50	2.30

ORGANIC FARMING VS. BIODYNAMIC FARMING

Biodynamic farming means biological dynamics. It is a method of organic agriculture, which considers farm as a living system and where one activity affects the other. Rudolf Steiner (1924) from Central Europe was founder of Biodynamic agriculture.

Agriculture has faced vast changes, challenges and difficulties in past years. Indiscriminate use of fertilizers, pesticides and herbicides has polluted our environment. Varied chemicals, which are applied to soil and plant systems were not completely degradable and further more appeared to have carcinogenic effects. Thus increase in the use of inorganic compounds has led to residual toxicity problem which in turn poisoned our food chain. Pesticide residue of DDT and BHC in lactating women is a typical example of the deleterious effect of pesticides in agriculture. Moreover continuous use of insecticides in particular synthetic pyretheroids created the problem of pest resurgence and resistance against concerned chemicals. The continuous usage of inorganic nitrogenous fertilizers resulted in the excess vegetative growth, which is not accomplished with plant strength, structure and resistance. Thus the biodynamic farming will be great boon to the farmers, since the system reduces the pesticides and chemical fertilizers load to the soil eco systems, which in turn reduce the cost of cultivation. This is a needful technology which helps in the production of quality products, which will fetch special prize in future.

The biodynamic farming is more than just another organic method. It stands for a truly scientific way of producing humus. It does not involve in the application of organic matter in a more or less decomposed form but the use of completely digested from of crude organic matter known as stabilized and stable humus. In this aim, the method differs from organic farming. IN the case of biodynamic farming the organic material to be used as basis for compost is transformed either by means of the biodynamic compost preparations or by means of the biodynamic compost starter.

PRINCIPLES OF BIODYNAMIC FARMING

1. Biodynamic farming aims to restore humus status of the soil ecosystem to hold its fertility and productivity.
2. Also it helps to restore the soil for a balanced funtioning of flora and fauna. Because soil is a living system wherein the microbes can be fully established and maintained.
3. The biodynamic farming does not deny the role and importance of mineral nutrients of the soil like nitrogen, phosphate, potash, calcium, magnesium etc., and it considers the skilful use of organic matter as the factor for soil life.
4. It involves skilful application of all the factors contributing to soil life and health. Because a plant grows under the influence of abiotic factors like temperature, oxygen, CO_2 light, water etc. And these energies are transformed in the plant system into chemically active energies by way of photosynthesis.
5. Biodynamic farming considers a plant as living entity whih consists not only of mineral elements (like N, P, K, Ca, Mg, Mn, Cl, Fe, etc.,) but also of organic matter such as protein, carbohydrates, cellulose, starch.
6. Biodynamic farming gives importance to enzymes and growth

substances.

7. To restore the soil fertility, biodynamic farming gives priority to proper crop rotation Soil exhausiting crops should be cultivated alternatively with fertility restoring crops.
8. It also provides recognition to green manuring and cover cropping.
9. To restore the soil environmental conditions, forests, wind protection and water regulation are important.
10. Biodynamic farming also gives importance to maintain the soil structure (physical characters like bulk density, pore space, water holding capacity and texture).

RULES FOR USING BIODYNAMIC AGRICULTURE

1. Use of biodynamic sprays to stimulate and retain soil fertility
2. Following crop rotation to combat pests and disease and to increase the beneficial micro flora
3. Recycling on farm wastes
4. Following prevention strategies against pest and disease by strengthening plant and animals in the farm.

The produce of this farming is guaranteed by presence of international certification trademark Demeter™ this assures for a buyer that produce is biodynamically grown. The Demeter™ appeared first in Germany around 1928. It is certification trademark of biodynamic farming.

FEW BIODYNAMIC PREPARATIONS

When a living organism/entity, leaves behind the dead mineral of some form, that is subjected to laws of decay. The life force held the living creature from decay. To sustain and foster these forces biodynamic preparations were developed. The concept behing this is, these preparations/mediates influence the planets of the solar system and livings in the ethenic formative forces into the heap.

Table 8.10: Few biodynamic preparations

Preparation	Herb/material	Elements	Results
502	Yarrow Flower *Achillea millifolium*	Sulphur (S) Potassium (K)	Permits plants to attract trace elements in extremely dilute Quantities for their best nutrition.
503	Chamomile Flower	Calcium (Ca)	Stabilizes Nitrogen (N) within the compost and increase soil life so as to stimulate plant growth.
504	Stinging Nettle *Urtica dioca*	Sulphur (S) Potassium (K)	Stimulate soil health, by
		Calcium (Ca) Iron (Fe)	providing plants with the individual nutrition; enlives the earth (soil).
505	Oak Bark	Calcium(Ca)	Provides healing forces (or qualities) to combat harmful plant diseases.

Preparation	Herb/material	Elements	Results
506	Dandelion Flower *Taraxacum officinale*	Silicon (Si) or Siliclic Acid Potassium (K)	Stimulates relation between Si and K so that the Si can attract cosmic forces to the soil.
507	Valerian Flower *Valeriana officinalis*	Potassium (P)	Stimulates compost so that Phosphorous component is properly used by the soil

CONCLUSION

The natural nutrient supply, the flows and the nutrients added needs proper management for achieving as high crop yields as possible under the climatic circumstances without avoidable environmental pollution.In this contest the concept of fertilization need to be widened into a more comprehensive concept of nutrient management being integrated with other production factors. This goal requires an optimum use of soil nutrient sources, a sustainable soil fertility, recycling of nutrients and a comprehensive fertilization in order to supply the crops with all nutrients in sufficient amounts.An improved nutrient management also includes the adaptation of crops to the nutrient supply available and better use of nutrients from previous crops in rotations. The heterogeneity of the nutrient supply in large fields, especially of N, should be compensated by precision fertilization.

Nutrient management should not only care for high yields, but also for good product quality, stress resistance and reduction of nutrient losses.

Different farming systems in different climatic regions require special nutrient management ranging from exploitation cropping to sustainable agriculture at low-medium or high yield level. The claim that the nutrient management of biofarming produces better food is not justified.

Optimum nutrient management requires not only good expertise, but also knowledge and engagement of the farmers as well as an appreciation of ht especial problems involved in primary production by the general public.

In the near future, several modifications of nutrient management are required for saving nutrients, at least those with a limited global supply, whereas, in a distant future the food production may have to be adapted to a sustainable nutrient use level.

CHAPTER 9

Perspectives

The nitrogen fixing biological systems such as blue green algae, *Azolla, Axospirillum, Acetobacter,* and *Azorhizobium* are compatible to different cropping systems and could fix considerable amounts of nitrogen.

The biomass of these biological systems decompose rapidly in soil and supply the nitrogen to various crops grown in the field. Besides nitrogen addition, these biofertilizers also contribute significant amounts of phosphorus, potassium, sulfur, zinc, iron, molybdenum and other micronutrients. The biological fertilizers were known to influence the total soil microbial population and soil enzymes activity, which directly reflect on soil fertility index.

The use of biological fertilizers would reduce the cost of chemical fertilizers involved in crop production. The effective utilization of these biological fertilizers for different cropping systems will not only provide economic benefits to rice farmers, but also improve and maintain the soil fertility and sustainability of the soil ecosystem.

Biological nitrogen fixation is an important process in rice farming systems because it is an inexpensive source of nitrogen for increasing the productivity of crops. The nitrogen fixing bio systems such as Azolla, cyanobacteria and legume green manures were able to adopt well under wetland rice field ecological conditions and fixing considerable amounts of nitrogen. The biomass of these bio systems decomposes rapidly in rice soil and supply the nitrogen to rice crop.

Besides nitrogen addition, these biofertilizers or bio-manures also contribute significant amounts of phosphorus, potassium, sulphur, zinc, iron, molybdenum and other micronutrients. The organic acids released during the mineralization process of the biomass of bio-manures would accelerate the phosphorus availability in rice soil.

These biological fertilizers or bio-manures were known to influence the total soil microbial population dynamics of nitrogen fixing free-living bacteria like, *Azotobacter, Azospirillum* and also the free-living symbiotic bacterium such as *Rhizobium*. It is evidently clear now that application of biological fertilizer enhanced the accumulation of soil enzymes, which directly reflected on soil fertility index. The use of biofertilizers would reduce the cost of chemical fertilizers involved in crop production. The effective utilization of biofertilizers for crops will not only provide economic benefits to the farmers but also improve and maintain the soil fertility and sustainability in natural soil ecosystem. Since these biosources represent a great diversity in chemical, physical and biological characteristics, their efficient use depends upon the particular agro-ecological environment and local availability.

Therefore, it is very much essential to develop strong, workable and compatible package of nutrient management through organic sources for various crops based on scientific facts, local conditions and economic viability.

Biofertiilzers are environment friendly, low cost agricultural input playing a significant role in improving nutrient availability to the crop plants. It is wide term including a diverse category of bi-inoculants such as nitrogen fixers (symbiotic and asymbiotic) phosphate solubilises and plant growth promoting rhizobacteria etc. Performance of biofertilizers is highly unpredictable due to their biological nature and susceptibility to biotic and abiotic stresses. Future research should focus development of effective competitive and stress tolerant strains. T eh potential of biofertilizers to supply micronutrients is yet to behamessed. Lesser used biofertilizers, such as phosphate solubilizing bacteria and phosphate modilizers like VAM are promising in supplying phosphorus and other micronutrients. In this direction culture of obligate symbiont (VAM) under laboratory conditions will enable testing of their performance in field conditions. Genetic basis for competitive superiority needs to be worked out. Selection of an economical synthetic carrier capable of maintaining high viable count and improvement in inoculation techniques to ensure establishment and persistence in soil are areas needing further research.

CONSTRAINTS IN BIOFERTILZER USE

Though the benefits derived from the use of biofertilizers are quite significant, they have not become popular due to various constraints at production, distribution, field and marketing level.

These constraints can be classified as:

1. Production Constraints
2. Marketing Constraints
3. Resource Constraints
4. Field Level Constraints

Production Constraints

Despite significant improvement/refinement in BF production technology over the years, the progress in the field of BF production is below satisfaction due to following constraints:

Unavailability of appropriate and efficient strains

Lack of region specific strains is one of the major constraints as biofertilzers are not only crops specific but soil specific too. Moreover, the strains selected for production should have various qualities particularly competitive ability over other strains, N fixing ability over a range of environmental conditions, ability to survive in adverse physical conditions; ability to multiply in broth and survive in inoculant carrier.

Strains having all these qualities are not available for various regions of the country.

Unavailability of a suitable carrier

Unavailability of suitable carrier due to which shelf life of biofertilizers is short, is a major constraint. Peat has been university recognized as the most suitable but its shelf life is about six months. Availability of good quality peat is major constraint. Being readily available, lignite is the most common carrier in the country. Biofertilizer development cannot be achieved to desired level unless a good carrier is evolved.

Mutation during fermentation

Biofertilizers tend to mutate during fermentation and thereby raising production and quality control cost. Presently there are no effective measures to eliminate mutation. Extensive research work on this aspect is urgently needed.

Table 9.1: Number of Biofertilizer production units located in different States/Union Territory

Sl No.	Name of the State/Union Territory	Total No. of BFPU	Total BFP capacity (tonne/yr/state)	Distributed pattern of BFPU
01.	Andhra Pradesh	4	165	SG(1) + AU(1) + P(1) + FI(1)
02.	Assam	2	90	AI(1) +AU (1)
03.	Bihar	2	20	AU (2)
04.	Delhi	1	75	IARI
05.	Gujarat	3	300	FI(2) + MFD (1)
06.	Haryana	2	125	GI(1) + AU (1)
07.	Himachal Pradesh	1	10	SG(1)
08.	Karnataka	6	150	AU(2) + SG(1) + GI(1) + P(2)
09.	Madhya Pradesh	4	450	GI(1) +AI(1) + MFD (1) + SG(1)
10.	Maharashtra	5	250	GI(1) +AI(2) + AU (1) + P(1)
11.	Manipur	1	50	GI(1)
12.	Orissa	2	125	GI(1) + AI(1)
13.	Punjab	1	75	GI(1) + (AI(1)
14.	Rajasthan	2	100	AI(1) + AU (1)
15.	Tamil Nadu	9	450	SG (4) + AU (1) + FI (3) + P(1)
16.	Uttar Pradesh	13	300	GI(1) +SG(10) + AU(2)
17.	West Bengal	4	150	AU (1) + P (3)
	Total	62	2885	

Abbreviations: AU = Agriculture University, AI = Agriculture Industries, BFPU = Biofertilizer Production Unit, BFP = Biofertilizer Production, FI = Fertilizer Industries, GI = Government of India, MFD = Marketing Federation, P = Private, SG = State Government. *Source*: Singh (1993).

The total estimated demand for various biofertilizers for different crops grown in India is as follows:

Rhizobium .... 35,730 tonnes
Azospirillum .... 4,82,000 tonnes
Azotobacter .... 1,62,610 tonnes
Blue Green Algae .... 2,67,720 tonnes
Azolla .... 20,380 tonnes
Phosphate solubilizers .... 2,75,500 tonnes

The demand is calculated based on the area under different crops and the dose of biofertilizer.

The present production capacities of different biofertilizers in India are as follows:

National Biofertilizer Development Centre (Central Sector)	375 tonnes
State Agricultural Universities	245 tonnes
State Agricultural Departments (Major Biofertilizer producing States include Tamil Nadu, Uttar Pradesh and Maharashtra)	665 tonnes
Agro Industries Corporations	1130 tonnes
Private Sector	195 tonnes

The annual requirement and production of different biofertilizers have clearly shown tremendous gap in this area. Thus a strategy for judicious combination of chemical fertilizers and biofertilizers will be economically viable and ecologicall useful. It should also be remembered that biofertilizers are not a substitute, but a supplement to chemical fertilizers for maximizing not only yield but also agro system stability.

Table 9.2: Potential demand of Biofertilizers

Biofertilizer	Potential demand (based on area in 1992-93) (tonnes)	Estimated demand in 1999-2000 (tonnes)
Rhizobium	4950	12500
Azotobacter	1987	5850
Azospirillum	9264	18528
BGA	91050	205250
Total	107251	242128

Source: Biotech Consortium India Limited (1992).

Lack of standards in packaging

There are inadequate standards available with regard to type of package material to be used. Moreover, size of packs are also not standardized.

Market Level Constraints

Though use of biofertilizer in India started in late 50's, consumption has not picked up to desired level due to various marketing and market related constraints.

Lack of awareness of farmer

Inspite of considerable efforts in recent years, majority of farmers in India are not aware of biofertilizers, their usefulness in increasing crop yields and soil productivity.

Inadequate and inexperienced staff

Because of inadequate marketing staff and that too not technically qualified who can attend to technical problems, acceptance of biofertilzers has been very limited in the country. Farmers are not given proper instructions about the application aspects, particularly the type and quantity of biofertilzers in different crops and cropping systems.

Lack of quality assurance

The sale of poor quality bio-innoculants through corrupt marketing practices results in loss of faith and as a result, product image is suffered. It may mentioned here that to regain the faith once lost is very difficult and challenging.

Seasonal and unassured demand

A small scale production unit solely dependent on biofertilizer is not economically viable because biofertilizer use is seasonal and both production and distribution is done only in a few months of year. Moreover, production units particularly in private sector are not sure of their demand. During 1970s, many private sector units had to close their units because of their inability to book orders in those states where the government was actively involved in production. The past experience have thus shown that when state governments have their own production units and placing of indent depends on them, it is very difficult for other producers to market their product in those states.

Limited scope for marketing

As biofertilzers are soil and crop specific, a large number of formulations of a microbial inoculant each having unique features are required to be produced by a manufacturer. This limits the market size for each type of biofertilizer by narrowly segmenting the markets.

Resource Constraints

In addition to technological and marketing constraints, resource constraint is also one of the main constraints in production of biofertilizers on commercial scale.

Limited resources generation for BF production

The investment in biofertilizer production unit is considerably low. But keeping in view of the risk involved largely because of short shelf life and no guarantee of off take of biofertilizers, the resource generation particularly from private sector is very limited. Although Government of India is offering an aid of Rs. 13 lakh for setting up biofertilizer units, the response is not that encouraging.

Limited risk taking capacity of farmers

Biofertilizer are considered as low cost input for the resource poor farmers. But so long as the crop response to biofertilzer is inconsistent, poor farmers would find it difficult to take the risk of using biofertilizers.

Field Level Constraints

Besides various production, marketing and resources constraints, there are a large number of field level constraints too in promoting Biofertilizer usage. There is low level of acceptance of biofertilizers at farmers level largely because the response is not immediately visible and many times it is erratic also.

Soil and climatic factors

Among soil and climatic conditions, high soil fertility status, unfavourable pH, high nitrate level, high temperature, drought, deficiency of P, Cu, Co, Mo or presence of certain elemens in toxic amounts affect the crop response.

Native microbial population

Antagonistic micro organisms already present in soil compete with microbial inoculants and many a times do not allow their effective establishment by out-competing the inoculated population.

Faulty inoculation techniques

What to talk of farmers, majority of marketing/sales personnel do not know proper inoculation techniques. Biofertilizers being living organisms required proper handling, transport and storage facilities otherwise these result in reduced shelf life and poor viability.

Crop management

Besides various soil and climatic factors, response to biofertilizer largely depends upon balanced use of other inputs and overall crop management. Poor crop management and use of poor agricultural practices make biofertilisers ineffective. It may be noted here that just making available biofertilizers to resources poor farmers is not going to benefit unless they are able to afford for afford for other crop production inputs particularly improved seed, fertilizer and water.

Quality control

Quality control of biofertilizer is very important. Quality standards have been set up for *Rhizobium* and *Azotobacter*. But, its effectiveness is ensuring the supply of quality biofertilzers at field level is doubtful. Distribution of substandard inoculant particularly those whose shelf life is over, is also being practiced. This is eroding the farmers belief in biofertilizers.

RESEARCH AND DEVELOPMENT NEEDS

Current annual use of biofertilizers at around 2200 tonnes (of which more than 1000 tonnes is of *Rhizobium* alone) is much below the projected requirement. Undoubtedly requirement calculated on the basis of crop acreage is not realistic. However, there is no difference of opinion that the country has potential to increase use of biofertilizers. To popularize biofertilizers' use, concerted efforts right from working out realistic demand estimates, delineation of areas which require inoculation on priority basis, development of suitable production technology, streamlining marketing and distribution channels and strengthening extension and training activities are needed. Some of the steps which need urgent attention for successful production, promotion and distribution of biofertilizer are mentioned.

Widening Research Base

The mechanism by which the introduced microorganisms benefit crop plants and soil is not fully understood. There is need to widen research base to understand the mechanism as these biological systems involve complex interactions among the host; rhizosphere micro fauna and flora; and other inputs used in crop production. Work on gene manipulation which is necessary to bring desirable properties in strains should be done more extensively.

Screening New Efficient Strains

Identification of factors governing competitive ability and efficiency of strains is the prerequisite for screening more efficient and competitive strains which are capable of multiplication and fixing of N under varying agro-climatic conditions. Strains which perform well under adverse conditions like soil acidity, salinity, moisture stress and insect resistance need to be selected, developed and tested for their performance. In order to harness potential benefits of biofertilizers, the survival and competitive ability of strains must be improved.

Improvement in Production Technology

At present the production of biofertilizers is being done by using septic conditions introducing acid producing contaminants which are not inductive to the growth of *Rhizobium.* There is urgent need of improving the present system of mixing and packaging biofertilizers. Search for suitable synthetic carrier materials which can be uniform, non-toxic, simple to use and sustain viability for longer period should be pursued vigorously.

Preparation of Biofertilizer Map

Like the soil maps for nutrient status, there is need to prepare biofertilizer map. Surveys should be conducted to quantify native soil microbial population which will help in determining the priority areas and inoculation requirement for different kind of biofertilizers.

Quality Assurance

Strict quality control mandatory is a must for all biofertiliser producers. Quality control should not only be ensured at the production level but at the sales points also. Proper storage facilities should be provided at the production unit, sale points and during transportation to maintain required level of viable cells. Appropriate legislation should be enacted to prescribe the required standards for good quality inoculants.

Improving Distribution System

Distribution system which is the key to supply quality products has not been effective in the past. Accessibility of farmers to biofertilizers needs to be ensured by expanding distribution network.

Improving Technical Support System

Technical support through skilled personnel should be provided to educate the farmers in storage and usage of biofertilizers. Intensive education of extension workers, dealers, farmers and exposition of policy makers through training, field demonstrations and audio-visual aids is needed to generate awareness about the importance of biofertilizers in sustainable agriculture.

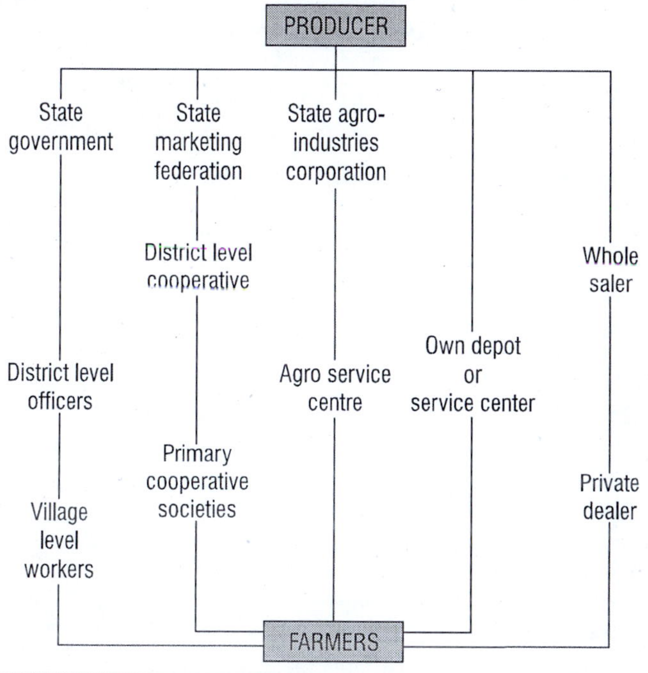

Fig. 9.1: Biofertilizer marketing channels in India.

Institutional Support for Research and Development

There has to be strong institutional support for ongoing research and development work. ICAR has AICRP on BNF with a few centers in different parts of the country. Ministry of Agriculture has sponsored the National Project on Biofertilizers with a few regional and sub-centres. Department of Biotechnology has established National Facility on Blue green algae and *Rhizobium* at IARI, New Delhi. Further, it is very much necessary that the Departments of Agricultural Microbiology in the State Agricultural Universities are strengthened to do more of fundamental research work. A few centers of excellence need to be prompted wherever some good work is being done. In each state, regional laboratories need to be established for isolation, testing and maintenance of cultures.

QUALITY CONTROL

Recently, the biofertilizer production technology has attracted much attention of the small and large producers because of its simple and cost effective nature. The current annual production of biofertilizers is in the range of 2000-2500 tonnes. Though there is lot of scope for scaling up the production level, the biofertilizer technology faces the problems of quality control, inconsistent field performances, poor transport and storage and lack of knowledge about these inputs. This has necessitated the implementation of better quality control mechanism to ensure the availability of quality inoculants to the farmers. The quality of biofertilizers depends on the use of specific and appropriate microbiological techniques for production and quality control of biofertilizers. The following points have to be taken into consideration while adopting quality control procedures for biofertilizers with the objectives of ensuring supply of quality inoculatns to the farmers.

1. Biofertilizer production unit should have adequate microbiological laboratory facilities to produce quality inoculum. The unit should be maintained by a qualified microbiologist. The production unit should have one quality control laboratory to check their own biofertilizer packets.
2. Each biofertilizer production unit should maintain crop specific and pre-tested efficient N_2 fixing microorganisms or phosphobacteria either collected from a reputed scientific organization or of their own. The maintenance of microorganisms should be as per the standard microbiological procedures and repeated sub-culturing should be avoided. If possible, lyophilized cultures should be used for each batch of biofertilizer production. The manufacturer should indicate the source of the culture in the biofertilizer packet produced.
3. A biofertilizer production unit should use fermentors so as to ensure high microbial counts at broth level itself.
4. The biofertilizer production unit should use good quality carrier material like peat or lignite or wood charcoal or any other suitable carrier and the biofertilizer products should indicate the name of the carrier.
5. The inoculum packet should contain 200 g carrier based culture packet in standard polythene or polyprophylene bags. The BGA biofertilizer packets should be in 1 kg, 4 kg, 10 kg pack size.
6. Recently the cost of 200 g pack size biofertilizer packet is fixed at Rs. 5 and therefore there is no need to enhance the cost at present.
7. The microbial inoculum must contain the specified population count at the time of packing and expiry. The packet should be crop specific and should give effective yield increase.The production centers and distribution centers should store the biofertilizer packets in cool temperature so as to ensure longer shelf life.

Table 9.3: Suggestions for Distribution Agencies

- Keep biofertilizer packets away from direct heat and sunlight. Store it in cool and dry palce.	- Don't store biofertilizer packets under heat and sunlight.
- Sell only biofertilizer packets which contain batch number, the name of the crop, the date of manufacture and expiry period.	- Don't sell biofertilizer packets after their expiry period is over.
- If the expiry period of the packet is over, discard it. It will not be effective.	- Don't prick holes into the packets or puncture them.
- Keep biofertilizer packets away from fertilizer or pesticide containers.	- Don't store heavy items above biofertilizer packets

8. Sample packets of biofertilizer should be drawn randomly for quality testing by the Central quality control authority appointed by the state, and out of 5 samples in a batch, at least 3 should pass the specified quality standards.
9. To ensure microbial inoculum quality control testing and quality assurance of the biofertilizers produced by different production units of the state, an exclusive Central Quality Control Laboratory with growth chambers and immunological laboratory facilities should be established. A certification unit should be included in the Quality Control Laboratory for certification of the inoculants. An amount of Rs. 50 as certification charge should be collected for each batch of biofetilizer produced.
10. A Central biofertilizer germplasm collections should be established with sub-centres at different agroclimatic zones so as to select efficient agriculturally useful microorganisms.
11. The government should make legislation for enforcement of quality standards for different biofertilizers produced in the state.

PROMOTIONAL STRATEGIES

1. The reason for inadequate popularity of biofertilizers is that biofertilizers cannot show instant and dramatic response like fertilizers.
2. In many cases, quality of culture supplied is poor.
3. Inadequate correct awareness among the farmers about inoculants.
4. Lack of strong promotion and extension work and insufficient publicity.
5. Lack of availability of the quality products in time in rural areas.

For successful launching of a biofertilizer programme it is necessary to equip the extension machinery at the grass-root level with the full technical back up of production as well as application technology. There is need to give emphasis to biofertilizers as supplementary, renewable and non-polluting sources of nutrients. Once the farmers are convinced that the biofertilizer are an inexpensive but effective way of keeping their soil fertility, they will accept them and this will lead to higher crop production and higher economic returns.

FUTURE THRUSTS

Developing strains of bacteria which can withstand high temperature and alkali, and are acid resistant and high pH tolerant; which produce antibiotics in the soil or any chemicals which impart the strain competitive edge over other strain is the need of the hour. Some of the genes responsible for production of plant hormones have been characterized and cloned. These genes could be introduced into *Rhizobium* strains. Strains that can fix nitrogen in the presence of added nitrogen and those that can increase nitrogen-fixing efficiency should be selected.

Efforts to mass multiply mycorrhizae in a large quantity using lighter carrier material should be intensified. The PCR techniques should be standardized to identify mycorrhizal colonization within plant roots using different species specific primers.

APPENDIX 1

Practical Methods for Estimation of Biological Activities

1. Isolation of Rhizobium from Root Nodules

Introduction

Rhizobium is a symbiotic nitrogen fixer and forms nodules on the roots of leguminous plants. Rhizobia can be isolated from the root nodules after cleaning and surface sterilizing the nodules. The nodules can be either crushed in small volume of sterile water or dissected and the nodule contents streaked on the plate containing yeast extract mannitol agar (YEMA) medium. To distinguish rhizobia from other contaminating bacteria incorporation of dye Congo red (10 ml of 1/400 aqueous solution) in YEMA medium is advantageous as rhizobial strains absorb the dye weakly whereas many other bacteria take it up strongly.

Requirements

Petridishes, sterile tap water, mercuric chloride (0.1%), ethyl alcohol (70%), test tubes containing 5 ml sterile water, scalpel, yeast extract mannitol agar (YEMA) medium with congo red, glass rod, forceps.

Procedure

1. Uproot healthy leguminous plant.
2. Wash the root system with water.
3. Select nodules which are healthy, pink and preferably on the tap root.
4. Separate the nodule by giving diagonal incisions from two sides of nodule so as to cut nodule along with root portion.
5. Wash the nodules thoroughly with running water.
6. Transfer nodule to sterile beaker and surface sterilize with 0.1% mercuric chloride and 70% ethyl alcohol for 3 min and 30 sec respectively.
7. Decant the alcohol and wash nodules with sterile water, atleast give six washings.
8. Place the nodule in small vial containing sterile water or on porcelain cavity plate.
9. Crush the nodule with sterile glass rod.
10. Plate the suspension on CRYEMA plates (See Annexure).
11. Incubate at 28°C till small, round colourless or while with central red dot and entire margin develops.
12. Transfer the isolated colony to fresh CRYEMA plate for purification.

13. Maintain the pure culture on YEMA slants till further use.

2. Assay of nitrogenase enzyme in free living bacteria and nodules by acetylene reduction technique

Introduction

Nitrogenase is a versatile enzyme capable of catalyzing the reduction of several substrates other than nitrogen, including acetylene (C_2H_2), azide, nitrous oxide, nitriles and isonitriles. The observations that acetylene is an inhibitor of dinitrogen (N_2) fixation, and is converted by the N_2 fixing enzyme nitrogenase (N_2ase) to ethylene (C_2H_2) provided the basis for the development of the first simple method for the estimation of N_2 fixation. The C_2H_2 reduction method provides a simple, inexpensive, highly sensitive and non-destructive procedure for measuring rates of N_2 fixation. It can be used to measure N_2 fixation in free living N_2 fixing bacteria and blue green algae, associative symbiosis and nodule bearing leguminous and non-leguminous species.

The reduction of N_2 and C_2H_2 by N_2ase is identical and may be designated as follows:

$$N_2 + 6H^+ + 6e^- \rightarrow 2NH_3$$

$$C_2H_2 + 2H^+ + 2e- \rightarrow C_2H_4$$

Thus the theoretical ratio of substrates reduced is 3:1, which is included in the equation used to quantify the rate of N_2 fixation based on C_2H_2 reduction

$$mgN = \frac{mMC_2H_2 \text{ X } 28}{3}$$

ARA technique is an indirect method of measuring N_2 fixation at a point of time. The technique involves incubation of nitrogenase containing system in a known atmosphere of C_2H_2 (5-20%) in a gas phase, and after an optimum time of incubation, C_2H_2 produced is measured by a gas chromatograph using a flame ionization detector (FID).

Requirements

Assay vials/containers with serum stopper, Compressed C_2H_2 cylinder (C_2H_2 can also be conveniently prepared from calcium carbide and water), Compressed C_2H_2 standard gas cylinder, Compressed N_2 and H_2 gas cylinders, Gas tight syringes (1,2,5 and 10 ml), Incubator, Gas chromatograph with Porapak N column and FID detector.

Proceduro for frcc-living bacteria

1. Prepare slants or use semi solid medium using appropriate medium. (*Azosprillum* – semi solid N – free malate medium) (*Azptobacter* – Jensen's N_2 free medium) (Explanta Nitrogenase activity of *Rhizobium* – CS7 medium)
2. Streak loopful culture on the slants.
3. Incubate at 28°C for 2-7 days.

4. Replace cotton plugs with air tight serum stoppers.
5. Remove 10% of the atmospheric air from the tube and inject equal volume of acetylene gas using a syringe.
6. Incubate tubes at 28°C for 16-24 h.
7. Draw 1 ml of gas sample from the tube and inject into gas chromatograph for ethylene estimation.
8. Similarly record values for 1 ml standard ethylene gas.
9. Estimate total protein of each sample tube as follows:
   a) Collect bacterial cells in 2ml of 2N NaOH.
   b) Keep cell suspension in boiling water bath for 10 min.
   c) Cool it, neutralize with 2ml, 2N HCl and estimate by Lowry's method or Bradford's method.
10. Calculate C2H4 produced using the formulae

$$\text{nmoles of } C_2H_4 \text{ produced } hr^{-1}mg^{-1} \text{ protein} = \frac{C \times Ps \times As \times V}{Pstd \times Astd. \times T \times P}$$

ARA of root nodule system

1. Uproot legume plant, shake gently to remove adhering soil (washing not advisable).
2. Detach the root system along with nodules.
3. Transfer the nodulated root to assay vials or bottle.
4. Insert serum stopper, replace 10% of air with equal volume of acetylene.
5. Incubate for 30 – 60 min. at 28°C.
6. Carry on gas chromatography as described earlier.
7. Determine the volume of air space in the assay vial (remove serum stopper, pour water in the assay vial and measure the volume of water (= volume of air space).
8. Detach nodules from the roots. Take fresh weight or dry them in oven at 80°C for 3 days and take dry weight.
9. Calculate C_2H_4 produced as follows

$$\text{nmoles of } C_2H_4 \text{ produced } hr^{-1}\, g^{-1} \text{ dry weight of nodule} = \frac{C \times Ps \times As \times V}{Pstd. \times Astd. \times T \times wt}$$

Terms C, Ps, As, Pstd., Astd, V and T are described earlier.

Wt. = fresh or dry weight of nodules (g)

*Calculation of ethylene concentration (C) in standard ethylene gas

Standard ethylene gas is supplied as compressed gas in air tight cylinders. Standard ethylene comes as a mixture with inert gases like argon. The volume of ethylene in gas mixture is mentioned on the cylinder and can be 105 vpm or 1050 vpm (volume per million), that is, 10^6 ml of gas mixture will have 105 or 1050 ml of ethylene.

To calculate ethylene concentration let us consider a standard ethylene cylinder with a volume of 105 vpm.

10^6 ml of gas mixture contains = 105 ml of ethylene gas

i.e. 1 ml of gas mixture contains $= \frac{105}{10^6}$ ml ethylene gas

According to Avagadro's law

22,400 ml of any gas at NTP = 1 mole

Or 1 ml of any gas at NTP $= \frac{1}{22{,}400}$ moles

Or $105/10^6$ ml of any gas at NTP $= \frac{1 \times 105}{22{,}400 \times 10^6}$ moles

$$= \frac{1 \times 105 \times 10^6}{22{,}400 \times 10^6} \text{ nmoles } (1 \text{ mole} = 10^6 \text{ nmoles})$$

The value 105/22,400 nmoles represents the concentration of standard ethylene in 1ml of gas mixture.

3. Strain Identification in *Rhizobium* using Intrinsic Antibiotic Resistance

Introduction

Strain identification is a major problem in field experimentation with strains *Rhizobium.* Where legumes are regularly grown soils invariably contain some rhizobia capable of nodulating the particular host. It is, therefore, desirable to develop simple techniques for identifying inoculant strains of *Rhizobium* spp. The use of genetic marker such as rapid method for strain identification. Microorganisms, in general, display an intrinsic resistance to the inhibitory or lethal effects of particular antibiotics. Such resistance may depend e.g. on the absence, or inaccessibility, of those structural and/or functional features against which the antibiotic is effective. The genes responsible for resistance to antibiotics are present either on the chromosome or on the plasmid. A plasmid, the intracellular presence of which confers on a cell the property of resistance may be gained through R factor include aminoglycoside antibiotics (gentamycin, kanamycin, neomycin, paromomycin, streptomycin and spectinomycin), chloramphenoicol, erythromycin, penicillins, sulphonamides and tetracyclines.

The intrinsic levels of resistance to host of antibiotics is developed by using different concentrations of antibiotics, and once the resistance pattern for a particular strain is developed it can be used to identify the strain.

Requirements

YEMA medium, Petri plates, Ahtibiotics, *Rhizobium* isolates

Procedure

1. Raise the broth culture of the strains to be tested in YEMA medium.
2. Prepare stock solutions (10 mg ml^{-1}) of the antibiotics in the appropriate solvent. Filter sterilize it (If filter membranes are not available, dissolve the antibiotics in sterile solvent and store it at 4°C.
3. Prepare agar based YEMA medium. Dispense 50 ml in 100 ml flasks.

Autoclave at 15 lb. psi.

4. Cool the medium to 50 – 60°C and add appropriate concentrations of antibiotics.
5. Pour two plates for each concentration. Allow it to solidify.
6. Streak or spot the *Rhizobium* strains on the antibiotic plate and control plate (no antibiotic).
7. Incubate the plates at 28°C for 2-5 days.
8. After incubation, compare the growth on antibiotic plate with control plates. Score the growth as positive or negative.
9. Concentration till which the growth occurs is taken as the resistance limit.
10. Prepare the intrinsic antibiotic resistance pattern of each strain. It can now be used for identification of a strain mixed culture, soil or nodules.

Table 1: Some antibiotics and their solvent

Antibiotics	Solvent system
Gentamycin	Distilled Water
Chloramphenicol	Methanol
Rifampicin	Ethanol
Trimethoprin	Distilled water + Lactic acid
Nalidixic acid	Distilled water + NaOH
Ampicillin	Distilled water
Neomycin	Distilled water
Streptomycin	Distilled water
Strectinomycin	Distilled water
Spectinomycin	Distilled water
Tetracycin	50% Ethanol
Kanamycin	Distilled water

4. Enumeration of *Azospirillum* by Most Probable Number (MPN) Method

Introduction

The MPN method permits estimation of population density without an actual count of single cells or colonies. It is sometimes called the method of ultimate or extinction dilution or end point dilution. The MPN technique is based on a determination of the presence or absence of microorganisms in several individual aliquots of each of several consecutive dilutions of soil. A prerequisite of the method is that microorganisms whose population is to be determined must be able to bring about some characteristic and readily recognizable transformation in the medium into which it is inoculated, or else the microorganism itself, after it has undergone multiplication, must be easily recognizable in the substrate. On the basis of probability theory, it is possible to calculate, from the numbers of positive and negative tubes receiving a certain quantity of inoculum, the most probable number of organisms in that quantity of inoculum. Multiplying the result by the appropriate dilution factor gives the MPN for the sample. The theoretical aspects of the theory is dealt.

Azospirillum is a microaerohilic nitrogen fixer. A semi-solid N-free malate medium with Bromothymol blue dye is used for enumeration. The pressure of *Azospirillum* in the tube is indicated by

a) change in dye colour from green to blue
b) formation of pellicle at the subsurface of medium
c) reduction of acetylene to ethylene

Requirements

Glass tubes, 90 ml water blanks, soil, pipettes (10 ml and 1 ml), Syringe, gas chromatograph, N free semi solid malate medium.

Procedure

1. Prepare glass tubes (25) each containing 5 ml of semi solid malate medium. Sterilize them at 121°C, 15 lb pressure.
2. Take 1 gm soil and prepare a 10 fold dilution series upto 10^{-6}.
3. Withdraw by sterile pipette 1 ml aliquots from dilutions 10^{-2}, 10^{-3}, 10^{-4}, 10^{-5} and 10^{-6} and dispense it in the medium containing tubes.
4. For each dilution 5 tubes have to be inoculated with 1 ml aliquot separately.
5. Incubate the inoculated tubes for 2 days at 30°C.
6. Examine each tube for colour change and pellicle formation. Record the growth in each individual tube as either positive or negative.
7. The visual observations can be confirmed by estimation of nitrogenase activity of each tube (as described earlier). The tubes +ve for *Azospirillum* should show nitrogenase activity.
8. Calculate the MPN count as follows.

Calculations

To calculate the MPN in the original sample.

Select as p_1: The number of positive tubes in the least concentrated dilution in which the greatest number of tubes is positive.

and let P_2 and p_3 represent the numbers of positive tubes in the next two higher dilutions.

Use the table of MPN for use with 10-fold dilution and 5 tubers per dilution. Then find the row of numbers in MPN table in which p_1 and P_2 correspond to the values observed experimentally. Follow that row of numbers across the table to the column headed by the observed value of p_3. The figure at the point of intersection is the most probable number of organisms in the quantity of the original sample represented in the inoculum added in the second dilution (p_2). Multiply this figure by the appropriate dilution factor to obtain the MPN for the original sample.

Example

Suppose following observations were made

Dilution.. +ve
10^{-3}.. 5

10^{-4} 3
10^{-5} 1
10^{-6} 0

In this series; $p_1 = 5 = 3$, $p_3 = 1$. For this combination of p_1, p_2 and p_3, the MPN table gives 1.1 as the most probable number of organisms in the quantity of inoculum applied in the 10^{-5} (p_2) dilution. Multiplying this number with dilution factor 10^5 gives 1.1 x 10^5 as the MPN for the original sample.

5. Enumeration of Rhizobia in Soil by Most Probable Number (MPN) Method

Introduction

The population of specific rhizobia in the soil is estimated by the indirect plant infection test following the most probable method. This method depends on the ability of a specific *Rhizobium* cell added to the test plant leads to a sufficient population in the root surroundings to cause nodulation. The number of tubes with nodulated plants can then be used to determine the likely number of rhizobia in the soil sample under test. Brockwell gave tables for estimation of most probable number of rhizobium in soil, both based on 6 dilution steps, for give fold and ten fold dilutions.

Requirements

Gibson tubes (200 mm x 25 mm), Non absorbent cotton, tube stands, growth chamber, acidified 1% $HgCl_2$ (1g $HgCl_2$ in 95 ml water + 5 ml HCl), sterile water blanks (4 ml and 90 ml), 70% ethanol, forceps, 2% water agar, seedling nutrient agar (SNA) medium.

Procedure

Preparation of agar tubes

1. Prepare SNA medium (full strength) and boil it to melt agar.
2. Add about 30 ml medium to Gibson tubes
3. Cover the mouth of the tube with aluminium foil and fix it with tight rubber band.
4. Autoclave the tubes at 121°C, 15 lb pressure
5. Set the tubs as slopes (the agar should touch the surface of aluminium foil).
6. Tubes are ready for use.

Preparation of seedlings of host to be used

1. Select undamaged, healthy seeds
2. Treat with 70% ethanol healthy seeds
3. Decant off alcohol
4. Treat with 1% $HgCl_2$ for 5 min
5. Decant off $HgCl_2$.
6. Wash thoroughly with sterile water (give atleast six changes).
7. Keep the seeds in sterile water for imbibition for 2 hrs.

8. Transfer imbibed seeds to petri plates containing 2% water agar.
9. Incubate for 24-28 hrs at 28-30°C or till germination occurs.

Planting, Watering and Inoculation

1. Take the Gibson tube with slant of SNA medium
2. Swab the surface of the aluminium foil with 70% alcohol.
3. Make a small hole with the help of needle at the side where the medium is touching the aluminium foil.
4. With the help of forceps transfer the seedling to the Gibson tube (The radicle of the seedling is inserted in the tube through the hole. Care should be taken that the radicle touches the surface of the medium).
5. Cover the planted tubes with specimen tubes (size 200 x 40 mm) containing a plug of moist cotton so as prevent desiccation of the young seedlings.
6. Maintain the tubes in growth chamber for 1 to 3 days so as to allow the seedling to lodge properly.
7. Make a small hole in the cap (Opposite the growing plant) near the side of the tube for watering and ventilation.
8. Add quarter strength sterile seedling solution through the hole
9. Add inoculum through the hole
10. Plug the hole with sterile cotton
11. Maintain the tubes in growth chamber under proper growth conditions
12. Water the tubes as and when required.

Experimental set up for MPN method

Following the method outlined the procedure described below is based on 6 dilution steps and for 5 fold dilutions.

1. Prepare initial suspension by adding 10g of soil to 90 ml of water blank and shake it for 30 minutes. Prepare the dilution as follows.
2. Dilutions and inoculations Dilution level

(a) 1 ml of initial suspension + 4 ml water blank 1 ml to each 4 plant tubes and 1 ml to 'b'	1
(b) 1 ml of 'a' + 4 ml water blank 1 ml to each of 4 plant tubes and 1 ml to 'c'	5
(c). 1 ml of 'b' + 4 ml water blank 1 ml to each of 4 plant tubes and 1 ml to 'c'	25
(d) 1 ml of 'c' + 4m water blank 1 ml to each of 4 plant tubes and 1 ml to 'e'	125
(e) 1 ml of 'd' + 4 ml water blank 1 ml to each of 4 plant tubes and 1 ml to 'f'	625
(f) 1 ml of 'e' + 4 ml water blank 1 ml of 'e' of 4 plant tubes	3125

3. Keep the inoculated tubes in a suitable growth chamber
4. Observe the seedlings after 3-4 weeks for nodulation.
5. Tubes +ve or –ve for nodulation for each dilution are recorded.
6. Using the MPN table, calculate the MPN of specific *Rhizobium* following the example given below:

    If a soil sample diluted initially 100 x gave the following distribution of +ve tubes in then further five fold dilution series

Dilution level	1	5	25	125	625	3125
+ve tubes	4	4	3	1	0	0

The characteristic is 443100 which, from table, corresponds to 164 rhizobia. The number per g of soil is, therefore: 164x102 (Since the initial dilution was 100x).

Composition of Growth Media

Seedling Nutrient Agar

$CaHPO_4$	1.0 g
K_2HPO_4	0.2 g
$K_2SO_47H_2O$	0.2 g
NaCl	0.2 g
$FeCl_3$	0.1 g
Water	1000 ml
Trace Element solution	1 ml

Trace element solution (Stock)

Bo	0.05%
Mn	0.05%
Zn	0.005%
Mo	0.005%
Cu	0.002%
PH	6.5-7.0

Yeast Extract Mannitol Agar (YEMA) medium

Mannitol	10.0 g
Yeast Extract	1.0 g
K_2HPO_4	0.5 g
$MgSO_4.7H_2O$	0.2 g
NaCl	0.1 g
Distl. water	1000 ml
Agar	18 g
pH	7.0

For isolation and purification of *Rhizobium* sterilized 10 ml (1/400 aqueous solution) of congo red is to be added to YEMA just before pouring of plates and the medium is referred as CRYEMA.

Modified Jensen's Agar medium

Sucrose	20.0 g
K_2HPO_4	1.0 g
$MgSO_4$	0.5 g
Na_2MoO_4	0.001 g
$FeSO_4.7H_2O$	0.01 g
$CaCO_3$	2.0 g
Agar	18.0 g
Dist. water	1000 ml
pH	7.0-7.2

CS7 medium

	mM		μM
KH_3PO_4	2.2	$MnSO_4.4H_2O$	58.0
$CaCl_2.2H_2O$	0.7	H_3BO_3	8.2
KCl	0.9	$ZnSO_4.7H_2O$	3.5
$MgSO_4.7H_2O$	0.14	KI	6.0
Glutamate	2.0	$CuSO_4.5H_2O$	0.8
Myoinositol	5.6	$Na_2MoO_4.2H_2O$	0.4
Sod. Succinate	25.0	$CoCl_2.6H_2O$	0.4
L-Arabinose	25.0	Na.EDTA	54.0
Difco Noble Agar	1.5%	Thiamine.HCl	15.0
pH	5.9	Nicotinic acid	41.0
		Pyridoxine.HCl	2.4

N-free Malate Medium

Malic acid .... 5 g
K_2HPO_4 .... 0.5 g
KOH .... 4.0 g
$MgSO_4$ .... 0.1 g
NaCl .... 0.02 g
$CaCl_2$ .... 0.01 g
$FeSO_4$ .... 0.05 g
Na_2MoO_4 .... 0.002 g
$MnSO_4$ .... 0.01 g
0.5% Alc. BTB .... 2.0 ml
Agar .... 1.8 g (semi-bold) 18.0 g (solid)
Water .... 1000 ml
pH .... 6.9 – 7.3

For growing *Azospirillum* in solid or liquid medium, NH_4Cl (1g/l) has to be added.

6. Isolation and Characterization of *Azotobacter chroococcum*

Azotobacter chroococcum (family: Azotobacteraceae) had been isolated. At present seven species of this bacterium are known Species of *A. chroococcum* most frequently occurring in different soils. Due to its multiple physiological attributes of broad-spectrum utility, the use of the *Azotobacter* is recommended for various area.

Characteristics

Pleomorphic, gram negative, often motile (polar or peritrichous flagella), non-spore forming relatively large rods or even yeast like appearance, showing variation in shape and size, mesophilic (optimum growth temperature 30°C), obligate aerobes, use carbohydrate as an energy source, catalase positive, macrocyst forming, cultures grow best with free nitrogen or simple forms of combined nitrogen, capable of fixing atmospheric nitrogend asymbiotically, widely distributed in soil (pH 6.0-7.5). There is only one species i.e. *Azotobacter paspalum*, which specifically associated with *Paspalum notatum*, a grass, which is classed as a case of associative symbiosis.

It has been observed that *Azotobacter chroococcum* displayed three fold effect as follows:

a) **Nutritional:** Helps in saving nitrogenous fertilizers (10-20 kg N/ha) and helps in phosphate solubilization.
b) **Stimulatory:** Secretes growth promoting substances (IAA, Gibberillins etc.) which helps in better seed emergence and expanded root system.
c) **Therapeutic:** Suppresses the growth of saprophytic and pathogenic microorganisms near the root system and reduces the damage to crop by plant diseases.

Requirements

Jensen's nitrogen free medium, soil samples (rhizosphere), petridishes, pipettes (1 ml, 10 ml), sterile water blanks (9ml, 90 ml) autoclave, U.V. Chamber, inoculation needle, B.O.D. incubator, slides and microscope etc.

The composition of modified Jensen's agar medium is given below.

Constituents	g/l
Sucrose	20.0
K_2HPO_4	1.00
$MgSO_4$	0.50
NaCl	0.50
Na_2MoO_4	0.001
$FeSO_4$	0.01
$CaCO_3$	2.00
Agar Agar	20.00
PH	7.0 – 7.2

(Autoclave at 15 lb psi at 121°C, for 15 minutes)

Procedure

1. Take 10.0 g of representative soil samples (rhizosphere) in 90 ml water blank.
2. Dilute the soil in sterile distilled water blanks of 90 ml, it will give 10^{-1}.
3. Like wise prepare serial dilutions upto 10-3.
4. Pipette out 1.0 ml soil suspension (as an inoculum) from 10-1 to 10-3 dilutions in stersilized peridishes separately.
5. Add molten Jensen's Agar medium (45-50°C), in the plates aseptically (in the inoculation room).
6. After solidification of the medium (present in the petridishes) put the plates for incubation at 28-30°C for 24-48 hrs in B.O.D incubator.
7. After incubation period colonies showing light showing light brown to black pigment on Jensen's medium should be selected for isolation.

After isolation the organisms is to undergo various morphological, biochemical and physiological test to prove its authenticity. Purification of

Azotobacter can be done by repeated streaking and through dilution-plate technique using Jensen's Agar medium as a nutritional source.

Table 1: Important characteristics of *Azotobacter chroococcum*

Cell shape	Oval or rod
Cell size (μm)	2.0-3.0 x 3.0-6.0
Cyst formation	Present
Motility	Motile especially young cultures grown in ethanol
Pigment formation time	With ageing
Pigment characteristics	Dark brown to black (water insoluble)
Utilization of citrate	+
Utilization of starch	+
Utilization of Mannitol	+
Utilization of sodium benzoate (0.5% and less)	+
Growth in Mannitol media with 1.0% benzoate	-
Utilization of rhamnose	-
M.R. test	-
V.P. test	-
Litmus milk reaction	Acidic
% (G+C) ratio	65-66
Habitat	Soil and Water

7. Production and Application of BGA Biofertilizer in Rice Crop

Introduction

Diazotrophy, or the ability to fix nitrogen biologically is widespread in blue green algae. This property alongwith their ability to photosynthesize and capacity to withstand adverse environmental conditions has made blue green algae play a pivotal role in improving agricultural production. Among BGA, some fix nitrogen with the help of specialized cells called heterocysts while others adapt various spatial and temporal strategies to overcome the oxygen sensitivity of the enzyme. Besides, their ability to independently fix N_2, their other qualities which are of agricultural significance are:

- Excretion of polysaccharides and other related compounds which help in improved aggregation, water holding, capacity and aeration of soil-thus improving soil structure.
- Liberation of growth promoting substances (Vitamin B12, Folic acid and auxins-IAA), sugars which help in improved growth of plant.
- Solubilization of insoluble phosphates-Mussorie rock phosphate, tricalcium phosphate – increased P availability.
- Ability to reclaim 'usar' or salilne-alkali land as a result of their ability to reduce sodium ions, soluble salts and other toxic compounds – thus giving the soil pH to neutrality.

Algalization trials in the past two decades has indicated that:

- BGA alone can contribute 25-30 kg nitrogen/ha/season besides enhancing crop yields by 5-30% in diverse agroclimatic regions.

- Application of BGA along with recommended levels of fertilizers and pesticides results in 10% increase in crop yields.

Production technology

The BGA production technology has been improved over the years and at present basically two technologies are available. These are:

- Open air soil based biofertilizer production
- Polyhouse/glasshouse straw based production.

Both these methods offer flexibility in size and quantity required and utilize the four strains –

Improved Open Air Production Technology

Depending upon the requirement, algal biofertilizer can be produced by individual farmers and or also on commercial basis.

At individuals level

1. Prepare shallow troughs, at least 2 or multiple of two, approximately 2 m x 1 m x 23 cm of galvanized iron sheet, permanent cement and mortar tanks or polythene lines pit on ground. All these structure should be in an open area so as to receive sunlight for the entire day. A light intensity of 1000 to 4500 lux is good enough for BGA growth.
2. Spread about 3 kg soil (1.4 kg m^2) per multiplication unit and add 200 g single super phosphate. For best growth of BGA soil pH should be between 7.0 to 7.5, however, growth to occur at higher or lower soil pH. To bring the pH about 7.5 soils can be amended with gypsum or lime depending upon initial soil pH.
3. Add water upto a height of 3-4 inches to the multiplication unit and mix the contents.
4. As soon as the suspended soil settles and water becomes clear, sprinkle/broadcast the starter culture containing *Anabaena* and *Nostoc* (mixture 1), and *Aulosira* and *Tolypothrix* (mixture 2) in two different troughs/multiplication units. This is essential to give equal opportunity to all the forms to grow and multiply equally.
5. To prevent insect breeding, add carbofuran (3% granules, 20 g/tray or unit) or malathion, BHC or any other suitable insecticide.
6. In hot summer months (temperature more than 30°C) BGA grows fast and in about seven to ten days a thick mat is formed. During this period if rate of evaporation is high, add water intermittently. Once the algal growth is thick, stop watering. Allow the algal mat to dry, dried algae form flakes alongwith soil.
7. Collect the dried algal flakes, mix from both the multiplication unit to make composite mixture and store in polythene lined bags in cool dry place.
8. Repeat the process till the requirement is met. From a single harvest about 3 kg dried material per unit is obtained.

Field scale production is scaled up process for the production in large quantity. It can be adopted in cooperative basis by a group of workers. This process is possible in soil where water remain standing after pudding.

1. Demarcate the area ($40m^2$), stubbles are removed and soil is well puddle to facilitate waterlogging.
2. Bund the area with strong 15 cm earth embankments.
3. Flood the area with water to a height of 1-2 inches.
4. Apply single super phosphate at the rate of 12 kg/$40m^2$.
5. To control the predators or algal grazers like snails, dephnids, mosquitoes larvae etc. apply carbofuran (3% granules) or ekalux (5% granules) at 250 g/$40m^2$.
6. Inoculate the plot with 5 kg composite inoculum.
7. In clay soil within about 2 weeks good growth of BGA takes place, whereas in loamy soils three to four weeks are required. In between if the plots get dry fill with water. As soon as good algal mat is formed, it is allowed to dry. Dry algal flaskes are collected and stored.
8. One can continuously harvest BGA from the same area by reflooding the plot and applying super phosphate and pesticides.
9. During summer months (April-June), the average yield of BGA per harvest ranges form 16-30 kg/$40m^2$.
10. This production procedure can be adopted at the time of sowing nurseries for rice seedlings and by setting BGA production unit side by side.

Storage of soil based BGA

- The sundried BGA can be stored for a long time (more than 2 years) in a dry state at normal room temperature in shade.
- Storage bags for BGA biofertilizers should be fresh and not the bags earlier containing urea or other chemical fertilizer and other agrochemicals.

Field application

- Broadcast the dry algal inoculum over the standing water in the rice field at a rate of 10 kg/ha. In clay soils water remain turbid for 2-3 days, hence BGA inoculum should be applied after 2-3 days of transplantation of rice seedlings. Whereas in loamy soils application should be done immediately after transplantation is completed. Application of excess BGA inoculum is not harmful and will accelerate the fast growth of BGA in the fields within 2-3 weeks.
- Recommended pest control measurers and other management practices do not normally interfere with the establishment and activity of the BGA in the field.
- BGA biofertilizers should be used in every rice crops as a kind of insurance to the crop yield as well as to give a stop to the deterioration of soil physico-chemical properties.
- Use of BGA fertilizers is mainly recommended for rice. However, it has given good results with crop which require large amount of water like sugarcane.

PRODUCTION TECHNOLOGY FOR COMMERCIAL PURPOSES

Production unit: Indoor production can be undertaken in polyhouse or glasshouse. The dimensions and number of such production units are variable depending upon the turnover. A smallest recommended size is 12 m x 5 m x 3.5 m (lxbxh). Covering material (polythene/glass) quality depends upon the cost input capacity.

BGA multiplication units: These units can be of either RCC, brick and mortar or polythene lines pits on ground as suggested for soil based production units. The life of polythene lined shallow pits on ground is maximum 6 months, whereas other structures are permanent. Possible size of polythene lines pits is 4m x 0.23 m (lxbxd). The smallest recommended pits of RCC or brick and mortar is 1.7m x 0.9m x 0.23m (lxbxd). In glasshouse with RCC pits one can go for double tier system.

A special formulated medium is to be filled in each pit/trough.

Only one form of BGA is to be inoculated per pit to avoid species competition and dominance during multiplication. The initial inoculum is 4 flasks with 100 ml actively grow in culture of about 7 days old equivalent to 1.2 g biomass.

BGA is allowed to grow for 3-8 days depending upon the ambient temperature, light intensity and humidity till final biomass of 14-15 g is attained.

Once fully grown, culture is mixed with carrier and sundried. There are various kind of carriers, some require pretreatment others not. Most suitable carriers are rice or wheat straw. Approximately 400 g carrier is needed per tank.

Once culture is sundried, it is packed in suitable size packets (polythene), sealed and stored for future use. All precaution are identical as for storage of soil based BGA inoculum.

Every time fresh medium is added and pits are freshly inoculated.

In a double tier system, illumination is achieved by providing suitable light source at the lower (bottom) surface of the upper tier so that the lower tier is illuminated.

When temperature is below 25°C heating provision for the glass house is made.

FIELD APPLICATION OF STRAW BASED BGA BIOFERTILIZERS

- BGA biofertilizer produce in a formulated medium in indoor culture facility contains 100,000 to 1,000,000 units (propagules) per gram carrier, hence 1 kg/ha (400g/acre) straw based BGA inoculum in enough.
- Broadcast 1 kg/ha in the fields immediately after transplantation of rice seedlings.
- Ensure that there is enough water atleast 1 week after BGA application.
- In case, problem of broadcasting 400 g material in a acre mix with dry soil from the field or ash or saw dust and then use for even distribution.

8. Isolation, Purification and Enumeration of Algae

Algae is a heterogenous group of organisms having ubiquitous distribution. However, being photoautotrophic in nature, they have restricted distribution along the soil profile. In fallow fields, algae are generally found upto about 2-3 cm depth but in cultivated fields, these can be found even upto 20 cm depth because of the turning of the soil during ploughing. To isolate and maintain algae in the laboratory, these are cultured in appropriate medium under congenial environmental conditions. Phosphate, nitrate, magnesium and calcium are the macronutrients. These best grow in a pH range of 6.5-8.0, light intensity of 1.5-2.5 k lux and temperature optima of 30-35°C.

Isolation and Purification

Cyanobacteria can be isolated both from the natural water bodies as well as from the soil. In former, case, suspend a loopful of the algal growth in 5 ml sterilized distilled water, homogenize and 0.1 to 0.5 ml is spread on plates containing the suitable medium and incubated. Isolated colonies observed through a binocular microscope after incubation are picked up, and transferred to agar slants.

A variety of culture media are available for growing algae in the laboratory. For cyanobacteria, generally, Fogg's medium without N is used for nitrogen fixing and with N (0.5 g KNO_3/litre) for non-nitrogen fixing BGA. The composition of Fogg's medium is: Potassium nitrate, 3.0 g; Potassium dihydrogen phosphate, 0.2 g; Magnesium sulphate, 0.2 g; Calcium chloride, 0.1 g; A_5 Micronutrient solution, 1.0 ml and Fe-EDTA Stock Solution, 1.0 ml. The volume is made to 1L and pH is kept at 7.0. A_5 Micronutrient solution contain Boric acid (2.86 g), Maganese chloride (1.81 g), Zinc sulphate (0.222g), Sodium molybdate (0.0177g) and Copper sulphate (0.079g) dissolved in one litre of distilled water. Fe-EDTA Stock solution is prepared separately by dissolving 26.1 g of ethylene diamine tetracetic acid (disodium salt) [EDTA] in 268 ml of 1 N potassium hydroxide solution and 24.9 g of ferrous sulphate is added. The volume is made to 1 litre with distilled water. The solution is aerated overnight to produce a stable complex marked by the change in colour to dark brown. The volume is again made to 1 litre. 1 ml of this stock solution in 1 litre medium will give 5 ppm iron. Volume of each ingredient is taken accordingly for one litre of the medium. The pH is adjusted before autoclaving the medium. For solid media, agar is added at 1.5 per cent (w/v). The glassware like petri dishes, pipettes plugged with cotton at sucking like petri dishes, pipettes plugged with cotton at sucking end and cotton plugged culture tubes are wrapped in a paper and sterilized in a hot oven at 180°C for 1 hr.

Isolation

The soil samples transterred aseptically to 100 ml medium. The flasks are shaken and incubated undisturbed in a growth room for 20-25 days. Flasks are observed daily for the algal growth and when visible algal growth appears, wet mounts from each flask are prepared for microscopical examination. Individual colonies are lifted using an inoculation needle suspended in 5 ml sterilized distilled water in test tubes, shaken vigorously to make a homogenous suspension. 0.5 ml of this suspension is spread on an agar plate, incubated in light. The plates are observed regularly and isolated colonies are picked up and examined under microscope. If found unialgal, another similar colony is picked up from the plate and transferred to agar slant. Contaminated culture can be made unialgal by pour planting or streak method. For making axenic cultures free from bacterial or fungal contamination, the contaminated culture is treated either with UV rays or with antibiotics. Unialgal cultures can be maintained on agar slants with appropriate medium contained in screw capped culture tubes.

Enumeration

Algae can be enumerated by direct count method or by the most probable number method.

I. Direct count by haemocytometer
II. Most probable number method:

In this method, graded dilutions of the soil sample are inoculated into culture tubes containing the appropriate medium and incubated. The tubes showing algal growth are scored. Based on the number of tubes showing algal growth in each dilution, probability is calculated with the help of standard tables.

9. Estimation of Phosphate Solubilising Capacity of Microorganisms

Introduction

Phosphorus is a macro-nutrient and essential for living systems. Phosphorus remains fixed in soil; even a major portion of applied phosphorus as fertilizers is also get fixed in soil and released very slowly. The solubilization is achieved by several group of microorganisms like fungi, bacteria and even cyanobacteria. This capacity of p-solubilization can be quantitative estimated by the preponderance of specific microorganisms as well as the process can be enhanced by enrichment of desired microorganisms.

Principle

In acid solution, soluble orthophosphate combine with molybdate to form double salts though the amount of phosphomolybdate present may be insufficient to form a precipitate upon reduction a coloured soluble complex is formed, the intensity of colour is proportional to the amount of phosphate present. Arsenates and silicates behave similarly. For quantitative determination it is necessary that final solution acidity should be adjusted. Too much acidity prevents or slow down development and too less acidity leads to reduction of molybdenum which also gives blue colour. The usual range of concentration is 0-50 μg in 50 ml of final solution.

Requirements

Conical flasks 250 ml, Volumetric flasks 50 ml, Pipette 10 ml, Spectrophotometer; HCl, Distt. water and Pikovaskya broth.

- Chlorostannous acid: 2.5 gm of SnCl2.2H2O dissolve in 10 ml of conc. HCl. Heat gently. After cooling make the volume to 100 ml with distilled water.
- Chloromolybdic acid: 15.0 gm of Ammonium molybdate dissolve in 400 ml of warm distilled water. Add 342 ml of 12 N HCl and cool. Make up the volume to one litre with distilled water.

Standard curve: For standard curve prepare 100 ppm phosphorus solution by dissolving 0.439 g of dried KH_2PO_4 in 400 ml distilled water. Add 25 ml of 7N H_2SO_4 and make the volume to 1 litre.

Procedure

- Prepare Pikovskaya broth with known amount of inert phosphorus source (Rock phosphate/ tricalcium phosphate).
- Inoculate it with test microorganism (*Aspergillus* sp.)
- Incubate at 28-30°C on shaker for 3-4 days.
- Filter/centrifuge the broth in case of fungi and bacterium respectively.
- Take aliquot (0.1-1.0 ml) from the supernatant/filtrate and add 10 ml of Ammonium molybdate.
- Shake well. Dilute the contents of the flask to 45 ml. Add 0.25 ml of Chlorostannous acid and immediately make up the volume to 50 ml.
- Measure the blue colour intensity of the solution at 600 nm or filter No. 61 using Hilger's absorption meter.
- The amount of soluble phosphorus is obtained with the help of standard curve.

Pikovaskya medium

Constituents	g/L
Glucose	10.0
Yeast extract	0.5
$NH_4(SO4)_2$	0.5
KCl	0.2
NaCl	0.2
$MgSO_4.7H_2O$	0.1
$FeSO_4.7H_2O$	traces
$MnSO_4$	traces
$Ca_3(PO_4)_2$ (Inert P-source)	5.0
pH	7.0 – 7.2

PRECAUTIONS

1. Make the reagent Chlorostannous acid just before use.
2. Store Ammonium molbdate in amber coloured bottle.

10. Plant Growth Promoting Substances Produced by Phosphate Solubilising Microorganisms

Introduction

Plant growth promoting substances *viz.* auxins, gibberellins like substances are elaborated by certain group of microorganisms. The production of these substances are of significance in the rhizosphere of plants as these substances favourable affect the plant growth. Microorganisms like *Rhizobium, Azospirillum* and phosphate solubilizing microorganisms are few examples synthesizing growth promoting substances.

Principle

A pink colour develops when a mineral acid is added to a solution containing indole acetic acid in the presence of ferric chloride. Different mineral acids, HCl, phosphoric acid, nitric acid, sulphuric acid, perchloric acid can be used for development of colour. FeCl-$HClO_4$ reagent is most sensitive and show least interference from other indole compounds *e.g.* Tryptophan, Skatole. Allyindole, indole, indole aldehyde show slight interference with this colorimetric estimation. Since Beer's law is not followed at high concentration of IAA, absorbances obtained are converted to IAA concentration by a standard curve. Such a curve does not vary appreciably with reagent stored in light at room temperature.

Requirements

Glassware: Test tubes, Conical flasks 250 ml, Pipettes, Wash bottle.

Chemicals: $FeCl_3$ (Ferric chloride), Perchloric acid (HClO4), Tryptophan, Ethanol, Distilled water, Pikovskya broth, Orthophosphoric acid.

- 1 ml of 0.5 $FeCl_3$ in 50 ml of 35% $HClO_4$
- **IAA stock solution:** 100 μg/ml in 50% ethanol.

Procedure

1. Inoculate loopful of culture in 25 ml broth of Pikovaskya. Amendment of medium with amino acid Tryptophan (precursor of IAA) is optional.
2. Incubate for 24 hr at 28°C on a rotary shaker.
3. Centrifuge at 10,000 rpm for 15 minutes.
4. Take 2 ml of supernatant and add 2-3 drops of O-phosphoric acid.
5. Add 4 ml of reagent to the aliquot.
6. Incubate the samples for 25 minutes at room temperature.
7. Read absorbance at 500 nm.
8. Record auxin quantification values by preparing calibration curve, made by using indole acetic acid (IAA) as standard (10-100 μg/ml).

Precautions

1. Use only log phase culture for auxin estimation.
2. Handle perchloric acid carefully.

11. VAM Fungi – Quantification in Plant Roots

Introduction

Vesicular-arbuscular (VA) mycorrhizal fungi has received importance in recent past year, because of its role in sustainable agriculture. If the benefits rendered by VA-mycorrhizal fungi are inconsistent, then quantification is needed to determine the degree and intensity of root colonization of the plans. Assay of VA-mycorrhizae is also needed to find out their establishment within the roots of plants and asses the treatments response. To enumerate and to study the morphological structures of fungi a common method employed.

Procedure

Collection of Root Samples

Roots can be collected at any time of the year. Since, fine terminal feeder roots are the primary sites for VA-mycorrhizal developments precaution must be adopted not to lose terminal roots, when plant are excavated. The terminal feeder roots can be used as sample, need not be immediately processed for staining, and are preserved with Formalin-Aceto-Alcohol (FAA). This solution is made in 50% alcohol with a v/v/v ratio of 18:1:1. In FAA solution samples can be preserved for 2-3 years. However, the samples are not suitable for nutritional and biochemical assay.

Clearing and staining of samples

- Wash the root specimens stored in FAA and place in a beaker and cover with 10% KOH solution.
- Place the beakers with specimens in an autoclave at 15 lb pressure for 10 min. If an autoclave is not available, heat the samples at 90°C for 1 hr in well-ventilated fume hood. The KOH solution clears the host cytoplasm and nuclei and facilitate the stain permeability.
- Pour off the KOH solution and rinse the samples three times with complete change of water or till brown colour disappears in the rinsed water.
- Cover the specimens in the beaker with alkaline H_2O_2 at room temperature for 10 to20 min or until the roots are bleached. Alkaline H_2O_2 is made by adding 3 ml of NH_4OH to 30 ml of 10% H_2O_2 and 567 ml of tap water. This solution should be made as and when required since its effectiveness is lost if stored overnight.
- Rinse the specimens in the beaker thoroughly using at least three complete changes of tap water to remove the H_2O_2.
- Cover the specimens in the beaker with 1% HCl and soak for 3 to 4 min and then pour off the solution.
- Cover the specimens in the beaker with 0.01% acid-fuchsin-lactic acid staining solution and autoclaves for 10 min at 15 lb inch 2 pressure. The lactic acid solution consist of 875 ml lactic acid, 63 ml glycerine, 63 ml of tapwater and 0.1 g of acid fuchsin.

- Decent excess of staining solution, remove the specimens from the beaker and place in petri plates for destaining and mycorrhizal assay. Do not rinse specimens after staining because the stain is readily removed from fungal structure.

Note: *Specimens can also be stained without heat. In this method the solutions and procedures remain the same as used in heat method. Time required for each step is greatly increased. If bleaching is required, upto 6 hr exposure of specimens is sufficient in unheated KOH. Fungal structures absorb the stain with the exposure of 2 to 4 hrs depending upon the nature of the specimens.*

Assessment of colonization by VA mycorrhizal fungi

Assessment of colonization after cleaning and staining can be achieved by several procedures. It includes – (i) visual assay, (ii) slide length (estimated length of colonized tissues in root segment of standard length maintained on slides), (iii) slide ± presence or absence of colonization in the same root segment, and (iv) grid line intersect. These assay procedures can be put into two categories: (i) Systematic, and (ii) Non-systematic.

Non-systematic Method

In all the studies, the main objective is to determine the percentage of roots colonized as well as the intensity of colonization within the roots. To accomplish, stained roots are spread uniformly in a glass petridish with help with help of dissecting needle. Susceptible feeder roots are counted to determine the percentage of roots colonization by VA-mycorrhizal fungi. A workable classification used at the Institute for Mycorrhizal Research and Development, USDA Forest Service, Athens Georgia, is as follows: class 1, 0.5%; class 2,6-26%; class 3, 26-50%; class 4, 51-75% and class 5, 76-100%.

Karmanik and others have reported three categories for evaluating intensity of colonization. Roots with small colonization sites widely scattered along the roots, larger colonization sites more uniformly distributed through the colonized roots, but rarely coalescing, feeder roots are almost solidly colonized with few easily identified isolated patches of colonization and 1, 2, 3 is given respectively.

SYSTEMATIC METHODS

Gridline intersect method

It can be used to estimate both the proportion of root length colonized and total root length in the sample. In this method root sample is spread out evenly in petridishes, having gridline markings (1.27 cm^2) on the bottom. Gridlines can also be made on circular piece of acetate paper and placed on the bottom of inverted plate top. The bottom portion of the plate containing the sample is then placed inside the top.

The gridline method serves as a tool for systematic selection of observation points. Vertical and horizontal gridlines are scanned and the absence of presence of colonization is recorded at each point where a root intersects a line. For better accuracy of percentage of root length colonized atleast 100 gridline intersects required to be tallied.

Slide method

Root segments (1cm long) are selected at random from a stained sample and mounted on microscopic slides in group of ten and 30 to 100 root segments are scanned under microscope, for better accuracy. Langth of cortical colonization is assessed (at 100 X and 250X) in millimeters for each segment, averaged for each of 10 segment in a group, and expressed as a percentage of root length colonized. This method gives as assay base on total root length and takes into consideration the percentage of root colonized, as well as the intensity of colonization.

Isolation of Vesicular-arbuscular mycorrhizae (VAM) fungal spores

Still Vesicular-arbuscular mycorrhizae (VAM) fungi are being collected from field and/or pot culture soils. Collection and production of VAM inoculum are time-consuming clamydospores or azygospores, soil borne vesicles and mycelium or infected root pieces form the propagules of VAM fungi. Since they occur in soil these propagules are termed as mixed inoculum as compared to the spores, separated from soil which are more purer.

Wet Sieving and Decanting methods

1. Take 50 g of soil sample in 200 ml of water and keep it undisturbed for few moments to settle down heavier particles.
2. Pour supernatant in a 30-60 mesh size sieve to remove large pieces of organic matter, collect the liquid in 1 L beaker. Wash the sieve with water to ensure that all small particles have passed through the sieve.
3. Resuspend the particles in the water which passed through the coarse sieve and allow heavier particles to settle for a while.
4. Pass this suspension through a sieve fine (100-300 mesh) enough to retain the desired spores.
5. Wash the material retained on the sieve to ensure that all colloidal material passes through the sieve.
6. Transfer small amount of remaining to a petridish and examine under dissecting microscope.

Isolation of spores through Density Gradient Centrifugation

- Add 10 ml of 50% sucrose solution to a clean 50 ml centrifuge.
- Add another layer of 25% sucrose solution on top of the 50% sucrose layer and further add water layer above the second layer of sucrose using hypodermic syringe with a curved glass tube along the glass surface of the tube.
- Add a suspension of sievings and centrifuge 5 min at 3100 rpm.
- Remove debris which collects in middle layer, transfer to a fine sieve and wash with water.

Isolation of spore through sucrose centrifugation

Place a 100 to 500 g of soil sample on a 20 mesh screen and wash with water to remove large debris.

Collect sieved soil and water in a container.

Mix the sieved soil by stirring and allow the mixture to stand for 30 sec. Decant through a 270 or 325 mesh sieve and collect the residue from the sieve in a beaker. Add water to the sieved soil and repeat decantation. Finally discard the remaining soil.

Transfer the collected residue in to dry 50 ml centrifuge tubes and centrifuge the contents for 4-5 min at 1750 rpm in a horizontal rotor.

Decant the supernatant liquid carefully and resuspend the pellet in a sucrose solution (45%). Again centrifuge for 0.5 to 1 min.

Pour the supernatant containing spores into a 325 mesh sieve and rinse with water to remove sugar.

Finally collect the spores

12. Estimation of Soil Microbial Metabolic Activity by Carbon Dioxide Evolution Method

Principle

The quantity and rate of carbondioxide evolution from soil is one of the best methods available to measure the activity of soil microflora. A part of the carbon assimilated is metabolized into their microbial biomass, a larger and relatively constant proportion of carbon assimilated is oxidized during cell respiration and released as carbon dioxide. Thus amount and rate of carbon dioxide released is an expression of metabolic activities and also of their interaction with environmental factor.

The carbon dioxide released is trapped in alkali (NaOH) forming corresponding carbonate ($NaCO_3$)

$$2NaOH + CO_2 \rightarrow Na_2CO_3 + H_2O$$

$$Na_2CO_3 + BaCl_2 \rightarrow NaCl + BaCO_3 \text{ (white precipitate)}$$

Amount of NaOH consumed by the evolved carbon dioxide is obtained by titrating against standard solution of HCl in presence of phenolphthalein indicator.

Solutions

Standard NaOH: Prepare standard NaOH by weighing gram equivalent weight of NaOH in 1L distilled water. For example for preparing N/2 NaOH 20.0 g of NaOH pellets were dissolved in 1L dist. water. Standardise this against N/2 Potassium hydrogen phthalate solution. N/2 potassium hydrogen phthalate was prepared by dissolving 10.21 g of it in 100 ml of distilled water.

Standard HCl: N/2 HCl was prepared by adding 45 ml of HCl (35 N commercial) to 955 ml dist. water. It was standardized against Standard NaOH solution using phenolphthalein as an indicator.

Procedure

1. Take 100 g moistened soil to moisture equivalent into a one litre flask.
2. Keep three empty flasks as atmospheric control.
3. Suspend a test tube with 15 ml of N/2 NaOH and stopper the flask.
4. Incubate for 24 hr.
5. Remove NaOH in a separate flask and add 10 ml saturated $BaCl_2$ solution followed by a few drops of phenolphthalein solution and back titrate with standard HCl.
6. Calculate the amount of CO_2 evolved as:
7. NaOH consumed (c) = Volume of NaOH taken (X) – volume of HCl consumed (Y)

Calculations

$$2NaOH + CO_2 \rightarrow Na_2CO_3 + H_2O$$

2 Moles NaSOH = 1 mole of CO_2
80 gm NaOH = 44 gm of CO_2
1L of N NaOH = 11 gm of CO_2

Control

Volume of NaOH taken = 15 ml
Vol. of HCl consumed = 13.9 ml
NaOH consumed = 15 – 13.9 = 1.1 ml
Test sample vol. of NaOH taken = 15.0 ml
Vol. of HCl consumed = 10.0 ml

13. Estimation of methane Emission from Paddy Fields

Production of Methane in Submerged Rice Fields

Biological methane formation is a geological process that occurs in most anaerobic environments where organic matter undergoes decomposition: swamps, lake sediments, the intestinal tract of animals, anaerobic sewage digesters. Reports have suggested that flooded paddy fields are one of the major contributors of atmospheric methane and nitrous oxide. Flooded rice soils are predominantly anaerobic due to the poor diffusion of oxygen through floodwater and consumption of available oxygen for various biochemical processes in soil layers below the soil-floodwater interface. The one of the organic carbon source for the decomposition in this anaerobic environment is the root exudates produced by the plant roots.

The fermentation process involving methane generation is purely a process involving successive proliferation of various physiological groups of microorganisms. Three group of microbes viz. hydrolytic, acid formers and methanogens are involved in the formation of methane. First stage involves species of fermentative bacteria which hydrolyse complex carbohydrates, proteins and lipid to the correspoinding monomeric units and then ferment them to fatty acids, hydrogen and carbondioxide. The second group called as the "Hydrogen producing acetogenic bacteria" produce acetate and hydrogen from the fatty acids generated in the first stage. The methanogenic bacteria appear in the third stage and utilize the products of the first two stages to produce methane and CO_2.

Collection of Gas Samples

Collection of gas sample is carried out by closed chamber technique. 50 cm × 30 cm × 70 cm chambers made of 6 mm thick acrylic sheets are used. An Aluminium channel is used with each chamber to make it air tight when placed in the field. One Mediflex three way stop cock (Eastern Medikit Ltd., India) is fitted at the top of the chamber to collect gas samples. Minimum of three replication should be taken from each plot of field. Head space inside the box is recorded which is used in calculations.

Analysis of Methane

Methane concentration in the gas samples collected from natural ecosystems is estimated using Gas chromatograph fitted with Flame ionization detector (FID), 6'X1.8" stainless steel column packed with Molecular sieve 5Å (60-80 mesh) maintained at 80°C. Carrier gas is Nitrogen and the flow rate is 15 to 20 ml/min. Injector and detector are kept at 130°C. 1 ml of gas is injected by air tight syringe. The retention time for methane is 1.9 to 2.2 minutes. The area under peak is measured by the means of microprocessor based integrator attached to the chromatograph. The concentration of methane in sample is determined from the standard curve obtained.

Flux Measurement and Calculation

To measure flux, the chamber is fixed at the experimenting site and the change in methane concentration in the chamber so formed, with time, is determined by taking replicate gas samples from the chamber head space by syringe and analyzing them in laboratory. The following precautions are taken for getting satisfactory results.

1. Gas samples should be drawn from the head space immediately after, placement and at equal time intervals there after over a period not exceeding 30 minutes.
2. A minimum of three samples should be drawn to check the linearity of the increase in concentration.
3. The chamber should be removed after drawing the final sample to minimize the disturbance to the environmental conditions with in the chamber.

Calculation of Flux

Cross sectional area of the chamber	–	m^2
Headspace (Height)	=	Hm
Volume of Headspace = AH m^3	=	1000 X AH 1
CH_4 concentration at 0 time	=	C_0 ppmV (ppm on volume basis)
CH_4 concentration at time t	=	C_t ppmV
Change in concentration in time t	=	$(C_1 - C_0)$ ppmV
	=	$(C_t - C_0)$ µ1 1^{-1}

Volume of CH_4 evolving in time t = $(C_t - C_0)$ µ1 1^{-1} X 1000 X A X H
= $(C_t - C_0)$ X AH ml

when t is in minute, then flux is:

F = $[(C_t - C_0)$ X AH] / (A Xt) ml m^{-2} min^{-1}
$[(C_t - C_0)$ X H] / t ml m^{-2} min^{-1}
= Y ml m^{-2} min^{-1} (say)

Now, 22.4 ml contains 16 mg methane at STP (Avogadro's Law)

F = Y X 16 / 22.4 mg m^{-2} min^{-1}
= Y X 16 / 22.4 X 60 mg m^{-2} hr^{-1}

Hence,

F = $[(C_t - C_0)$ /t] X H X 42.857 mg m^{-2} hr^{-1}

This working formula was used to calculate methane flux in all the cases.

14. Recycling of Organic Waste for Soil Enrichment and Nutrient Uptake through Microbial Activity

Introduction

The annual production of crop residues in India accounts to about 770 m tonnes. These crop residues are either burned in fields or incorporated into soil. These crop residues are decomposed in soil by a variety of microorganisms. Organic waste decomposition is most closely tied to ecosystem processes and soil fertility. The largest fraction of organic carbon entering the soil is that contributed by plant residues, which contain variable concentration of hemicellulose, cellulose and lignin. Cellulose, the most abundant constituent of plant residue is nearly always associated with hemicellulose and lignin. Cellulose constitute 13 to 51% of the dry weight of plants. Among the various constituents, lignin is the most resistant plant material.

A wide range of microorganisms can attack cellulose. The fungus *Trichoderma* probably is the most studied for cellulose decomposition. When plant residues are incorporated into soil, normally their decomposition is slow, however, the process can be hastened if the residues are pre-fermented with celulolytic fungus, *Trichoderma* can be incorporated in the soil, thus reducing the time of its decomposition in the soil. Simultaneously *Trichoderma* can also serve as a biological control agent for soil borne plant pathogens. Partially degraded plant residues form the soil organic matter (SOM), which plays a major role in the soil structure and thus has great impact on water infiltration, root development and erosion resistance. It also is the store house of nutrients such as N, S, P and many minor nutrients, which can be mobilized to plant through higher microbial activity in organic matter rich soils.

This phenomenon can be demonstrated through an experiment which can be divided in the following parts:

1. Pretreatment of plant residue with *Trichoderma viride.*
2. Incorporation of fermented residue in soil.
3. Demonstration of cellulolytic and lignolytic activity of *Trichoderma* by qualitative methods.
4. SOM and its effect on density of soil microflora.
5. Determination of organic matter, N and P content in organic matter amended and unamended soil

Pretreatment of plant residue with *Trichoderma viride*

1. Take wheat straw (1 kg) in a basket and treat with 2% NaOH (w/w) for 48 hr. Then wash the alkali with excess water several times till pH reaches to neutrality.
2. Remove the excess water by filtration compression.
3. Add 4% urea to the straw (at 40% moisture level).
4. Compress the material in the polythene bag and close the bag tightly, leave at room temperature for a period of 15 days for absorption of ammonia.
5. Then open the bag and add 3% (V/W) inoculum of *Trichoderma viride,* mix uniformly and incubate for solid state fermentation (SSF) at 28-30°C in horizontal position for 15 days.
6. Pack the bag very loosely with holes in polythene bag for proper aeration for growth of *T. viride.*

Incorporation of Fermented residue in soil

After solid state fermentation incorporate the fermented residue at the rate of 80 Q/ha in soil. Maintain a control where no amendment is provided in the soil.

Demonstration of cellulolytic and lignolytic activity of *Trichoderma* by quantitative methods

Cultural test for cellulolytic activity

Prepare the medium with following composition.

KH_2PO_4	2.0 g
$(NH_4)2SO_4$	1.4 g
Urea	0.3 g
MgSO4.7H2O	0.3 g
$FeSO_4.7H_2O$	8.0 mg
$MnSO_4$	1.6 mg
$CoCl_2$	2.0 mg
O-phosphoric swollen cellulose*	10.0 g
Distilled water	1 lt.
pH	5.0
Agar Agar	18.0

*For preparing swollen cellulose, suspend 10 g of cellulose powder in 85% O-phosphoric acid for 2 hr at 4°C, remove O-phosphoric acid by filtration followed by addition of 1% (w/w) $Na_2\ CO_3$ solution. Remove the alkali by repeated washing, finally with distilled water till it become neutral and filter through an ordinary filter paper.

Procedure

- Dispense the above mentioned medium into test tubes to a height of 4 cm and sterilize at 15 lb psi pressure for 30 min.
- Cool the contents, thus vertical column of solid medium will be obtained.
- Aseptically transfer an agar piece with the fungal mycelium from the growing edge of fungal colony (15 days old culture of *Trichoderma*) to 5 test tubes.
- Incubate at 29±1°C. After 30 days of incubation measure the depth of clearance zone in mm.

Test for detection of lignolytic activity

The preferential degradation of lignin component by fungi can be tested by method. Prepare the medium with following composition:

$NH_4H_2PO_4$	0.50 g
KH_2PO_4	0.15 g
K_2HPO_4	0.10 g
Malt exctract	900 ml
Distilled water	900 ml
pH	5.0
Agar	18.0 g

1. Dissolve all the chemicals in 900 ml of distilled water and adjust the pH to 5.0 sterilize the media at 15 lb for 15 min after adding 18.0 g agar.
2. Dissolve 240 mg of Tannic acid in 250 ml capacity Erlenmyer flask containing 100 ml distilled water and autoclave at 15 lb pressure for 15 min.
3. Mix both medium and tannic acid solution before pouring into plates.
4. After solidification of the medium inoculate the petri plate with mycelial clump taken from a 15 day old culture of *T. viride*.
5. Incubate the plates at 29±1°C for 4-10 days.
6. Observe for radial growth of fungi and brown colour formation (on the back side of plate) after incubation. Brown or yellow colour formation indicate presence of lignolytic enzymes.

Soil organic matter and its effect on density of microflora

Prepare medium for population count of bacteria, fungi and actinomycetes.

Soil extract agar medium for bacteria

Glucose	1.0 g
K_2HPO_4	0.5 g
Agar	18.0 g
Soil extract (stock*)	100 ml
Tap water	900 ml

pH..6.8

One kg of sieved garden soil is mixed with 1 L of tap water and steamed in an autoclave for 30 min. A small amount of $CaCO_3$ is added and the whole is filtered through a double filter paper.

Rose Bengal medium for fungi

Glucose..10.0 g
Peptone..5.0 g
KH_2PO_4..1.0 g
$MgSO_4.7H_2O$..0.5 g
Streptomycin..30.0 g
Agar..15.0 g
Rose Bengal..0.035 g
Distilled water..1 L

The antibiotic solution was filter sterilized and added aseptically to the sterilized medium.

Inorganic salt starch medium for Actinomycetes

Starch*..10.0 g
$(NH_4)SO_4$..2.0 g
K_2HPO_4..1.0 g
$MgSO_4.7H_2O$..1.0 g
NaCl..1.0 g
$CaCO_3$..2.0 g
Agar..18.0 g
Distilled water..1 L
pH..7-7.5

*Make a paste of the starch with a small amount of cold distilled water and bring to a volume of one litre. Then add other ingredients and autoclave the media at 15 lb pressure for 15 min.

- To compare the density of microflora in SOM rich soil with control, take 10 g soil from both the treatments.
- Dispense the 10 g soil in 90 ml sterilized water blanks. This will provide 10^{-1} dilution.
- Prepare serial dilution from 10^{-1} to 10^{-6}.
- After preparing the dilution, pour 1 ml suitable diluted sample into sterilized plates. Some sample dilutions can be used for enumeration of bacteria, fungi and actinomycetes. Therefore, prepare 9 plates each for 10^{-2} to 10^{-6} dilution.
- Now dispense or pour approximately 20 ml liquefied medium in the plates. Prepare 3 plates each for bacteria, fungi and actinomycetes.
- After solidification of medium, incubate the plates at 30°C in inverted position for 2 days.
- Take the counts and record the observation.
- Calculate the population of bacteria, fungi and actinomycetes per gram

of soil.

- Find the difference in population density in organic matter amended and unamended soil.

Determination of organic matter, N and P content in organic amended and unammended soil

Organic matter content in soils can be determined

Requirement

1. 1N $K_2Cr_2O_7$: Dissolve 49 kg $K_2Cr_2O_7$ in 1 L distilled water to prepare 1 N $K_2Cr_2O_7$ solution, Conc. Sulphuric acid, Erlenmyer flasks 250 ml, Spectrophotometer.
   a) Take 1 g air dried and powdered soil.
   b) Mix the soil with 10 ml of 1N $K_2Cr_2O_7$ solution in 250 ml Erlenmeyer flask.
   c) Add 20 ml conc. H_2SO_4 with rapid speed.
   d) Swirl the mixture quickly for 10 sec and allow to stand for 10 min.
   e) Then add 100 ml distilled water and allow to stand for 4 hr so that the supernatant liquid becomes clear.
   f) Determine the absorbance of the supernatant on Spectrophotometer at 645 nm against a reagent blank.
   g) Prepare a standard curve with sucrose (0 – 80 mg) and calculate the concentration of organic matter from standard curve.
2. Quantification of soil N – will be done by Kjeldahl method
3. Quantification of P will be done by method as described in earlier exercises.

15. Immunological Techniques to Study Microbial Ecology

This study productive branch of classical ecology however, has been virtually closed to the microbial ecologist. This is mainly due to the small size of microorganisms, their non-descriptive morphology, and the extreme complexity of their microenvironments. Looking under the microscope at a small portion of the natural environment can reveal fascinating interrelationships between the microorganisms and their microhabitat. But it is virtually impossible to identify the various species of the microbial assemblage in situ. In study of microbial ecology, particularly in soil the microorganisms are to be identified in the complex environment having mixed population of various species and strains of microorganisms. The conventional plate culture method, due to its ease and simplicity, has become the most widely used technique in microbial ecology. However, it supplies indirect, artificial and often erroneous information about the actual conditions in nature. Though microorganisms are nondescript in morphology, they do have unique chemical, physiological and genetic features which can be used to identify genera, species, and even strains of a particular microorganisms. Such features as antigenic (immunological) characteristics, bacteriophage susceptibility and antibiotic resistance have been used in ecological studies as markers for the identification of specific microorganisms.

Immunological techniques, since their introduction by Border in 1885 have widely been used in the diagnosis of human and animal pathogens and have also been adopted for the identification of several microflora including the phytopathogens and soil microorganisms. The technique is based on the principle that a wide spectra of foreign substances including pathogenic and non pathogenic microorganisms (usually having high molecular weight, more than 10,000 Dalton) (proteins and lipoproteins) when introduced into an animal, cause the formation of specific proteins (the immunoglobulins), which are commonly called "antibodies. The causative substance is called "antigen" or "immunogen". These antibodies are found circulating in the blood and react specifically with the antigen. All microbes, except non-spore forming fungi and viroids possess, as a part of their structure, specific antigenic moieties. In case of fungi, the germinating spores have successfully been used as antigen.

The antigen-antibody reaction can be examined in vitro as well as in situ. Positive reaction confirm the diagnosis of the organism. There are a variety of techniques to examine the antigen-antibody reaction in vitro and in situ for the identification and localization of microbes in tissues. Our aim here is to describe the principle, relative merits and applicability of the techniques, currently used to study microbial ecology, particularly the inoculated strains in soil.

Antigen

The antigens of soil microorganisms, particularly of nitrogen fixing bacteria (*Rhizobium, Azospirillum*) have been prepared and used in various ways, viz. whole cell preparations and heat treated cells for somatic antigen formalin treated cells for flagellar antigens, sonicated or disrupted cells for intra-cellular antigens lipopolysaccharide fractions of cells, extracellular polysaccharides and capsular polysaccharides.

Antiserum

One of the best species available for antibody production is the rabbit. The route of injection varies depending upon the antigen preparation. Agglutination, immunofluorescence (IF), and the enzyme-linked immunosorbent essay (ELISA) deal with the particulate and structural components of the bacterial cell (somatic antigens). The intravenous (IV) route is the one choice for these preparations and results in a rapid antibody response. However, it is not sustained over a long period. Therefore, to maintain the response, the subcutaneous (SC) route is used for booster injections. For economical reasons and also to conserve the time, each rabbit can be injected first with a somatic antigen preparation of a bacterial strain for use in (IF, ELISA or agglutination reactions). After sufficient quantities of this antiserum have been obtained, the same rabbit can then the injected with soluble antigens of the same bacterial strain to obtain antisera for immunodiffusion studies.

The most common methods of preparing antisera agains *Rhizobium* have been intravenous injections of antigens on four successive days, intravenous injections spread over several weeks with rest period between injections, and intramuscular injection with adjuvant followed one month later by one or more intravenous injections. Various modifications of these have been used, however, the mina aim of each is to obtain high titer (not less than 3200) specific antibodies.

Antigen-antibody Reaction

Agglutination techniques

The agglutination method is the most common procedure, since its introduction. This technique is based on the reaction between the particulate antigen and serially diluted polyclonal antibody serum, either in small vials or micro-titer trays or on glass slides. In case of vials, the mixture is incubated at a constant temperature between 37°C to 52°C for 2-4 hr, and the observations are taken for agglutinated particles, which scintillate near the wall of the tube to the bottom, and the clearing of the suspension. Reaction is always compared with the serum-free antigen and antigen-free serum. Doubtful reactions are further observed after keeping the tube in refrigerator for overnight. The micro-titer trays are kept at 37°C and after 24 h are observed for agglutination, which appears as the wrinkled skin at the bottom of the wells. The slide agglutination reveals clumping of antigen-antibody complex within a few minutes and is used for preliminary detection. The method has well been tested with strains of *Rhizobium*, *Azospirillum* and *Pseudomonas*.

Gel-diffusion

The basic principle underlying this test is that the antigen and antiserum are placed opposite to each other at certain distances in soft gel (0.75 to 1%) of agar or agarose. The soluble antigen and antibodies diffuse through the gel, and at the place of meeting form antigen-antibody or complex, which appears as a precipitin line. Since the diffusion rate of different antigens may be different, each reacts with its homologous antibody different distance and forms a complex. The gel is prepared either in physiological saline water (0.85% NaCl) or in phosphate buffer saline. For controlling the growth of contaminants, sodium azide at a concentration of 0.01% is incorporated before solidifying the gel. The double diffusion test may be performed in small tubes even capillaries, on microscope slides or in petridishes. He was the first to use immunodiffusion technique of Ouchterlony for *Rhizobium*. Since then, it has been in extensive use in the identification of various strains of rhizobia as well as in defining different antigens of cells components.

Immunoelectrophoresis

Immunoelectrophoresis is a powerful analytical method widely exploited for examining mixture of antigens or for differentiating related antigens. The antigens are first separated by electorphoresis, normally in purified agar or agarose, then detected by diffusion against antiserum subsequently placed in troughs, cut parallel to the direction of the electrophoresis. Like agar diffusion plates, the gel slide in incubated at constant temperature in electrophoretic bath and the formation of preciption lines is observed. Organisms have minor antigen differences can be distinguished. However, only a few antigen samples can be analysed at a time and a large amount of antiserum is required. In case of *Rhizobium* also, it has been successfully used.

Immunofluorescence

The immunoflorescence technique is known since 1942 and others a convenient means to study the ecology of microorganisms directly in the soil. It was first demonstrated with soil fungi and later with bacteria. Technical considerations in the preparation of fluorescent antibody conjugates have been illustrated, and subsequently its success in the study of ecology of *Rhizobium* has been confirmed by many.

Antibody can be labeled (conjugated by chemical combination with fluorescent dyes without material effect on the immunological properties. The labeled or conjugated antibodies are used to react with the antigen in antigen-containing tissue or the extract leachate fixed on microscopic slide. Inoculum as fixed either by heat or the fixatives like acetone of Kirkpatrick's fixative. The staining is performed at a constant temperature for a fixed time. The preparation is washed to remove the unreacted antibodies, mounted with buffered glycerine and examined for the fluorescence under fluorescent microscope. The preparation is illuminated by ultraviolet or ultra-violet blue light, which is stopped above the object by filter, designed to transmit only the visible fluorescence emission. Advantage of this technique is that the antigen even in mixed form or in tissue can be identified and located. There are many fluorochromes but fluorescein isothicynate (FITC), rhodamine, 1-dimethy-amino phenopthalene-5-fluorescence (epifluorescence) microscopy, visualization or weak fluorescence has also been improved considerably. The epifluoresence attachment is installed on a good bright field transmitted microscopic. By using labeled anti-species antibody, the conjugation of primary antibody (homologous antibody) can be avoided. The techniques is highly sensitive and requires very little labeled antibody, however, many negative and positive controls during the assay procedure are essential.

Enzyme-Linked Immunosorbent Assay (ELISA)

The ELISA procedure is relatively simple, rapid, very sensitive and requires small amounts of antiserum. It is being effectively employed for detecting most of the viruses and bacteria. The techniques allows simultaneous handling of a large number of samples and can be subjected to automisation. There are two major types of enzyme immunoassays i.e., (i) homogenous assays, which, in general, are restricted to molecules of low molecular weight (drugs, haptens, hormones, etc.) and (ii) heterogenous assay, which are suitable for detecting macromolecules, and plant or animal pathogens. The heterogenous assays are characterized by the requirement for a separation step to distinguish bound labels from free ones.

Das-ELISA

Of the various kinds of heterogenous assay, the most commonly used for detecting plant associated microorganisms including seed-borne pathogens is the soluble antibody sandwich procedure, popularly known as DAS-ELISA. Specific antibodies (immunoglobulins) are absorbed to a solid surface, which are the wells of microtiter plates, generally made of polystyrene. A sample, suspected to contain the antigen is added and incubated. Antigens recognized by the antibodies are serologically bound. Unreacted sample is removed by washing. Enzyme-labeled specific antibodies (immunoglobulins) are added. The enzyme, generally employed for labeling is alkaline phosphatase or reddish peroxidase. The labeled antibodies bind to the antigen already bound to the coating antibodies. Then p-nitro-phenyl phosphate which is the substrate for alkaline phosphatase is added and development of colour is observed at 405 nm using multiscan spectrophotometer. The colour change in the substrate is proportional to the amount of enzyme present, which in turn is proportional to antigen concentration.

Indirect ELISA

Apart from DAS-ELISA there are numerous variations in the order in which various antibodies and antibody derivative are reacted with the antigen. Indirect ELISA systems employed F(ab')2 fragments and antibody from different animal species have been used successfully. Recently, an indirect ELISA procedure employing direct antigen (DAC) followed by the addition of specific immunoglobulins as crude antisera and detection of antigen-antibody complexes with enzyme-conjugated protein-A was described. Another indirect system using an initial protein A coating (PAC) for binding and orientation of specific antibodies added in the subsequent step was used. The protein-A binds to immunoglobulins at the Fc region, leaving the F(ab')2 region available for antigen binding. The bound antigen is then detected by another layer of antibodies, which are in turn detected by enzyme conjugated protein-A. The system was termed PAS-ELISA, acronym for protein-A sandwich. The indirect forms of ELISA have the advantages over the DAS method, because of the use of crude antiserum instead of purified immunoglobulins and a single enzyme conjugate with all detection systems. These features make the method preferable to the DAS-form in developing countries, where limited resources and facilities make simplified procedures particularly valuable. We have found that even the low-agglutination titer sera against the immunological recalcitrant strains of *Rhizobium* could be used to distinguish closely related strains of *Cicer-rhizobia* by using indirect ELISA.

Dot Immunobinding Assay (DIA)

This technique of dot immunobinding assay (DIBA) is also known as dot-ELISA, EIBA (enzyme linked immunoblot assay), NC-ELISA (nitrocellulose-

ELISA) or FiPSA (filter paper sero-assay).

In this technique a few microlitres of extract of samples are blotted on to a nitrocellulose sheet (membrance) which is then submerged in primary (crude antiserum) antibody. The precipitated specific antibody is then detected with enzyme-labeled protein A or second antibody. Phosphatase or peroxidase are convenient labels, since these are substrates for these enzymes, which yield dark-coloured precipitates. Using single seeds or seed samples or seed flours, DIBA has been successfully used to detect various strains of virus. In case of *Rhizobium* also it has been found very sensitive technique.

Immunodiagnostics to *Rhizobium* Ecology

Wright was probably first to differentiate immunologically the strains from a collection of 156 isolates of soybean rhizobia from the soils of wide geographical distribution. Kriss did not observe serological relationship between strains isolated from five Arctic species of legumes and those from the legumes common to the temperate regions. However, others observed that the strains from diverse geographical origin may or may cross react or be similar in their flagellar and somatic antigens. Usually the place of origin seems to have not much bearing on the immunological characteristics in the case of rhizobia of clover, medic and chickpea. However, in case of *Cicer*-rhizobia, though immunological heterogencity, some immunological unity among strains isolated from different agroclimatic and soil types have been observed. Strains isolated from grey and brown soils in western Rajasthan have been found to belong to a separate serogroup not found in other soil types in India. These observations could be mainly attributed to the localization of such strains to areas, where they could have arisen from single parent *Rhizobium*.

Distinctly different population of *Bradyrhizobium japonicum* in the nodules of plants grown in each of the six different soils, (b) essentially the same distribution of serological groups from plants in a given field and plants grown in the green house on soil from the same field, (c) distribution of serological types in the nodules similar if not identical and (d) each soil was devoid of one or more serological groups but had one or two other groups. In the subsequent studies, it was found that the distribution of serogroups can vary from year to year and within seasons. Soil temperature, host genotype, soil factors such as pH, soil type, organic matter, nutrient status, Fe content, physical properties and previous cropping have also been found to influence the distribution of the serotypes of rhizobia.

In the study of the strains of Cicer-rhizobia, it has further been observed that the two strains isolated from Northern Negav region of Israel and India did not show may any cross reaction for somatic antigens with the antiserum prepared against the *Rhizobium* strains of American origin. None of the serotypes were found to share somatic antigens with their American origin. Similarly with clover rhizobia isolated from different hill country areas in Otago (New Zealand) 10 serotypes were identified. Among strains of *Azospirillum*, isolated from different cultivars of sorghum, maize and barley in India, 9 serotypes were distinguishable and these were different form a strain of Brazalian origin.

Intra-strain competition among Rhizobium strains

The efficiency of an inoculated strain of *Rhizobium* mainly depends upon, among other factors, its competitive ability to occupy maximum nodulation sites on root system of the host. Immunological techniques have opened the way by which the organism (native or introduced) responsible for nodule formation could be identified. In the initial studies, the isolations were made from the nodules formed, and the identity was established by agglutination with antisera developed against each of the strains. By using this method, it was possible to show the influence of the legume host, of the *Rhizobium* inoculum size and strains on the proportion of nodules formed by each of them.

Means evolved a quick serological method of identifying the introduced strains of *B. japonicum* directly in nodules. By comparing agglutination reactions with nodule antigens as well as with pure cultures, it was shown that nodule antigen could directly be used for typing the *Rhizobium* in the nodules. Using either nodule antigens or isolates from the nodules formed or the rhizosphere soil, numerous studies on the survival, persistence and establishment of introduced strain in a particular soil have been followed in different parts of the World. In these studies, diffusion, immunofluorescence. ELISA or immunoblot have been used.

The studies have revealed that the success of the applied strain in forming nodules was predominantly strain specific, but was also many factors like the quality of the inoculum, nature of the native population of *Rhizobium*, pre-treatment of the seeds, soil properties, persistence of the introduced strain in soil and host genotypes.

16. Enzyme assay

Introduction

Soil is a living system where all biochemical activities proceed through enzymatic processes. Enzymes are present in soil as free enzymes, such as exoenzymes released from living cells or due to autolysis after cells death and may or not bound to cell constituents such as cell fragments. Most of the enzymes added to soils by decaying microbial tissues and by plant and animal residues are likely to be degraded by soil protease and remains incorporated with humus. Therefore, soil can be looked on as a system of humus and minerals containing both immobilized enzymes and occluded microbial cells. All these enzymes helps in the availability of nutrients to the microbial cells. All these enzymes helps in the availability of nutrients to the plants in soil, either through decomposition, mineralization or other activities.

Dehydrogenases

Biological oxidation of organic compound is generally a dehydrogenation process there are many dehydrogenases which are highly specific. The overall process for dehydrogenation may be presented as follows:

$$XH_2 + A \rightarrow X\text{-}AH_2$$

Where XH_2 is an organic compound (hydrogen donor) and A is hydrogen acceptor. The dehydrogenases play a significant role in the oxidation of soil organic matter as they transfer hydrogen from substrate to acceptors. Many different specific dehydrogenase systems are involved in the dehydrogenase activity of soils; these systems are an integral part of the microorganisms. Therefore, the result of the assay of the dehydrogenase activity would show the average activity of the active population. The most widely used method for determination of activity is that of Lenhard. This method involves colorimetric determination of 2,3,5-triphenyl formazan (TPF) produced by the reduction of 2,3,5 – triphenyltetrazolium chloride (TTC) by soil microorganisms. Tetrazolium salts are representatives of a unique class of compounds. These compounds have combinations of desirable properties. They are quaternary NH_4^+ salts and, as such, possess a high degree of water solubility.

Requirements

Test tubes (16 × 150 mm), Spectrophotometer Funnels, Filter paper discs. Calcium carbonate ($CaCO_3$) AR, 2,3,5- Triphenyltetrazolium chloride (TTC) 3% (Dissolve 3 g of TTC in about 80 ml of water and make up the volume to 100 ml), Methanol AR, Triophenyl formazain (TPF) standard solution (Dissolve 100 mg of TPF in about 80 ml methanol and make up the volume to 100 ml with methanol).

To prepare standard curve, dilute 10 ml of TPF standard solution to 100 ml with methanol (100 μg of TPF ml^{-1}). Pipette 5, 10 15 and 20 ml aliquots in to 100 ml volumetric flask (500, 1000, 1500 and 2000 μg of TPF 100 ml^{-1}). Make up the volumes with methanol and mix thoroughly, measure the intensity of red color of TPF as described for samples, plot the absorbance readings against the amount of TPF in 100 ml standard solutions.

Procedure

- Take 20 g air dried soil and 0.2 g of CaCO3 and mix thoroughly. Place 6 g of this mixture in test tubes in triplicate.
- To each tube add 1 ml 3% aqueous solution of TTC and 2.5 ml distilled water.
- Mix the contents thoroughly and incubate at 37°C for 24 hrs.
- After 24 hrs add 10 ml methanol in each tube, shake thoroughly and filter through Whatman No. 5 filter paper and collect the filtrate. Wash the soil with methanol until no more color could be extracted and pool with earlier extracts in a 100 ml volumetric flask and finally make up to volume with methanol.
- If suspended soil particles are present in tubes, centrifuge the filtrate at 14,000 rpm for 10 min before reading absorbance.
- Determine the absorbance at 485 nm.

- Calculate the amount of TPF produced by referring to the calibration graph prepared from TPF standards.

Urease

Urease is a hydrolytic enzyme that attacks the nitrogen and carbon bond in amide compounds such as urea and forms the alkaline end produce ammonia. Urease is detectable when the organisms are grown in a urea broth medium containing the pH indicator phenol red. Urea is split into its products, the presence of ammonia creates an alkaline condition which turns phenol red to a deep pink.

Requirements

Cultures of *Escherichia coli, Proteus vulgaris* (24 hr to 48 hr), Urea broth concentrate (filter sterilized solution) 10.0 ml, Sterile distilled water, 90.0 ml (Aseptically add the urea broth concentrate to the sterilized and cooled distilled water). Under aseptic conditions, dispense 3 ml into sterile tubes, Burner and inoculation loop.

Procedure

Using sterile technique, inoculate each experimental organisms into its appropriately labeled tube by means of inoculation needle.

- Incubate cultures 24 to 48 hr at 37°C.
- Examine all urea broth cultures for colour and record the results.

Nitrate Reductase

Some aerobic and facultative anaerobic microorganisms reduce nitrate in the absence of molecular oxygen. In these, anaerobic respiration is an oxidative process whereby the cell uses inorganic substances such as nitrates or sulfates to supply oxygen that is subsequently utilized as a final hydrogen acceptor during energy formation.

$$NO_3^- + 2H^+e^- \underset{Enzyme}{\overset{Nitrate}{\rightleftharpoons}} NO_2^- + H_2O$$

Some organisms act further on nitrites to reduce to ammonia (NH_3^+) or molecular nitrogen.

$$NO_2^- + NH_3$$

$$2NO_3^- + 10e^- + 12H^+ \rightarrow N_2 + 6H_2O$$

Nitrate reduction can be determined by cultivating organisms in nutrient broth supplemented with 0.1% KNO_3 as the substrate. Also, medium is made semi solid by the addition of 0.1% agar. The semisolid nature of medium favours anaerobic requirement for nitrate reduction. After incubation, an organisms's ability to reduce nitrates to nitrites to nitrites is determined by addition of two reagents: Solution A, is sulphanilic acid and solution B, is alphanaphthylamine. Following reduction, the addition of sol. A and B will immediately produce cherry red colour.

Cultures not producing a colour change suggest either (i) nitrates were not reduced by the organism or (ii) the organism possessed such potent nitrate reductase enzyme that nitrate were rapidly reduced beyond nitrites to ammonia or even molecular nitrogen. To determine whether or not nitrates were reduced past the nitrite state, a small amount of zinc powder is added to the basically colourless cultures already containing solutions A and B. Zinc reduces nitrates to nitrates. The development of red colour verifies that nitrates were not reduced and if still there is no colour change, the nitrates in medium were reduced beyond nitrites to ammonia or nitrogen gas.

Requirement

- Cultures of Escherichia coli, Pseudomonas aeruginosa 24 to 48 hr old.
- Trypticase nitrate broth (pH 7.2) (Trypticase 20 g, Disodium phosphate 2g, Dextrose 1.g, Agar 1.g, Potassium nitrate 1.0 g).
- Solution A Sulphanilic acid, Solution B (a-naphthylamine), zinc powder.
- Bunsen burner and inoculating loop etc.

Procedure

- Using sterile conditions, inoculate each experimental organism into its appropriately labeled tube by means of a inoculation loop.
- Incubate all cultures for 24 to 48 hr at 37°C.
- Add 5 drops of soil. A and then five drops of sol. B to all nitrate broth cultures, observe development of red colouration if any, and record the observation.
- Add a little quantity of zinc to the culture tubes in which no red colour developed, observe and record the change in colouration if any, now develops in each culture.
- On basis of observations, determine whether each organism was capable of nitrate reduction or not. Identify end products (NO_2^- or NH_3/N_2).

Phosphatases

Phosphatases is a group of enzymes that catalyze the hydrolysis of both esters and anhydrides of H_3PO_4. A considerable part of soil phosphorus is organically bound. The forms of organic phosphorus can vary with soil. The largest fraction of it appears to be in the form of phytin andits derivatives. Phosphatases are adaptive enzymes and their activity increase when plants are deficient in phosphorus.

Phosphomonoesterases are the most widely studied phosphatases and are classified into either acid phosphatase (EC 3.1.3.2) or alkaline phosphatase (EC 3.1.3.1.) The enzymes are classified acid and alkaline phosphatases because they show their optimum activities in acid and alkaline ranges. The general equation of the reaction catalyzed by acid and alkaline phosphatases is:

Requirement

Bacterial culture (24 hr old), 100 ml conical flaks, Spectrophotometer, Centrifuge, Tris-ctirate buffer

Procedure

- A loopful of culture of *Azospirillum* sp (grown in malate medium) inoculated int to Nutrient broth (50 ml)
- Incubate 37°C for 24 hrs
- Centrifuge 10,000 rpm (20 min, 4°C)
- Resuspend the pellet in sterile dist. water
- OD at 620 nm (adjust to 0.2)
- Take 1 ml of culture
- Adjust pH 8.8 (Tris-citrate buffer)
- Add 2 ml P-nitrophenyl phosphate
- Incubate 37°C for 1 hr
- Add 0.2 ml of 2 N NaOH
- Centrifuge
- OD at 420 nm.

Soil phosphatase activity

- Take 1 g soil in 100 ml conical flask
- Add 4 ml modified universal buffer, 0.25 ml toluene and 1 ml p-nitrophenyl phosphate
- Incubate at 37°C for 1 hr
- Now add 1 ml 0.5 M Calcium chloride and 4 ml 0.5 M sodium hydroxide, mix
- Filter through Whatman No. 42 filter paper
- Collect filtrate and read OD at 400 nm

Malate medium

Costituents	**g/1**
Malic acid	5.0
KOH	4.0
KH_2PO_4	0.5
$FeSO_4.7H_2O$	0.05
$MnSO_4.H_2O$	0.01
$MgSO_4.2H_2O$	0.10
NaCl	0.02
$CaCl_2$	0.01
Na_2MoO_4	0.002
Bromo thymol blue	2.0 ml (0.5% alcohol)
NH_4Cl	1.0
pH	6.6-7.0
Distilled water	1 L

APPENDIX 2

Biofertilizers Statistics and Production Estimates

All India Capacity and Production of Biofertilisers

Year	Capacity	Production tonnes
1992-93	5400.5	2005.0
1993-94	6125.5	3084.0
1994-95	8114.5	5800.5
1995-96	10680.4	6692.3
1996-97	12647.0	7406.6
1997-98	N.A.	7104.6
1998-99	16446.0	8010.1

Statewise Production of Biofertilisers – 1999-2000 & 2000-01

Zone/State	Production	
	1999-2000 (Estimated)	2000-01 (Likely)
East	**242.0**	**361.0**
Assam	38.1	55.0
Bihar	34.9	35.0
Orissa	71.5	166.0
West Bengal	97.5	105.0
North	**207.2**	**119.0**
Haryana	54.0	
Himachal Pradesh	2.4	2.5
Uttar Pradesh	147.4	110.0
Punjab	3.4	6.5
South	**4491.6**	**6117.5**
Andhra Pradesh	48.2	45.0
Karnataka	1534.5	3152.5
Kerala	1.0	1.0
Tamil Nadu	2870.0	2879.0
Pondicherry	37.9	40.0
West	**5098.2**	**6429.9**
Gujarat	630.8	800.0
Madhya Pradesh	1812.9	2045.5
Maharashtra	2434.0	3173.0
Rajasthan	220.5	411.4
All-India	**10038.9**	**13027.4**

Source: State Departments of Agriculture.

Product-wise Distribution of Biofertilisers – All India 1992-93 to 1998-99

(tonnes)

Product	1992-93	1993-94	1994-95	1995-96	1996-97	1997-98	1998-99
Rhizobium	916.31	1180.23	1467.24	1330.16	1392.61	1249.54	1247.39
Azotobacter	208.05	646.95	921.52	1160.98	1036.42	1089.18	1188.62
Azospirillum	200.61	323.70	702.46	1131.27	757.38	640.12	788.41
Blue Green Algae	0.03	0.03	2.78	2.51	0.74	1.43	2.44
Acetobacter	-	-	110.25	119.35	75.66	66.91	67.04
Phosphate Solubilise	275.01	763.46	1784.65	2544.04	3332.53	3069.17	3281.56
Total	**1600.01**	**2914.37**	**4988.90**	**6288.32**	**6681.44**	**6295.63**	**6700.27**

Unitwise, Productwise Sale Prices of Biofertilisers 1996-97 to 1998-99

Name of the unit	Products	Packing Size	(Price (Rs./kg.)		
			1996-97	1997-98	1998-99
Ajay Bio-tech (India), Pune (Maharashtra)	Ajay Rizo (*Rhizobium*)		22.00	22.00	-
	Bio phos (PSB)		37.25	37.25	37.25
A.V.S. Agro Products, Sangamner (Maharashtra)	PSB		30.00	29.00-30.00	30.00
Agricultural Bacteriology Section, College of Agriculture, Pune (Maharashtra)	*Rhizobium, Azotobacter*, PSB		40.00	40.00	40.00
	Azospirillum		60.00	60.00	60.00
	BGA		10.00	10.00	10.00
Bharat Laboratory & Biological House for Agriculture, Dhule (Maharashtra)	*Rhizobium, Azotobacter, Azospirillum*	250 gm	20.00	20.00	20.00
	PSB	500 gm	20.00	30.00	30.00
BAIF Laboratories, Wagholi, Pune (Maharashtra)	*Rhizobium, Azotobacter, Azospirillum*, PSB		N.A.	N.A.	N.A.
Bio Agro Fertilizers, Pune (Maharashtra)	*Rhizobium, Azotobacter*, PSB		-	-	50.00
Bio-Fertilizer Production Unit, Dept. of Agriculture, Pudukkottai (Tamil Nadu)	*Rhizobium, Azospirillum*, PSB		25.00	25.00	25.00
Bio-Fertilizer Production Unit, Dept. of Agriculture, Salem (Tamil Nadu)	*Rhizobium, Azospirillum*, PSB	200 gms	25.00	25.00	25.00
Bio Science Laboratories, Salem (Tamil Nadu)	Agro bium (Nitrogen fixer), Agro phos (PSB)		25.00	25.00	25.00
Deptt. Of Agriculture, Hawthornvilla, Shimla (Himachal Pradesh)	*Rhizobium*		60.00	60.00	60.00
Deptt. of Agriculture, Kotnur, Gulbarga (Karnataka)	*Rhizobium*		*	*	*
Ecosense Labs (I) Pvt. Ltd., Mumbai (Maharashtra)	PSB		-	-	14.50

Name of the unit	Products	Packing Size	(Price (Rs./kg.)		
			1996-97	1997-98	1998-99
Godavari Fertilizers & Chemicals Ltd. Secunderabad (Andhra Pradesh)	*Rhizobium, Azospirillum*, PSB		35.00	35.00	32.00
Gujarat State Co-operative Marketing Federation Ltd., Kankaria (Gujarat)	*Rhizobium, Azospirillum*, PSB		20.00	20.00	20.00
G.S.F.C., Vadodara (Gujarat)	*Rhizobium, Azotobacter*,	250 gms. – Gujarat	24.00	24.00	32.00
	Azospirillum, PSB	250 gms. – Other states	32.00	32.00	40.00
	Azotobacter,	1 kg. – Gujarat	22.00	22.00	28.00
	Azospirillum, PSB	1 kg. – Other states	28.00	28.00	32.00
HMG Biotech Pvt Ltd, Shimoga (Karnataka)	*Rhizobium, Azotobacter, Azospirillum*, PSB		37.50	37.50	30.00
Hindustan Fertilizer Corporation (All units)	*Rhizobium, Azotobacter, Azospirillum*, PSB		26.25	26.25 (Upto Sept. 97) 26.00 (from Oct. 97)	36.00
Institute of Natural Agriculture (INORA), Pune (Maharashtra)	BGA		-	-	10.00
	PSB		-	-	40.00
K – Ferts Lab, Nanded (Maharashtra)	*Rhizobium*,		30.00	30.00	30.00
	Azotobacter, PSB		32.00	40.00	40.00
Karnataka Bio Fertilizers, Bijapur (Karnataka)	*Rhizobium, Azotobacter, Azospirillum*, PSB		33.50	37.50	40.00
Kisan Agro Chem, Dhanegaon (Maharashtra)	*Rhizobium, Azotobacter, Azospirillum*	250 gms	22.00	22.00	22.00
	PSB	"	30.00	30.00	29.00
KRIBHCO, Hazira (Gujarat)	*Rhizobium, Azotobacter*,		22.50	30.00	30.00
	PSM		25.00	30.00	30.00
Kumar Krishi Mitra Bio Products (I), Pune (Maharashtra)	BIOPLIN (AzotobacterGenus)	For seed & soil appln.	35.00	35.00	35.00
	VITORMONE (AzotobacterGenus)	For foliar appln.	44.00	44.00	44.00
	PHOSFERT (PSB)	For seed & soil appln.	25.60	25.60	25.60
	PHOSFERT PSB (Pre sterile Lignite based)	For seed & soil appln.	30.00	30.00	30.00

Name of the unit	Products	Packing Size	(Price (Rs./kg.)		
			1996-97	1997-98	1998-99
Lakshmi Bio-Techs, Cuddalore (Tamil Nadu)	*Rhizobium, Acetobacter*,		16.00-	16.00-	16.00-
	Azospirillum, PSB		24.00	24.00	24.00
Madras Fertilizers, Manali (Tamil Nadu)	*Rhizobium, Azospirillum*, PSB		37.00	40.00	40.00
Magnam Associates, Chennai (Tamil Nadu)	*Azospirillum*, PSB		30.00	30.00	30.00
Maharashtra Bio-tech Industries, Pune (Maharashtra)	*Azotobacter*, PSB		30.00	30.00	30.00
Main Bio Control Research Laboratory (A Unit of Tamil Nadu Cooperative Sugar Fedn.) Chengalpattu (Tamil Nadu)	*Acetobacter*, PSB		25.00	25.00	25.00
Marathwada Agricultural University, Pharbhani (Maharashtra)	*Rhizobium, Azotobacter*,		20.00	20.00	40.00
Micro Bac India, Shyamnagar (W.B.)	*Rhizobium, Azotobacter*, PSB		35.00	35.00	35.00
Micro Biological Laboratory (Soil Testing Laboratory),, Pattambi (Kerala)	*Rhizobium*		10.00	10.00	10.00
Monarch Bio-Fertilizers and Research Centre, Chennai (Tamil Nadu)	*Rhizobium, Azotobacter, Azospirillum*, PSB		18.00	25.00	28.00
M.A.I.D.C. (Rhizobium Culture Prodn. Unit) Chinchwad, Pune (Maharashtra)	*Rhizobium, Azotobacter*,		22.00	22.00	22.00
	PSB		30.00	30.00	30.00
MLN Farmers Training Institute, (IFFCO) Phulpur (U.P.)	*Rhizobium, Azotobacter*, PSB		30.00	30.00	22.00
M.P. State Agro Industries Dev. Corpn., Bhopal (M.P.)	*Rhizobium, Azotobacter, Azospirillum*	150 gms	53.33	53.33	53.33
	PSB	250 gms	40.00	40.00	40.00
M.P. State Co-Op. Oil-Seed Growers Federation, (Bio tech Research Centre) Dhar (M.P.)	*Rhizobium, Azotobacter*		53.33	53.33	53.33
	PSB		40.00	40.00	40.00
Nafed Biofertilizer, Indore (M.P.)	*Rhizobium, Azotobacter*		-	-	8.00
	PSB		-	-	12.00
National Bio fertilizer Development Centre, Ghaziabad (U.P.)	*Rhizobium Azotobacter, Azospirillum*, PSB		20.00	20.00	20.00
Regional Biofertilizer Development Centre Hissar (Har.), Imphal (Manipur) Nagpur (Maharashtra) Jabalpur (MP)	" " "		20.00 20.00 20.00	30.00 30.00 30.00	35.00 40.00 35.00

Name of the unit	Products	Packing Size	(Price (Rs./kg.)		
			1996-97	1997-98	1998-99
	Rhizobium, Azotobacter, Azospirillum, PSB	200 gms	20.00	30.00	40.00
	BGA		2.00	2.00	4.00
Nav Maharashtra Chakan Oil Mills Ltd., Pune (Maharashtra)	*Rhizobium, Azotobacter*,		30.00	30.00	30.00
	PSB		28.00	28.00	28.00
Niku Bio-research Lab, Pune (Maharashtra)	*Rhizobium, Azotobacter*, PSM		-	50.00	50.00
Nitrofix Laboratories, Calcutta (W.B.)	*Azophos* (Composite culture of *Azotobacter* and PSM)		8.70	10.87	33.55
	Rhizophos (Composite culture of *Rhizobium* and PSM)		13.08	6.06	9.30
Nodule Research Laboratory, BCKV, Mohanpur (West Bengal)	Rhizobium, Azotobacter, Azospirillum, PSB		18.50	18.50	18.50
Pyrites, Phosphates & Chemicals, Amjhore (Bihar)	Rhizobium		50.00	50.00	50.00
Rashtriya Chemicals & Fertilizers Limited Chembur, Mumbai (Maharashtra)	*Rhizobium, Azotobacter*,		28.00	28.00	28.00
	PSB		45.00	45.00	45.00
Rashtriya Chemicals & Fertilizers Limited Bangalore, Mumbai (Karnataka)	*Rhizobium, Azotobacter*,		27.50	27.50	27.50
	PSB		45.00	45.00	45.00
Regional Agri-Testing & Demonstration All Units (U.P.)	*Rhizobium, Azotobacter*, PSB		20.00	20.00	20.00
Regional Soil Testing Laboratory (Bacterial Culture Production Laboratory) Rajendra Nagar, Hyderabad (AP)	*Rhizobium*		5.00	6.00	6.00
Rhizobium Scheme, Deptt. of Agriculture, Durgapura (Rajasthan)	*Rhizobium, Azotobacter*,		16.25	16.25	16.25
	PSB		30.00	30.00	30.00
Rhizobium Culture Prodn. Laboratory, Deptt. of Agriculture, Hebbal, Bangalore (Karnataka)	*Rhizobium, Azotobacter, Azospirillum*, PSB		*	*	*
Samarth Bio Tech Limited, Hubli (Karnataka)	*Rhizobium, Azospirillum*,		35.00	35.00	35.00
	PSB		32.00	32.00	30.00
Samruddhi Agrotech, Pune (Maharashtra)	*Rhizobium, Azotobacter*, PSB		13.00-	13.00-	13.00-
			30.00	30.00	30.00
SPIC, Bio Bioproducts, Porur, Chennai, (Tamil Nadu)	*Rhizobium, Acetobacter, Azospirillum*, PSB		28.00	28.00	28.00

Name of the unit	Products	Packing Size	(Price (Rs./kg.)		
			1996-97	1997-98	1998-99
Tejasvi Biofert, Pune (Maharashtra)	*Azotobacter*		25.00	25.00	25.00
	PSB		30.00	30.00	30.00
Terra Firma Biotechnologies Ltd.	Nitrogen fixer, PSM		32.00	32.00	40.00
The SIMA Cotton Dev. & Research Association, Coimbatore (Tamil Nadu)	*Rhizobium, Azospirillum*, PSB		33.00	33.00	33.00
T.N. Agricultural University, Dept. of Agril. Microbiology, Coimbatore (Tamil Nadu)	*Rhizobium, Azospirillum*, PSB		25.00	25.00	25.00
University of Agricultural Science, (Bio Fertiliser Scheme) G.K.V.K., Bangalore (Karnataka)	*Rhizobium, Azotobacter, Azospirillum,* PSB		For demonstration purposes		
University of Agricultural Sciences, Dept. of Agrl. Microbiology, Dharwad, (Karnataka)	*Rhizobium, Azospirillum,* PSB		30.00	30.00	30.00
Vasantdada Sugar Institute, Pune (Maharashtra)	*Azotobacter, Azospirillum,* PSB		20.00	20.00	20.00

* = Distributed free of cost to the farmers.
** = Farmers of Coop. & Public sector Sugar Mills in Tamil Nadu are availing 50% subsidy through the Sugar Mills (i.e. Rs. 12.50 per kg.).

Ready Reckoner on Biofertilizer Production (Tentative)

Total production capacity (1 packet = 200 gm)	75 tonne/year (375,000 packets)
Number of working days	240 days (8 months)
Average daily production capacity	312 kg (1560 packets)
Number of shifts	1 (8 hour)
Number of daily wages labour required - Unskilled - Semi-skilled	 3 (@ labour/500 pkt) 1
Requirement of mother culture	1%
Requirement of broth per day	104 litre
Requirement of carrier per day	208 kg
Broth required for 1 tonne (40% moisture basis)	400 litre
Carrier required for 1 tonne (on 40% moisture basis)	600 kg
Sievability of carrier	100 to 72 mesh (150 to 212)
Mixing ratio (carrier: both)	1.5:1.0
Time required for sterilization - carrier - broth (depending on quantity)	 4 hours 45-60 minutes or more
Carrier required for 1000 packets	120 kg
Requirement of chemicals for production Rhizobium broth/1000 pkts a. Mannitol b. Yeast extract c. Magnesium sulphate	 800 g 80 g 16 g

d. Dipotassium hydrogen phosphate e. Sodium chloride	40 g 8 g
Weight of corrugated box containing 40 packets	8 kg
Corrugated box ex required for 1000 packets	25

Source: "Biofertilizer Technology, Marketing and Usage – A Sourcebook-cum-Glossary" by Dr. M.R. Motsara, Dr. P. Bhattacharyya and Dr. Beena Srivastave (1995), Fertilizer Development and Consultation Organization, New Delhi.

Chemical and physical characteristic of common carriers

Parameters	**Peat (Indian)**	**Lignite**	**Charcoal**
Chemical			
Appearance	Dark brown	Brownish black	Black
pH	Acidic, needs neutralization	Acidic, needs neutralization	Neutral
Odour	None	None	None
Carbon (%)	54.20	72.79	84.00
Hydrogen (%)	5.70	4.74	2.30
Oxygen (%)	31.60	19.60	10.70
Nitrogen	1.50	0.98	0.01
Ash (%)	7.00	7.00	3..00
Physical			
Water holding capacity (%)	149.30	198.90	200.00
Moisture (%)	40-50	40-50	20-30
Bulk density (g/cm3)			
Particle density (g/cm3)	2.18	1.66	1.62
Porosity (5)	53.16	34.79	73.39
Total surface area (m2/g)	646.60	556.40	9870.90
Availability	5.5 million ton	3 million ton	Plenty
	Nilgiri Valley	a. Neyveli (TN)	Everywhere
	(Tamil Nadu)	b. Singrauli (MP)	
		c. Palana (Rajasthan)	

*Tilak and Subha Rao (1978), Source: "Biofertilizer Technology, Marketing and Usage – A Sourcebook-cum-Glossary" by Dr. M.R. Motsara, Dr. P. Bhattacharyya and Dr. Beena Srivastave (1995), fertilizer Development and Consultation Organization, New Delhi.

List of equipments needed for biofertilizer production (75 tpa)

No.	**Name**	**Quantity**
Main Equipments		
1.	Horizontal autoclave	1
2.	Vertical autoclave	4
3.	Double dacker rotary shaker (200 litre capacity)	1
4.	Fermentor (100 litre capacity)	2
5.	BOD incubator	1
6.	Laminar air flow	1
7.	Hot air oven	1
8.	Air conditioner	2
9.	Refrigerator	2
10.	Lyophiliser	1

No.	Name	Quantity
11.	Microscope	1
12.	Balance: Physical Chemical	2 2
13.	Powder mixer	1
14.	Sealing machine	1
Supporting Equipments		
15.	pH meter	1
16.	Colony counter	1
17.	Distillation apparatus	1
18.	Colorimeter	1
19.	Centrifuge	1
20.	Leonard Jar	100

Source: "Biofertilizer Technology, Marketing and Usage – A Sourcebook-cum-Glossary" by Dr. M. R. Motsara, Dr. P. Bhattacharyya and Dr. Beena Srivastave (1995), Fertilizer Development and Consultation Organization, New Delhi.

Space required for a biofertilizer production facility of 75 tonnes per annum (tpa)

Name of the Unit	Minimum size	Purpose	Facilities of provided
INOCULUM WAY			
Storage room	7.5 x 4 m = 30 sq. m.	For storage of chemicals glassware & polythene bags.	Different racks, cabinet lab. bench, sink, balance vertical autoclave, hot air oven, pH meter,
Media preparation room	9 x 4 m = 36 sq. m	For preparation of media for biofertilizer	Distillation plant deioniser. Lab. bench, sink, laminar air flow, BOD
Bacterio logical room	7.5 x 4 m = 30 sq. m	For isolation, incubation	Incubator, refrigerator
Fermentation shaker room	9 x 4 m = 36 sq. m.	For multiplication of cells.	Fermenters/shakers (rotary), Lab, bench air conditioner Lab. bench,
Quality room	7.5 x 4 m = 30 sq. m.	For checking quality	Sink, microscope, pH meter, colony counter, balance, spectrophotometer, BOD incubator.
Growth room	4 x 4 m = 16 sq. m.	For nodulation study	Slotted angle rack with illumination, Leonard jar, Gibson tube air conditioner, timer, thermograph.
CARRIER WAY			
Storage room for carrier	7 x 4 m = 28 sq. m.		
Carrier sterilization room	6 x 4 m = 24 sq. m.		Horizontal autoclave, trolley, sink Lab. bench, trays
Mixing room	7 x 4 m = 28 sq. m.	Mixing of both with carrier	Tray, bench
Curing room	5 x 4 m = 20 sq. m.		Air conditioner, temp. & humidity controller.
Packing room	5 x 4 m = 20 sq. m.		Sealer, corrugated box.

Name of the Unit	Minimum size	Purpose	Facilities of provided
VISITORS – CUSTOMERS WAY			
Office room	20 x 4 m = 80 sq. m.		
Biofertilizer storage room	8 x 4 m = 32 sq. m.		Air conditioner
Selling room	2 x 4 m = 8 sq. m.		
Net house	12 x 4 m = 48 sq. m.	For observation of the impact of biofertilizer	Bench, sink earthen pot
Garage	8 x 4 m = 32 sq. m.	For vehicle	
Total	**498 sq. m. or 500 sq. m.**		

Source: "Biofertilizer Technology, Marketing and Usage – A Sourcebook-cum-Glossary" by Dr. M.R. Motsara, Dr. P. Bhattacharyya and Dr. Beena Srivastava (1995), Fertilizer Development and Consultation Organization, New Delhi.

Properties of packing materials used for biofertilizer

Property	LDPE	HDPE	Polypropylene
Molecular weight		1500-7000	25000-50000
Thickness (mm) 0.038-0.08	0.08-0.1	0.05-0.09	
Appearance	Translucent	Translucent to Opaque	Transparent to
Melting point (°C)	80-90	105-120	120-130
Autoclavability (during sterilization)	Poor	Moderate	Good
Stability towards irradiation	Good	Good	Fair
Transmission of water vapour CC's (STP) cm2/sec/cm Hg at 25°Cx10^{-8}			
Permiability to gasses CC's (STP) cm2/sec/cm Hg at 25°Cx10^{-8}			
CO_2	28	6.5	6
N_2	7.5	1.6	2
O_2	2.5	0.5	0.4
Gas exchange property	Good	Moderate	Moderate

Source: "Biofertilizer Technology, Marketing and Usage – A sourcebook-cum-Glossary by Dr. M. R. Motsara, Dr. P. Bhattacharyya and Dr. Beena Srivastava (1995), Fertilizer Development Consultation Organization, New Delhi.

SPECIFICATION OF RAW MATERIALS

- **Wood Charcoal:** 50-200 mesh, good quality, pH 6.5-7.5, 120-140% water holding capacity, moisture percentage, 40% (w/v), ash content 5-10%.
- **Raw Lights Power:** 150-200 mesh, 35% moisture percentage, finely powdered, duly packed in HDPE bag of 40-50 kg each.
- **Gum Arabic:** 150-200 mesh dry powder.
- **HDPE Polythene Bag:** made of high quality, 1st grade milky white

poluthene tube having thickness of 300 guage and sealed from one end. Individual bag printed on both sides (Flexo printing) as per the design and size. Bag sizes range from 8 ×16 cm to 30 ×40 cm to contain 40 to 2,880 g inoculant respectively. In India the standard packet size is 15 ×25 cm for 200 g biofertilizer.

- **Instafix Gummed Plastic Tape:** 5 cm wide tape for fixing and sealing the corrugated card board boxes.
- **LDPE Plastic Strips:** For packing the cardboard made of first grade LDPE 5 cm wide thick in 200 m rolls.
- **Iron Clips:** For packing plastic strip of 5 cm sizes.
- **Aluminium Foil:** 0.03 mm thickness.

Source: "Biofertilizer Technology, Marketing and Usage – A Sourcebook-cum-Glossary" by Dr. M. R. Motsara, Dr. P. Bhattacharyya and Dr. Beena Srivastava (1995), Fertilizer Development and Consultation Organization, New Delhi.

APPENDIX 3

List of Biofertilizers Units in India and Abroad

LIST OF BIOFERTILIZER UNITS IN INDIA

1. Agriculture Bactroriology
   Section College of Agriculture
   Pune 411005
2. Acharya N.G. Ranga Agricultural University
   Agricultural Research Station
   Amaravathi 522020
   District Guntur
   Andhra Pradesh
3. Ajay Bio-Tech (India) Ltd.
   22658, Village Khalad
   District Pune 412301 (MS)
4. Assam Agro Industries Corpn.
   Agricultural Chemistry Research Stn.,
   Guwahati, Assam
5. Azo Fertilizers (Pvt.) Ltd.
   1063/89, 1st Main, 7th Cross
   Vidyaranyapuram
   Mysore 570008, Karnataka
6. Arvee Biotech
   134, East Car Street
   Chidambaram 608001
   District South Arcot Vallalar
   Tamil Nadu
7. BAIF Laboratories Ltd.
   Briahnagar
   Off Pune Nagar Road
   Wagholi 412207
   District Pune, Maharashtra
8. Balaji Biotech
   Village Sathanur
   Post Vazhuvur
   Vandhavas 604405, District T.S.,
   Chennai
9. Bidhan Chandra Krishi Vishwa Vidyalaya
   Nodule Research Laboratory
   Mohanpur
   Nadia Dist. 741252
   West Bengal
10. Biofertilizer Production Unit
    Govt. of Tamil Nadu
    Kukumiamalal
    District Pudukkottai
    Tamil Nadu 622104
11. Bio-Fertilizer Plant
    The Madhya Pradesh State
    Agro Industries Development Corpn. Ltd.
    Agro Complex, C – Sector Indrapuri, Bhopal
    462022, M.P.
12. Department of Agriculture
    Government of Himachal Pradesh
    Hawthomvilla, Shimla 4 (HP)
13. Biofertilizer Production Unit
    Department of Agriculture
    Government of Tamil Nadu
    83-E, Kannan Kurchi Main Road
    Salem 636007, Tamil Nadu
14. Department of Soil Science
    and Agricultural Chemistry
    Faculty of Agriculture
    Birsa Agricultural University
    Ranchi 834006, Bihar
15. Esvin Bio-Systems Limited
    Esvin House, Perungudi
    Chennai 600096 (TN)
16. Farm Organics Indian Pvt. Ltd.
    Khole Niwas
    Gurudwara Road
    Kopergaon 423601 Maharashtra
17. Godavari Fertilizers and
    Chemicals Limited
    Vaninilayam
    50, Sebastian Road
    Secunderabad 500003, A.P.
18. Gujarat State Cooperative
    Marketing Federation Ltd.
    Sahakar Bhavan 5351316
    Relief Road
    Ahmedabad 380001, Gujarat

19. Gujarat State Fertilizers & Chemicals Limited
P.O. Fertilizernagar 391750
District Vadodara
Gujarat
20. Hans Biogenics
14-146, Gandhipuram 2
Rajahmundry 533103
A.P.
21. Harayana Agricultural University
Department of Microbiology
Hissar 125001
Haryana
22. Hindustan Fertilizer Corporation Ltd.
Fertilizer Promotion & Agricultural
Research Division
52 A, Shakaspeare Sarani
Calcutta 700001
23. Hindustan Fertilizer Corporation Ltd.
Fertilizer Promotion &
Agricultural Research Division
Durgapur 713212
24. Indian Agricultural Research Institute
Division of Microbiology
New Delhi 110012
25. Indian Organic Chemicals Ltd.
Khopoli 410203
District Raigad
Maharashtra
Regd. Office
Chemtex House
26. 2nd Floor, Main Street Sector-12
Hiranandani Garden
Powai 400076
Mumbai
27. Kadur Agro
R.V. Vidyaniketan P.O.
Mysore Road
Mylasandu
Bangalore 560059
28. K. Ferts Lab.
25, First Floor
Gurunanak Market
Nanded 431601
Maharashtra
29. Karnataka Department of Agriculture
Rhizobium Culture Prodn. Laboratories
Hebbal Bangalore 560024
Karnataka
30. Karnataka Department of Agriculture
Soil Health Centre
Gulberga
Karnataka
31. Kerala Department of Agriculture
Vikas Bhawan
Thirurananthapuram 695035
Kerala
32. Kisan Agro Chem
Bhatia Complex
Behind Mundada Hospital
Near Gujrathi High School
Vazirabad
Nanded 431601
33. Krishak Bharati Coop. Ltd.
Marketing Division
A-10, Sector-1
Noida 201301
District Ghaziabad
Noida (U.P.)
34. Kumar Krishi Mitra BioProducts
(I) Pvt. Ltd.
12, Ganeshwadi Fergusson College Rd.
Pune 411004
Maharashtra
35. Lakshmi Bio-Techs
No. 32 Rajunaicken Street
West Mambalam
Chennai 600033
36. Nagam Associates
No. 11 T.T.K. Road
Royapettah
Chennai 600014
37. Marthwada Agricultural University
Parbhani 431402
Maharashtra
38. M.P. State Agro Industries
Corporation Limited
Biofertilizer Plant
Agro Complex, C-Sector
Indrapuri
Bhopal 462022
M.P.
39. M.P. State Coop. Oilseed
Growers 'Fedn. Ltd.
H.O.:
1, Arera Hills, Behind Govt. Press
Bhopal 462011
Madhya Pradesh
Factory:
Biotech Research Centre
Gulab Chokkar
Dhar 454001
Madhya Pradesh
40. Mahatma Phule Krishi Vidyapeeth
Deptt. of Agricultural
Bacteriology
College of Agriculture
Shivajinagar
Pune 411005
Maharashtra

41. Madras Fertilizers Ltd.
Post Box No. 2
Manali
Madras 600068
Tamil Nadu
42. Main Bio Control Research Laboratory
26393 (A Unit of Tamil Nadu Cooperative
Sugar Fedn.)
2E/1, Rajeshwari Vedachalem Street
Chegalpattu 603001
Tamil Nadu
H.O.:
Tamil Nadu Cooperative Sugar Fedn. Ltd.
474, Anna Salai
Periyar EVR Building
5th Floor, Nandanam
Chennai 600035
Tamil Nadu
43. Micro Bac India
OGG Road, Noapara
P.O. Shyamnagar 743127
Dist. 24 Parganas (N)
West Bangal
44. MLN Farmers Training Instt.
(Cooperative Rural Development Trust)
IFFCO Biofertilizer Production Unit
Ghiya Nagar, Phulpur
Allahabad 212404
U.P.
45. Monarch Bio-Fertilizers
and Research Centre
Sidco Industrial Estate
Thirumazhisai
Chennai 602107
Tamil Nadu
46. Nafed Biofertilizer
(A Unit of National Agricultural In
Coop. Mktg. Fedn. of India Ltd.)
Plot No. 51/A, Sector'F
Sanwer Road
Indore 452003
M.P.
47. National Biofertilizer
Development Centre
Dept. of Agriculture & Cooperative
Ministry of Agriculture (GOI)
C.G.O. Complex II, 204-B Wing
Kamla Nehru Nagar
Ghaziabad 201002
U.P.
48. National Fertilizers Limited
Area Office – Indore
First Floor, Vikas Tower
Indira Complex Navlakha
Indore 452001
M.P.
49. Nitrofix Laboratories
25, Bansdroni Avenue
Calcutta 700070
West Bengal
50. Orissa State Oilseed Growers'Federation
95, Satya Nagar
Bhubaneswar 751007
Orissa
51. Punjab Agricultural University
Ludhiana 141004
Punjab
52. Punjab Agro Industries Corpn.
2A, Sector 28-A, Madhya Marg
657398 Chandigarh 160002
Punjab
53. Pyrites, Phosphates & Chemicals Ltd.
4558893
12-A, Sector-24
Noida 201301 (U.P.)
54. Rajasthan State Agro Ind. Corpn. Ltd.
0-1, Subhash Nagar 71228
Jotwara Road
Jaipur 302016
Rajasthan
55. Rashtriya Chemicals & Fertilizers Ltd.
Priyadarshini Eastern Express
Highway, Sion
Mumbai 400022, Maharashtra
56. Regional Agricultural Testing
Demonstration Centre
Azamgarh 276001, U.P.
57. Regional Agricultural Testing &
Demonstration Centre
P.O. Industrial Estate
Varanasi 221100, U.P.
58. Regional Biofertilizer
Development Centre
CAU Campus, Iroisemba
P.B. No. 7
P.O. Lamphepat
Imphal 795004
59. Regional Biofertilizer
Development Centre
Deptt. of Agriculture & Cooperation
Ministry of Agriculture (GOI)
34, IInd Main Road
Hebbal
Bangalore 560024
Karnataka
60. Regional Biofertilizer
Development Contro
Deptt. of Agriculture & Cooperation
Ministry of Agriculture (GOI)
18, Sirsa Road
Near Sweak Sabha Hospital
Hissar 125001
Haryana

61. Regional Biofertilizer
Development Centre
VCA Complex, Civil Lines
Nagpur 440001
Maharashtra
62. Regional Biofertilizer
Development Centre
Plot No. B, Industrial estate
Near Shobhapur Railaway Crossing,
Adhartal
Jabalput 482004 M.P.
63. Regional Biofertilizer Development
A-158, Shahid Nagar
Bhubaneshwar 751007
Orissa
64. Regional Research Station
(Rhizobium Culture Lab.)
Deptt. of Agriculture
Govt. of Rajasthan
Durgapura
Jaipur 302006
Rajasthan
65. Regional Soil Testing Laboratory
Department of Agriculture
Government of Andhra Pradesh
Rajendra Nagar
Hyderabad 500030
A.P.
66. Soil Testing Laboratory
Deptt. of Agriculture
P.O. Mele Pattambi
Dist. Palakkad
Kerala
67. Southern Petro Chemical Industries
Corpn. Ltd.
8921 SPIC In
Spic Bio-Products
Agro Industrial
Chettiar Agaram Road, Perur
Channai 600116
Tamil Nadu
68. T. Stanes and Company Ltd.
to 213518
8/23-24, Race Course Road
TSCO In
Post Box No. 3709
Coimbatore 641018
Tamil Nadu
69. T. N. Department of Agriculture
Bacterial Culture Prodn. Unit
Post Box No. 3709
Coimbatore 641018
Tamil Nadu
70. Tamil Nadu Agricultural University
Deptt. of Agricultural
Microbiology
Coimbatore 641003
Tamil Nadu
71. The Maharashtra Agro
Industries Development Corpn. Ltd.
RCL Unit, Sugars Factory
Plot No. 52-2-D
MIDC Industrial Estate
Chinchwad
Pune 411019
72. Maharashtra
Regd. Office
Rajan House
3rd Floor, Prabhadevi
Mumbai 400025
Maharashtra
73. The Orissa Agro Industries Corpn. Ltd.
(A Govt. of Orissa Undertaking)
95, Satya Nagar
Bhubaneswar 751207
Orissa
74. The SIMA Cotton Development &
Research Association
'Shanmugha Manram'
P. O. Box No. 3871
Race Course
Coimbatore 641018
75. The University of Agricultural Sciences
Department of Agril. Microbiology
Krishi Nagar
Dharwad 580005
Karnataka
76. University of Agricultural Sciences
P. O. Bag No. 2477
Hebbal
Bangalore 560024
Karnataka
77. University of Agricultural Sciences
Dept. of Agricultural Microbiology
G.K.V.K. Campus
Bangalore 560065
Karnataka
78. U.P. Department of Agriculture
Soil Testing Laboratory
Badaun 243001
U.F.
79. U.P. Department of Agriculture
Regional Soil Testing Laboratory
Alambagh
Lucknow 226001
U.P.

80. U.P. Department of Agriculture
Regional Soil Testing and
Demonstration Centre
Etawah 206001
U.P.
81. U.P. Department of Agriculture
Regional Soil Testing Laboratory
Civil Lines
Jhansi 284001
U.P.
82. U.P. Department of Agriculture
Regional Agricultural Testing and
Demonstration Centre
B-2, Adarsh Nagar
Meerut 250001
U.P.
83. U.P. Department of Agriculture
Regional Agricultural Testing and
Demonstration Centre
Collectri Farm
Varanasi 221001
U.P.
84. U.P. Department of Agriculture
Soil Testing Laboratory
Civil Lines
Banda 210001
U.P.
85. U.P. Department of Agriculture
Soil Testing Laboratory
Malladiya Nala
Sultanpur 228001
U.P.
86. U.P. Department of Agriculture
Soil Testing Laboratory
Ramghat Road
Allgarh 202001
U.P.
87. Green Fibldo Agrotech
Vivek V-Gangwar
20/1, Vrmila Society
Dhankawadi
Pune 411043
88. Niku Bio-Research Lab.
613, Nanapeth
Pune 411002
89. Inora
Division of Know-Row Foundation
c/o Resoon 43/1, Karve Road
Pune 411038
90. 3, ACE Bio-Tech
43/2, Drandawana
Off. Karve Road
Behind Resoon
Pune 411038
91. Bio-E, a Technologies
B-15, Corporation on Building
1st Floor, Link Road, Sadar
Nagpur 440001
92. Biodynamic Products
4, Oriental Apartment
Lane No. 2
Dahanukar Colony
Kothrud
Pune 411029
93. A.V.S. Agro Products
At. Post Nimgaon Jali
Tal, Sangamner
Dist. Ahmednagar 413714
94. Nooscute Labe (I) Pvt. Ltd.
11/B, Tiwari Industries Estate
Ram Mandir Road
Goragaon (W.)
Mumbai – 400104
95. Tajasvi Bio-fert.
At – Dhumewadi
Pune – Saswad Road
Dist. Pune
96. Bharat Laboratories & Technologies
for Agriculture
Jambhale Complex
Near Rajkamal Theatre
Distt. Dhule
97. Nav Maharashtra Chakan Oil Mills Ltd.
Nav Maharashtra House
43, Shaniwar Peth
Post Box 571
Opp. Shaniwarwada
Pune 411030
98. Nomin Agro Bio Pvt. Ltd.,
K-3, Flat No. 267/D/1
Pt. Jawaharlal Nehru Industrial Estate
Tal. Mulahi
Dist. Pune
99. Nirantan Biotech
45, Vinak Girija Society
NIT College Road
Pune 411030
100. Nirmal Seeds Ltd.
Dhadgaon Road
At Post Tal. Pachora
Dist. Jalgaon 424201
101. Krishival Biofertilizers
NIDC lalampur
Flat No. D-14
Tal. Walva
Dist. Sangli
102. ECO Tek Agro
3 Prashanti
7th Lane
Prabhat road
Pune 411004

103. Maharashtra Biotech Industries
171, Kisan Kunj
Wanwadigaon
Pune 411040
104. Associate Director (Research)
College of Agriculture
Latur
Marathwada Agriculture University
Parbhani
105. Innovative Pest Control Laboratory
No. 2552, Yashashaswini
13th Main, Kumaraswamy Layout
II Stage
Bangalore 560078
106. Green Field Organics Pvt. Ltd.
1257, Dalai Street
Nipani 591237
107. HMG Biotech Pvt. Ltd.
197, II Cross, 7th Main
I Block, Koramangala
Bangalore 560034
108. T. Stanes & Co. Ltd.
13/2, 4th Cross
Nagappa Street
Palace Nagar
Bangalore 560003
109. Zuari Industries Ltd.
Jubilee Building IIIrd Floor
45, Museum Road
P.B. No. 5151
Bangalore 560001
110. Kadur Agro
R.V.V. Post
Mylasandra
Bangalore 560059
111. Madras Fertilizers Ltd.
3/1, Deewan N. Madhav Rao Road
Basavanagudi
Bangalore 560004
112. Fertilizer and Chemicals
Travancore Ltd.
28/1, Ulsoor Road
Bangalore 560042
113. Southern Petro Chemicals
Industries Corporation Ltd.
67, Ulsoor Road
Bangalore 560042
114. Karnataka Antibiotics and
Pharmaceuticals Ltd. 3321350
Dr. Rajkumar Road, I Block
Rajaji Nagar
Bangalore 560010
115. Rashtriya Chemicals &
Fertilizers Ltd.
17/7, Ali Asker Road
Bangalore 560052
116. Samarth Bio Tech Industries
D-302 & B-500, Industrial Estate
Airport Road
Hubli 580030
117. Vimpex Agencies
No. 7, Balaji Nilaya
I Cross, Swamming Pool
Extension (Pipe Line)
Malleswaram
Bangalore 560003
118. West Coast Hebochem Pvt. Ltd.
105/B Industrial Suburb, II Stage
III Cross
Goruguntepalay
Tumkur Road
Bangalore 560022
119. Terra Firma Biotechnologies Ltd.
104, Ranka Pert
5, Lalbagh Road
Bangalore 560027
120. Mardia Chemicals Ltd. (GNV FEL)
No. 3698 9th Cross Service Road,
HAL IInd Stage
13th D Main
Bangalore 560008
121. Krishak Bharati Coop. Ltd.
101/102, Swiss Complex
35, Race Course Road
Bangalore 560001
122. Associate Professor
Popularization of Biofertilizer
Scheme, Dept. of Agriculture
Microbiology, UAS, GKVK
Bangalore 560065
123. Hubli Biotechnology Pvt. Ltd.
Sumangala Complex
Lamington Road
Hubli
Karnataka
124. Karnataka Bio-fertilizers
Kiran Cottage Compound
Near Railway Bridge
Sindagi Road
Bijapur 586104
125. Eugenic Bio-Derviatives
Off. 12/2 & 12/3
Bellad Building
Vidyapeetha Road
Fort, Dharwad-8
Karnataka
126. Micro-Tech
D. No. 1208/77, 6th Cross
Taralabalu Extn.
Davangere 577005

127. Regional Director
Regional Bio-fertilizer Develpt. Centre
Hebbal
Bangalore 560024
128. Professor & Head
Dept. of Agricultural Microbiology
University of Agril. Sciences
Krishinagar
Dharwad 580005
129. Senior Scientist
Agri. Research Station (UAS)
Kotnur
Gulbarga
130. Biofertilizer Production Unit
Motilal Nehru Farmers Training Institute
Cordet
Phulpur
Allahabad (UP)

FOREIGN

1. Agrproyectos
Rivadavia y Obispo
Gelabert 3000
Sante Fe
Argentina
2. Dra. Haydee Lackner
Crawford, Keen y Cia. S.A.
Florida 681, 3er, Piso
CEP 1375, Buenos Aires
Argentina
3. INTA
Departmento de Microbiologia
C.C. 25 – 1712
Castelar
Argentina
4. Centro de Investigaciony
Desarrollo de Fermentaciones
Industriaies (CINDEFI)
Calles 47 y 115 (1900)
La Plata
Argentina
5. Fac. De Ciencias Exactas y Naturales
Avda. Uruguay No. 151
6300 Santa Rosa
La Pampa
Argentina
6. Bio-Care Technology, Pty. Ltd.
95-99 Carlingsford Road
Sefton
New South Wales
P.O. Box 8
Regents Park 2143
Australia
7. Inoculant Services
Teangi
Bethanga
Victoria 3691
8. Bio-Care Technology Pty. Ltd.
84 Rawson Road
P.O. Box 367
Woy Woy
New South Wales 2256
Australia
9. Queensland Inoculants
P.O. Box 1052
Toowoomba QLD 4350
Australia
10. University of Queensland
Dept. of Agriculture
Queensland 4072
Australia
11. Verband der Landwirtschaftlicher
Genossenschaften
Sailergasse – A – 1010
Vienna
Austria
Importer of Inoculant from:
Hellebiogri Gaiba, Italy
12. Biolink
c/o Department of Soil Science
Bangladesh Ag Univ. (BAU)
Mymensingh 2202
Bangladesh
13. Soil Microbiology Laboratory
Bangladesh Institute of Nuclear Agriculture
P. O. Box 4, Mymensingh 2200
Bangladesh
14 Centro de Investigation Agricola
Tropical (CIAT)
Av. Ejercito Nacional No. 131
Casilla 247
Santa Cruz de la Sierra
Bolivia
15. Agro-Quimic Planlto Ltda
R. Carlos Alberto Deninica 294
P. Fundo – RS
CEP: 99100
Brazil

16. Bio-Soja LTDA
Caixa Postal 87
Rua Soa Joaquim da Barra
SP 14600
Brazil
17. Centro de Energia Nuclear en Agricultura (CENA)
Av. Martins Francisco 119
Caixa Postal 96 CEP 13400
Piracicaba, Sao Paulo
Brazil
18. Empresa Brasilera de Pesquisa Agropecuaria (Embrapa)
Rodovia Carlos Joao Strass
Acesso Orlando Amaral
Caixa Postal 1061,
86001, Londrina, Parana
Brazil
19. IRFA
Av. Hulio de Castilhos
342 – 15q Andar
Porto Alegre, RS900300
Brazil
20. Nitral Ltda
Rua Rio Piquiri, 650
Pinhais, Piraquara
PR 83340
Brazil
21. Nitroar, Ind. Com. Prod. Agric.
Caixa Postal 70.008
Planaltina, DF 73300
Brazil
22. Cento de Recursos Microbiologicos en Rhizobium
MIRCEN
Fac. De Agronomia, UFRGS
Av. Bento Gonzalves 7712
91500 Porto Alegre, R.G.S.
Brazil 83420
23. Facotrigo Central-Sul-Cruz alta Brazil
Industria Bio-Soja de Inoculantes LTDA
Via Anhaguera Km 383 Marginal
esquerda, n! 1.000
Caixa Postal n! 87
CEP: 14.600
Sao Joquim da Barra – SP
Brazil
24. Leivas Leite S/A
Rua Benjamin Constant 1637
Pelotas, RS 96010
Brazil
25. Nitral S.A. Industria e Comercio de Inoculantes
e Products Agropecuarios
Rua Luiza Verone Picolli,
594 – Bacacheri
Caixa Postal, 4.162
80.00 Curitiba PR
Brazil
26. Solo-Fix
Av. Candido Abreu. 526
Conj. 407-A
Centro Civico
Curitiba, PR 80530
Brazil
27. Turfal
Caixa Postal 7410
CEP: 80.000 Curitiba PR
Brazil
28. Turfal, Ind. Com. Prod. Quim. LTDA
Rodovia BR 116 km 79
Quatro Barass, PR
Brazil
29. Agricultural Research Institute
Plant Pathology Division
Yezin
Burma
30. ISABU (Burundian Institute of Agronomic Sciences)
Bujumbura, Burundi
31. Candian Seed Coaters, Ltd.
P. O. Box 219
Brampton, Ontario L6V 2L2
Canada
32. First Line Seeds Ltd.
RR # 2
Gueip, Ontario
N1H6H8
Canada
33. Lamorna Enterprises Ltd.
2896 West 38th Avenue
Vancouver, B.C.
V6N 2X1
Canada
34. ESSO Chemical Canada,
Ag Biologicals
402-15 Innovation Blvd.
Saskatoon, SK
Canada S7N 2X8
35. Institut Rosell, INC.
8480 Boulevard St. Laurent
Montreal, Quebec
H2P 2M6
Canada

36. McIver Scientific, INC.
343 A Elgin St.
Cambridge, Ontario
N1R 7H9
Canada
37. Microbio Rhizogen Corporation
818 48th St.
East Saskatoon, Sask STK3R3
Canada
38. Rhizogen Corp.
Bay 5, 116
103rd Street East
Sask, Saskatchewan S7N 1Y7
Canada
39. Philom Bios, INC.
Mall 2, # 104-110 Research Dr.
Innovation Place
Sask, Saskatchewan S7N 3R3
Canada
40. University of British Colombia
Dept. of Plant Science
Suite 248, 2357 Main Mall
Vancouver B.C. V6T
Canada
41. Campax
Casilla 87
Temuco
Chile
42. Probical Ltda.
Alcalde Pedro Alarcon 754
Santiago
Chile
43. Microagro Ltda.
Transversal 22 No. 61 – 17
Bogota
Colombia
44. Universidad de Costa Rica
Centro de inestigaciones
Agron'micas
San Pedro, Montes de Oca
San Jose
Costa Rica
45. Estaclon Experimental la Renee
Quivican
Ministerio de Agricultura
La Habana
Cuba
46. Soils and Water Research Inst.
Department of Agricultural Microbiology
Agricultural Research Centre
Giza
Egypt
47. International Livestock Centre for Africa
(ILCA)
P.O. Box 5687
Addis Ababa
Ethiopia
48. Bioprox
6, Rue Barbes
B.P. 177
F – 92305 Levallois-Perret Cedex
France
49. INRA
F – 86600 Lusignan
France
50. Lipha
34 rue St. Romain
B.P. 8481
F – 69539 Lyon Cedex 08
France
51. GEP
P.O. Box 63
F – 41004 Bloiz Cedex
France
52. Laboratoire de Microbilogie des
Sols
INRA, BP 1540
21034 Dijon Cedrex
France
53. Liphatech
10 Avenue de Lattre de Tassigny
69330 Meyzieu
France
54. Elomestari Ltd.
Partala
51900 Juva
Finland
55. Nat. Ag. Research Inst. (NARI)
Soil Microbiology Laboratory
Mon Repos
East Coast Demerara
Guyana
56. Biotecnologia de Honduras
P.O. Box 1881
San Pedro Sula
Honduras
57. Esuela Agricola Penamericana
P.O. Box 93
Tegucigalpa
Honduras
58. Rhizogin Indonesia
Ventura Building
Third Floor Suite 303
Jl. Jagorwi-Cilandak
Outer Ring Road No. 26
Jakarta Selatan, 12430
Indonesia
59. Agricultural Research Institute
Tokachi Federation of Agricultural
Cooperatves (Tokachi Nokyoren)
KITA – 1 NISHI – 24
Hokkaido
Japan

60. Mircen Laboratory
c/o Head
Dept. Soil Sci/Fac. of Ag.
Box No. 30197
Nairobi
Kenya
61. Chitedze Agricultural Research Station
P.O. Box 158
Lilongwe
Malawi
62. Empresa Nitragin S.A.
Rio Tizapan 1716
Colonia Atlas
Guadalajara, Jalisco
Mexico
63. Institute of Agronomy & Veterinary
Hassan II
BP 6202
Rabat
Morocco
64. PTG
Postbus 8
2670 AA Naaldwijk
The Netherlands
65. IPO
Postbus 9060
6700 GW Wageningen
The Netherlands
66. Mintech (NZ) Ltd.
Agricultural Division
P.O. Box 440
Nelson
New Zealand
67. Rhizocoat Coated Seed Ltd.
P.O. Box 11 – 007
Christchurch
New Zealand
68. University of Lfe
Dept. of Plant Science
Ile – Ife Oyo State
69. Klonar, Paraquay S.R.L.
Biotechnolgia Aplicada
Yegros 437 – Edificio
San Rafael – Piso 19
Asuncion
Paraquay
70. Universidad Nacional de Cajamarca
Laboratorio de Rhizobiologia
Giudad Universitaria
Cajamarca
Peru
71. Universidad de San Cristobal de Huamanga
Laboratori de Rhizobiologia
Portal Independencia 57
Ayacucho
Peru
72. Universidad Nacional de San
Aqustin de Arequipa
Departmento de Microbiologia
Av. Alcides Carrison s/n,
La Pampilla, Arequipa
Peru
73. Los Banos Biotechnology Corporation
Suite 504, F&M Building
Legaspi St.
Legaspi Village, Makati
Metro Manila
Philippines
74. Republic of the Philippines
Department of Agriculture
Regional Soils Laboratory Region I
San Fernando
La Union 2500
Philippines
75. University of the Philippines Los Banos
Department of Soil Science
College Laguna
The Philippines
76. Zaklad Przemyslu Rolnego
Bydgoska 41
78-600 Walcz
Poland
77. Oil and Protein Seed Centre
Private Bag X1251
Potchefstroom
South Africa
78. Stimuplant
Republic of South Africa
79. Project Engrais FAO
GCPE/RWA/006/DEN
BP 337 Butare
Rwanda
80. Centro Inv. y Des. Agrario
"Las Torres"
CTRA, Rinconada – Alcald Apartado Official
41.200 – Alcola del Rio (Sevilla)
Spain
81. Central Laboratory for Biological Nitrogen
Fixation
Institute of Environmental and Natural
Resources Research
The National Centre for Research
P.O. Box 6096 Peoples' Hall
Khartoum, Sudan
82. National Taiwan University
Dept. of Agricultural Chemistry
Taipei
Taiwan
83. Bangkok Seeds Industry Co.
36 Soil Yeshit, Chan Road
Yannava
Bangkok 10120
Thailand

84. Entite, Industrial & Commercial
Enterprises, Inc.
Burkres Sokak 19/6
Cankaya – Ankara
Turkey
85. Makerere University
Inoculum Production Unit
Soil Science Department
P.O. Box 7062
Kampala
Uganda
86. Agricultural Genetics Company Ldt.
154 Science Park
Milton Road
Cambridge, Cambs
CB4 4G6
United Kingdom
87. Agricultural Genetics Co. Ltd.
Micro Bio Division
Church Street (0763) 208881
Thriplow, Nr Royston
Heris SG 877RE
United Kingdom
88. Biotal Ltd. & Biotechnica Ltd.
United Kingdom
Institute of Arable Crops Research
Crop Management Department
Rothamsed Exper. Station
Harpenden, Herts
AL5 2 JQ
United Kingdom
89. Abbott Laboratories
1 Abbott Park Road
Abbot Park llinois 60064
USA
90. Agbiochem Inc.
3 Fleetwood Court
Orinda, California 94563
USA
91. Agracetus
8520 University Green
Middletown, Wisconsin 53562
USA
92. Bentech Laboratories Inc.
14424 S.E. Industrial Way
Clackamas, Oregon 97015
USA
93. Clagene
1920 Fifth Street
Davis, California 95616-4918
USA
Contact: CEO: Roger Salquist
94. Liphatech Inc.
3103 West Custer Avenue
Milwauke, WI 53209
USA
95. NifTAL Center
1000 Holomua Road
Paia, HI 96779
USA
96. Hi-Bred International Inc.
P.O. Box 38
7300 Northwest 62nd Ave
Johnston, Iowa 50131
USA
97. Bentech Laboratories Inc.
14424 S.E. Industrial Way
Clackamas, Oregon 97015
USA
98. Celpril Industries
251 Oak Street
Manteca, California 95336-1937
USA
99. Microlife Technics
Box 3917/1833
57th Street
Sarasota, Florida 33578
USA
100. Pioneer Hi-Bred Intl., Inc.
700 Capital Square
400 Locust Street
Des Moines, Iowa 50309
USA
101. Trace Chemicals
839 Brenkman Drive
Pekin, IL 61554
USA
102. Urbana Laboratories
P.O. Box 1393
3105 Third St.
St. Joseph, Missouri 64502
USA
103. Yellowstone Valley Chemical
361 Rothiemay Rd.
Ryegate, Montana 59074
USA
104. Laboratorio de Microbiologia de Suelos y
Latinamerican Society of Rhizobiology
Control de Inoculantes del
Plan Agropecuario
Bulevar Artigas 3802
Montevideo
Uruguay
105. Miguel Castiglioni
ENZUR, S.A.
Azara 3787, Montevideo
Uruguay
106. Instituto Venezolano de Investigaciones
Cientificas (IVIC)
Aptdo 21827
Caracas 1020 A
Venezuela

107. Venezolano de Inoculantes C.A.
Av. La Anunciaciou Residencias
Grinoco
Piso 8
Apto 8-A
San Antonio de Los altos 1204
Edo Miranda
Venezuela
108. Universidad de Los Andes
Facultad de Ciencias
Dept de Biologia
Merida edo Merida
Venezuela
109. Mt. Makulu Cen Res. Sta.
Soil Micro/noculum Prod
Private Bag No. 7
Chilanga
Zambia
110. Grasslands Research Inst.
Soil Productivity Laboratory
Private Bag No. 3757
Marondera
Zimbabwe

APPENDIX 4

Edapho-Climatic Suitability of Nitrogen Fixing Trees

General climate requirements of important NFT species by mean annual temperature (altitude) and rainfall.

Over 24.5°C (0-500 m)	19.5-24.5°C (500-1000 m)	Below 19.5°C (over 1500 m)
Mean annual rainfall of 250-400 mm		
Acacia saligna	*Acacia saligna*	*Acacia Senegal*
Acacia Senegal	*Acacia Senegal*	*Prosopis alba/chilensis*
Acacia tortilis	*Prosopis alba/chilensis*	*Prosopis juliflora*
Parkinsonia aculeate	*Prosopis pallida/juliflora*	*Prosopis tamarugo*
Prosopis alba/chilensis		
Prosopis cineraria		
Prosopis pallida/juliflora		
Mean annual rainfall of 400-650 mm		
Acacia nilotica	*Acacia nilotica*	*Acacia senegal*
Acacia saligna	*Acacia saligna*	*Albizia lebbeck*
Acacia senegal	*Acacia senegal*	*Prosopis pallida/juliflora*
Albizia lebbeck	*Albizia lebbeck*	
Casuarina cunninghamiana	*Casuarina cunninghamiana*	
Casuarina glauca	*Casuarina glauca*	
Dalbergia sissoo	*Dalbergia sissoo*	
Prosopis cineraria		
Mean annual rainfall of 650-1000 mm		
Acacia nilotica	*Acacia mearnsii*	*Acacia mearnsii*
Acacia saligna	*Acacia nilotica*	*Albizia lebbeck*
Albizia lebbeck	*Acacia saliana*	*Casuarina equisetifolia*
Albizia procera	*Albizia procera*	*Leucaena diversifolia*
Albizia saman	*Albizia saman*	
Casuarina equisetifolia	*Casuarina equisitifolia*	Temperate zone
Dalbergia sissoo	*Dalbergia sissoo*	
Leucaena diversifolia	*Leucaena diversifolia*	*Alnus rubra*
Leucaena leucocephala	*Leucaena leucocephala*	
Mean annual rainfall of 1000-1600 mm		
Acacia auriculiformis	*Acacia auriculiformis*	*Acacia mearnsii*
Acacia mangium	*Acacia managium*	*Alnus acuminata*
Acacia nilotica	*Acacia mearnsii*	*Mimosa scabrella*

Over 24.5°C (0-500 m)	19.5-24.5°C (500-1000 m)	Below 19.5°C (over 1500 m)
Albizia lebbeck	*Albizia lebbeck*	
Albizia procera	*Albizia procera*	Temperate zone
Albizia saman	*Albizia saman*	
Paraserianthes falcataria	*Paraserianthes falcataria*	*Alnus rubra*
Cakkuabdra calothursus	*Calliandra calothyrsus*	*Robinia pseudoacacia*
Casuarina equisetifolia	*Casuarina equisitifolia*	
Gliricidia sepium	*Gliricidia sepium*	
Leucaena diversifolia		
Leucaena leucocephala		
Sesbania grandiflora		

General climate requirements of important NFT species by mean annual temperature (attitude) and rainfall (continued).

Over 24.5°C (0-500 m)	19.5-24.5°C (500-1000 m)	Below 19.5°C (over 1500 m)
Mean annual rainfall of more than 1600 mm		
Acacia auriculiformis	*Acacia auriculiformis*	*Acacia mearnsii*
Acacia mangium	*Acacia mangium*	*Acacia melanoxlym*
Calliandra calothyrsus	*Alnus acuminata*	*Alnus acuminata*
Gliricidia sepium	*Calliandra calothyrsus*	*Alnus nepalensis*
Paraserianthes falcataria	*Gliricidia speium*	
	Paraserianthes falcataria	Temperate zone
		Alnus nepalensis
		Alnus rubra
		Robinia pseudoacacia

Adapted from: MacDicken, 1988.

NFTs tolerant of problem soils

Species name	General environmental requirements
NFTs tolerant of saline soil	
Acacia nilotica	Arid, semiarid, subhumid
Acacia saligna	Arid, semiarid
Albizia lebbeck	Semiarid, subhumid
Casuarina equisetifolia	Humid, subhumid
Casuarina glauca	Humid, subhumid
Intsia bijuga	Humid
Parkinsonia aculeate	Arid, semiarid
Pithecellobium dulce	Arid, semiarid
Prosopis cineraria	Arid, semiarid
Prosopis tamarugo	Arid
NFTs tolerant of acid soil	
Acacia auriculiformis	Humid
Acacia mangium	Humid
Acacia mearnsii	Humid, tropical highlands
Acacia saligna	Arid, semiarid
Albizia lebbeck	Semiarid, subhumid

Species name	General environmental requirements
Albizia procera	Semiarid, Subhumid
Calliandra calothyrsus	Humid, subhumid
Casuarina glauca	Humid, subhumid
Paraserianthes falcataria	Humid, subhumid
Robinia pseudoacacia	Tropical highlands, temperate
NFTs tolerant of alkaline soil	
Acacia auriculiformis	Humid
Acacia tortilis	Arid
Casuarina equisetifolia	Humid, subhumid
Casuarina glauca	Humid, subhumid
Prosopis cineraria	Arid, semiarid
NFTs tolerant of poor drainage	
Acacia auriculiformis	Humid
Acacia nilotica	Arid, semiarid, subhumid
Acacia saligna	Arid, semiarid
Alnus rubra	Humid temperature
Casuarina equsetifolia	Humid, subhumid
Dalbergia sissoo	Semiarid, subhumid
Sesbania grandiflora	Humid, subhumid
Sesbania sesban	Semiarid, arid
NFTs tolerant of long dry seasons (> 3-4 months)	
Acacia auriculiformis	Humid, subhumid
Albizia lebbeck	Semiarid, subhumid
Albizia procera	Semiarid, subhumid
Albizia saman	Humid, subhumid
Calliandra calothyrsus	Subhumid, arid, semiarid
Dalbergia sissoo	Humid, subhumid
Gliricidia sepium	Humid, subhumid
Leucaena leucocephala	Humid, subhumid
NFTs tolerant of prolonged drought	
Acacia nilotica	Arid, semiarid
Acacia polycantha	Arid, semiarid, subhumid
Acacia saligna	Arid, semiarid
Acacia senegal	Arid, semiarid
Parkinsonia aculeate	Arid, semiarid
Prosopis alba/chilensis	Arid, semiarid
Prosopis cineraria	Arid
Prosopis pallida/juliflora	Arid, semiarid
Prosopis tamarugo	Arid

Selected Bibliography

Adil, M.L. and Katre, R.K. (1992). Blue green algae in rice soil ecosystem. In: National Seminar on Organic Farming. Eds. Rai, M.M. and Verma, L.N., Ravi Printing Press, Jabalpur, 200-210.

Alexander, M. (1977). "Introduction to Soil Microbiology." 2nd ed. John Wiley & Sons, New York.

Alexander, M. (1995). How toxic are toxic chemicals in soil? *Env. Sci. Technol.* 29, 2713-2717.

Allison, F. E. (1973). "Soil Organic Matter and Its Role in Crop Production." pp. 1-637. Elsevier Scientific Publishing, Amsterdam.

Am. Soc. Agric. Eng. 1977. Maure production and characteristics. ASAE Yearb. *D.* 384m St, Joseph, Mich.

Ange, A.L. 1995. Development of land use and plant nutrition practices during the last 30 years. In *Integrated Plant Nutrition Systems.* FAO Fertilizer and Plant Nutrition Bulletin 12, pp. 21-47.

Anonymous (1978). Annual Report, Joint Director, State Department of Agriculture, Bilaspur (M.P.). 1-52.

Anonymous (1987). Biofertilizers for sugarcane. Sugarcane breeding Institute Coimbatore. Extension Publication No. 20: 1-6.

Arora, B.K. (1984). Suspended and burning of rice husk – recent trend. In: Proceedings Bioenergy Society, First Convention and symposium, Bioenergy Society of India, New Delhi – 233-237.

Babbitt, H.E. and Baumann, E. R. 1958. Sewage and Sewage Treatment. John Willey & Sons, New York.

Bajpal, P.D. Gupta, B.R. and Ram, B. (1978). *Indian J. Agric. Res.* 12: 39-41.

Barea, J.M. 1991. Vesicular-arbuscular mycorrhizae as modifiers of soil fertility. *Advances in Soil Science* 15: 2-40.

Bartha, R. (1971). Fate of herbicide-derived chloroanilines in soil. *J. Agric. Food Chem.* 19, 385-387.

Bartha, R., and Bordeleau, L. (1969). Cell-free peroxidases in *Soil Biol. Biochem.* 1, 139-143.

Batistic, L., and Mayaudon, J. (1970). Stabilisation biologique dans le sol de l' acide ferulique ^{14}C, de. l'acide vanillique ^{14}C et de l' acide *p*-coumaric ^{14}C. *Ann Inst. Pasteur* 188, 199-206.

Bollag, J. M. (1983). Cross-coupling of humus constituents and xenobiotic substances. *In* "Aquatic and Terrestrial Humic Materials" (F. R. Christman and E. T. Gjessing, eds). pp. 127-141. Ann Arbor Sci. Publ., Ann Arbor. MI.

Bollag, J. M. M., Liu, S. Y., and Minard, R. D. (1980). Cross-coupling of phenolic humus constituents and 2,4-dichlorophenol. *Soil Sci. Soc. Am. J.* 44, 52-56.

Bollag, J. M., and Liu, S., Y. (1985). Copolymerization of halogenated phenols and syringic acid. *Pest. Biochem. Physiol.* 23, 261-272.

Bollag, J. M., and Loll, M. J. (1983). Incorporation of xenobiotics into soil humus. *Experientia* 39, 1221-1231.

Bollag, J. M., Chen, C. M., Sarkar, J., and Loll, J. M. (1987). Extraction and purification of a peroxidase from soil. *Soil Biol. Biochem.* 19, 61-71.

Bollag, J. M., Minard, R. D., and Liu, S. Y. (1983). Cross-linkages between anilines and phenolic humus constituents. *Env. Sci. Technol.* 17, 72-80.

Bollag, J. M., Sjoblad, R. D., and Liu, S. Y. (1979). Characterization of an enzyme from *Rhizoctonia praticola* which polymerizes phenolic compounds. *Can. J. Microbiol.* 25, 229-233.

Bourquelot, E., and Bertrand, G. (1895). Le bleuissement et le noireissement des champignons. *C.R. Soc. Biol.* 47, 582-584.

Buffle, J., and Leppard, G. G. (1895). Characterization of aquatic colloidx and macromolecules. I Structure and behavour of colloidal material. *Env. Sci. Technol.* 29, 2169-2175.

Buresh, R.J. 1995. Nutrient cycling and nutrient supply in agroforestry systems. In *Integrated Plant Nutrition Systems.* FAO Fertilizer and Plant Nutrition bulletin, Vol. 12, pp. 155-164.

Burgos, W. 1995. Effect of supply of organic inputs on soil fertility on western semi-arid African savannahs. In *Integrated Plant Nutrition Systems.* FAO Fertilizer and Plant Nutrition Bulletin. Vol. 12, pp. 129-138.

Burns, I. G., Hayes, M. H. B., and Stacey, M. (1973). Some physico-chemical interaction of paraquat with soil organic materials and model compounds. II Adsorption and desorption equilibria in aqueous suspensions. *Weed Res.* 13, 79-90.

Calderbank, A. (1989). The occurrence and significance of bound pesticide residues in soil. *Rev. Env. Contam, Toxtcol.* 108, 71-103.

Capriel, P., Beck, T., Borchert, H., and Harter, P. (1990). Relationship between soil aliphatic fraction extracted with supercritical hexane, soil microbial biomass, and soil aggregate stability. *Soil Sci. Soc. Am. J.* 54, 415-420.

Carmer, C. O., Converse, J. C., Tenpas, G. H. and Schlough, D. A. 1971. The design of solid manure storage for dairy herds. Am. Soc. Agric. Eng. (Winter Meeting) Paper 71-910.

Carringer, R. D., Weber, J. B., and Monaco, T. J. (1975). Adsorption-desorption of selected pesticides by organic matter and montmorillonite. *J. Agric. Food Chem.* 23, 568-572.

Clark, C. E. 1965, Hog waste disposal by lagooning. J. Sanit. Eng., Div. Am. Soc. Civ. Eng. 91 (SA6) 27-42.

Cooke, G.W. 1989. The intercontinental transfer of plant nutrients. In *Nutrient Balances and the Need for Potassium.* Proc. IPI Congress. Vol. 13, pp. 267-287. International Potash Institute, Bern.

CORT (1994) Changing Villages. 13(2). April/June, New Delhi India.

Dash M.C, Senaptai. B.K and Mishra P.C (Edited), (1986) Verms and vermicomposting. Shri Artantrana Rout for Five Stra Printing Press, Burla, India

Date, R.A. (1974). Legume inoculant production. Proc. INSA. 40 (B). 667-686.

Davis, J.G. 1994. Managing plant nutrients for optimum water-use efficiency and water conservation. *Advances in Agronomy* 53: 85-120.

Day, J.M. (1991). Inoculant production in the U.K. Paper presented in Expert Consultation on Legume Inoculant Production and Quality Control, FAO, Rome, 75-85.

De Datta, S.K. and Buresh, R.J. 1989. Integrated nitrogen management in irrigated rice. *Advances in Soil Science* 10: 143-169,

Dec. J., and Bollag, J. M. (1988). Microbial release and degradation of catechol and chlorophenols ound to synthetic humic acid. *Soil Sci. Soc. Am,. J.* 52, 1366 1371.

Dec J., and Bollag, J. M. (1994). Dehalogenation of chlorinated phenols during oxidative coupling. *Env. Sci. Technol.* 28, 484-490.

Dec. J., Haider, K., Benesi, A., Rangaswamy, V., Schaffer, A., Plucken, U., and Bollag, J. M. (1977a). Analysis of soil-bound residues of the ^{13}C-labeled fungicide cyprodinil by NMR spectroscopy. *Env. Sci. Technol.* 31, 1128-1135.

Dec. J., Haider, K., Rangaswamy, V., Schaffter, A., Fernandes, E., and Bollag, J. M. (1977b). Formation of soil-bound residues of cyprodinil and their plant uptake. *J. Food Agric. Chem.* 45,514-420.

Dec. J., Haider, K., Schaffer, A., Fernandes, E., and Bollag, J. M. (1977c). Use of the silylation procedure and ^{13}C-NMR spectroscopy to characterize bound and sequestered residues of cyprodinil. *Env. Sci. Technol.* In press.

Dinel, H., Schnitzer, M., and Mehuys, G. R. (1990). Soil lipids: Origin, nature, content, decomposition, and effect on soil physical properties. *In* "Soil Biochemistry"(J. M. Bollag and G. Stotzky, eds.), Vol. 6 pp. 397-429. Marcel Dekker, New York.

Dornbush, J. N. and Anderson, J. R. 1964. Lagooning of livestock wastes in South Dakota. Proc. Purdue Ind. Waste Conf., W. Lafayette, Ind., 19, 317-325.

Dubach, P., and Mehta, N. C. (1963). The chemistry of soil humic substances. *Soil Fert.* 26, 293-300.

Dudal, R. 1989. Vertisols of the subhumid and humid zone. In *Managing of Vertisols for Improved Agricultural Production.* ICRISAT. Proceedings IBSRAM Workshop, pp. 55-60.

Edwards C. A and Lofty J. R. (1977), Biology of earthworms, Chapman and Hall Lond. 333 pp

Edwards C. A and Neuhauser E. F. (1988) Earthworms in waste and Environment Management. (Edited), SPB Publishing. The Netherlands. P4. pp 392

Edwards, C.A. 1990. *Sustainable Agricultural Systems.* Soil and Water Conservation Society, Ankeny, Iowa, U.S.A.

Enders, C. (1943). Uber den Chemismus der Huminsaurebildung unter physiologichen Bredingungen. *Biochem. Z.* 313, 352-361.

Engebretson, R. R., Amos, T., and Wandruszka, R. v. (1996). Quantitative approach to humic acids association. *Env. Sci. Technol.* 30, 990-997.

Engebretson, R. R., and Wandruszka, R. v. (1994). Microorganization in dissolved humic acids. *Env. Sci. Technol.* 28, 1934-1941.

FAO 1984. *Fertilizer and Plant Nutrition Guide.* Fertilizer and Plant Nutrition Bulletin No. 9, Rome.

Felbeck, G. T., Jr. (1971). Chemical and biological characterization of humic matter. *In* "Soil Biochemistry" (A. D. McLaren and J. Skujins, eds.). Vol. 2, pp. 36-59. Marcel Dekker, New York.

Filip, Z. K., Flaig, W., and rietz, E. (1977). Oxidation of some phenolic substances as influenced by clay minerals. *In* "Isot. Radiat. Soil Org. Matter Stud.," Pt. II. Pp. 91-96. IAEA Bull., IAEA. Vienna.

Finck, A. 1992. *Dinger and Dungung* (Former edition: Fertilizers are Fertilization, 1982). VCH.

Finck, A. 1995. From the fertilization of crops to the management of plant nutrients in crop rotations and farming systems – an overview. In *Integrated Plant Nutrition Systems.* FAO Fertilizer and Plant Nutrition Bulletin Vol. 12, pp. 67-82.

Finck, A. and Venkateswarlu, J. 1982. Chemical properties and fertility management of vertisols. *Transactions 12th International Congress Soil Science.* Delhi: Symposium on Vertisols. pp. 61-79.

Flaig, W., Beutelspacher, H., and Rietz, E. (1975). Chemical composition and physical properties of humic substances. *In* "Soil Components. Vol. I. Organic Components" (J. E. Gieseking, ed.). pp. 1-211. Springer-Verlag, New York.

Francis, C.A. 1989. Biological efficiencies in multiple-cropping systems. *Advances in Agronomy* 42: 1-42.

Fuhr, F. (1987). Non-extractable pesticide residues in soil. *In* "Pesticides Science and Biotechnology. Proc. 6th Intl. Cong. Pestic. Chem." (R. Greenhalgh and T. R. Roberts, eds.), pp. 381-389. IUPAC, Blackwell Scientific Publ., London.

Fusteck-Mathon, E., Jambu, P., Joly, G., and Jacquesi, R. (1977). Analyse et role des bitumes dans lessols sableux acides. *In* "Proc. Symp. IAEA and FAO. Vol. 2. Soil Organic Matter Studies." Pp. 105-114, IAEA, Vienna.

Galstyan, A. S. (1958). Determination of comparative activity of peroxidase and polyphenol oxidase in soil. *Doklady Akademii Nauk Armyanskoi SSR* 26, 285-288.

Gaur, A.C. (1985). Phosphate solublising microorganisms and their role in growth and crop yields. Proc. National Symposium in Soil Biology, Hissar, 125-133.

Gaur, A.C. and Gaind, Sunita (1992). Role of phosphorus solubilisign microorganisms in crop productivity and enriched organic manure. In: National Seminar on Organic Farming, Eds, Rai, M.M. and Verma, L.N., Ravi Printing Press, Jabalpur, 134-143.

Gaur, A.C. and Singh, G. 1995. Organic and biological plant nutrient sources: potential, methods for reducing the bulk and improving the availability of nutrients. In *Integrated Plant Nutrition Systems.* FAO Fertilizer and Plant Nutrition Bulletin, Vol. 12, pp. 85-112.

Gaur, A.C., Mathur, R.S. and Sadasivam, K.V. (1980). Effect of organic materials and phosphate dissolving culture on the yield of wheat and green gram. *Indian J. Agron.* 25 (3): 501-503.

Gnanamani, A. and R. Kasturi Bai. 1992. Influence of biodigested slurry on rice-gram cultivation. Bioresource Technology 41:217-221.

Goring, C.A.I. 1990. The economic impact on sustainable agriculture on the agricultural chemical industry. In *Sustainable Agricultural Systems* (Ed. C.A. Edwards), pp. 495-503, SWCS, Iowa, U.S.A.

Greig-smith P.W; Becker, H. (1992) Ecotoxicology of earthworms. Edwards. P. J. and Heimbach F. (Edited). Intercept Ltd. UK pp. 269

Grub, W., Albin, R. C., Wells, D. M. and Wheaton, R. Z. 1969. The effects of feed, design and management on the controlof pollution from beef cattle feedlots. Proc. Agric. Waste Manage. Conf. Cornell Univ., Ithaca, N. Y.

Gupta, S.B. Rawal, A.K. and Khare, A.K. (1988). Effect of pre and post inoculation seed treatment with fungicides on nodulation and grain yield of soybean. Legume Research: 11 167-172.

Haider, K. (1976). Microbial synthesis of humic materials. *In* "Bound and Conjugated Pesticide Residues" (D. D. Kaufman, G. G. Still, G. D. Pasulson and S. K. Bandal, eds.). pp 244-257 ACS Symp. Series 29 Washington, D.C.

Haider, K. (1992). Problems related to the humification processes in soils of temperate climates. *In* "Soil Biochemistry" (G. Stozky and J. M. Bollag, eds.). Vol. 7, pp. 55-94. Marcel Dekker, New York.

Haider, K., and Martin, J. P. (1981). Decomposition in soil of specifically ^{14}C-labeled model and cornstalk lignins and coniferylalcohol over two years as influenced by drying, rewetting, and additions of an available C substrate. *Soil Biol. Biochem.* 13, 447-452.

Haider, K., Martin, J. P., and Filip, Z. (1975). Humus biochemistry. *In* "Soil Biochemistry" (E.A. Paul and A. D. McLaren, eds.), Vol. 4 pp. 195-244. Marcel Dekker, New York.

Haider, K., Spiteller, M., Wasis, A., and Fild, M. (1993). Evaluation of the binding mechanisms of anilazine and its metabolites in soil organic matter, *Intern, J. Env. Anal. Chem.* 53, 125-137.

Haider, R. J. (1969) The adsorption of linuron, atrazine, and EPTC by model aliphatic adsorbents and soil organic preparations. *Weed Res.* 9, 108-113.

Harmsen, K. 1995. Integrated phosphorus management. In *Integrated Plant Nutrition Systems.* FAO Fertilizer and Plant Nutrition Bulletin, Vol. 12, pp. 293-3065.

Hart, S. A. and Turner, M. E. 1965. Lagoons for livestock manure. J. Water Pollut. Control Fed. 37, 1579-1596.

Hatcher, P. G., and Spiker, E. C. (1988). Selective degradation of plant biomolecules. *In* "Humic Substances and Their Role in the Environment" (F. H. Frimmel and R. F. Christman, eds.), pp. 59-74. Wiley New York.

Hatcher, P. G., Bortiatynski, J. M., Minard, R, D., Dec. J., and Bollag, J. M. (1993). Use of high-resolution ^{13}C NMR to examine the enzymatic covalent binding of ^{13}C-labeled 2,4-dichlorophenol to humic substances. *Env. Sci. Technol.* 27, 2098-2103.

Hatzinger, P. B., and Alexander, M. (1995). Effect of aging of chemicals in soil on their biodegradability and extractability. *Env. Sci. Technol.* 29, 537-545.

Hayes, M. H. B. (1970). Adsorption of triazine herbicides on soil organic matter, including a short review on soil organic matter chemistry. *Res. Rev.* 31, 131-174.

Hayes, M. H. B. (1991). Concepts of the origins, composition, and structures of humic substances. *In* "Advances in Soil Organic Matter Research: The Impact on Agriculture and the Environment" (W. S. Wilson, ed.). pp. 3-22. Royal Soc. Chem., Cambridge, UK.

Haynes, R.J. and Williams, P.H. 1993. Nutrient cycling and soil fertility in the grazed pasture ecosystem. *Advances in Agronomy* 49: 119-199.

Hedge, D.M. and Diwidi, B.S. 1993. Integrated nutrient supply and management as a strategy to meet nutrient demand. *Fertilizer News* 38: 49-59.

Helling, C. S., and Krivonak, A.E. (1978). Biological characteristics of bounds dinitroaniline herbicides in soils. *J. Agric. Food Chem.* 26, 1164-1172.

Hill I. R. (1976). Degradation of the insecticide pirimicarb in soil-Characterization of "bound" residues. *In* "Bound and Conjugated Pesticide Residues" (D. D. Kaufman, G. G. Still, G. D. Paulson and S. K. Bandal, eds.). pp. 358-361. ACS Symp. Series 29, Washington, D.C.

Hsu, T. S., and Bartha, R. (1974). Interaction of pesticide-derived chloroaniline residues with soil organic matter. *Soil Sci.* 166, 444-452.

Hsu, T. S., and Bartha, R. (1976). Hydrolyzable and nonhydrolyzable 3,4-dichloroaniline humus complexes and their respective rates of biodegradation. *J. Agric. Food Chem.* 24,118-122.

Huang, P. M. (1990). Role of soil minerals in transformation of natural organics and xenobiotics in soil. *In* "Soil Biochemistry" (J. M. Bollag and G. Stotzky, eds.). Vol. 6, pp. 29-115. Marcel Dekker, New York.

Huang, P. M. (1991). Ionic factors affecting the formation of short-range ordered aluminosilicates. *Soil Sci. Soc. Am. J.* 55, 1172-1180.

Jenkinson, D. S., and Rayner, J. H. (1977). The turnover of soil organic matter in some of the Rothamsted classical experiments. *Soil. Sci.* 123, 298-305.

Jeyabal, A., K. Thanunathan, G. Kuppuswamy, and A.R. Lakshmanan. 1990. Effects of biogas slurry and other organic manures on root nodulation, growth and yield of groundnut cv. VRI 2 in vertisol. In Abstr. 30th Ann. Conf. Assoc. Microbiologists of India. Dep. Microbiol., Annamalai Univ., Annamalainagar, Tamil Nadu. P. 172.

Johar, K.S. (1992). Organic farming – a strategy. In: National Seminar on Organic Farming Ed. M.M. Rai and L.N. Verma, Ravi Printing Press, Jabalpur, 1-4.

Johnston, A.E. and Syers, J.K. 1998. *Nutrient Management for Sustainable Crop Production in Asia.* IMPHOS Conference, Bali 1997, CAB International. Wallingford, Oxon (in print).

Kaiser, J. P., and Bollag, J. M. (1990). Microbial activity in the terrestrial subsurface. *Experientia* 46, 797-805.

Katan, J., Fuhremann, T. W., and Lichtenstein, E. P. (1976). Binding of ^{14}C-parathion in soil: A reassessment of pesticide persistence. *Science* 193, 891-884.

Kazano, H., Kearney. P. C., and Kaufman, D. D. (1972). Metabolism of methylcarbamate insecticides in soil. *J. Agric. Food Chem.* 20, 975-979.

Khan, S. U. (1974). Adsorption of bipyridylium herbicides by humic acid. *J. Env. Qual.* 3, 202-206.

Khan, S. U. (1982). Bound pesticide residues in soil and plants. *Res. Rev.* 84, 1-25.

Khare, A.K., Vaishya, U.K. and Jain, H.C. (1989). Effect of cultural practices and herbicides on nodulation in soybean. *Geobiose,* 16: 272-274.

Kirk, T. K., and Farrell, R. L. (1987). Enzymatic combustion: The microbial degradation of lignin *Ann. Rev. Microbiol.* 41, 465-505.

Kogel-Knabner, I. (1993). Biodegradation and humification processes in forest soils. *In* "Soil Biochemistry" (J. M. Bollag and G. Stotzky, eds.). Vol. 8. pp. 101-135. Marcel Dekker, New York.

Kogel-Knabner. I., Zech, W., Hatcher, P. G., and De b euw, J. W. (1991). Fate of plant components during biodegradation and humification of forest soil: Evidence from structural characterization of individual bio-macromolecules. *In* "Advances in Soil Organic Matter Research: The Impact on Agriculture and the Environment". (W. S. Wilson, ed.), pp. 61-70. Royal Soc. of Chem, Cambridge, UK.

Kononova, M. M. (1966). "Soil Organic Matter. Its Nature, Its Role in Soil Formation and Soil Fertility." Pergamon Press, Oxford.

Kozlov, K. A. (1964). Enzymatic activity in the rhizosphere and soil in the East Siberia area. *Folia Microbiol.* 9, 145-149.

Krasil'nikov, N A., Terekhov, O. S., and Zenova, G. M. (1971). Effect of tyrosine stereoisomers on formation of the melanin-like pigments of actinomycetes. *Mikrobiologiya* 40, 481-484.

Kumada, K., and Kato, H. (1970). Browning of pyrogallol as affected by clay minerals. *Soil Sci. Plant Nutr.* 16, 195-200.

Kuppuswamy, G., A. Jeyabal and A.R. Lakshmanan. 1992. Effect of enriched biodigested slurry and FYM on growth and yield of rice. Agric. Sci. Digest. 12(2): 101-104.

Kuppuswamy, G., A. R. Lakshmanan and A. Jeyabal. 1991. Effect of gypsum-enriched biogas sludge and farmyard manure on rice yield. Int. Rich Res. News. 16 (4): 18.

Kyuma, K., and Kawaguchi, K. (1964). Oxidative changes of polyphenols as influenced by allophane. *Soil Sci. Soc. Am. Proc.* 28, 371-374.

Ladd, J. N., and Butler, J. H. A. (1975). Humus-enzyme system and synthetic, organic polymer-enzyme analogs. *In* "Soil Biochemistry" (E. A. Paul and A. D. McLaren, eds.), Vol. 4 pp. 143-194. Marcel Dekker, New York.

Lal, R. 1991. Myths and scientific realities of agroforestry as a strategy for sustainable management for soils in the tropics. *Advances in Soil Science* 15: 91-137.

Larson, R.R., and Hufnal, J. M., Jr. (1980). Oxidative polymerization of dissolved phenols by soluble and insoluble inorganic species. *Limnol. Oceanogr*, 25, 505-512.

Lee K.E. (1985) Earthworm: Their ecology and relationships with soil and land use. Academic Press. 411 pp.

Little, J. F. Agriculture and the Prevention of River Pollution as Experienced in West of Scotland. Inst. Sewage Purif. J. Proc. 65: 452-454.

Liu, S. Y., and Bollag, J. M. (1985). Enzymatic binding of the pollutant 2,6-xylenol to a humus constituent. *Water, Air, Soil Poll.* 25, 97-106.

Liu, S. Y., Freyer, A. J., Minard, R. D., and Bollag, J. M. (1985). Enzyme-catalyzed complex-formation of amino acid esters and phenolic humus constituents. *Soli. Sci. Soc. Am. J.* 49, 337-342.

Liu, S. Y., Minard, R. D., and Bollag, J. M. (1981). Oligomerization of syringic acid, a lignin derivative, by a phenoloxidase. *Soli Sci. Soc. Am. J.* 45, 1100-1105.

Loehr, R. C. 1974. Agricultural Waste Management: Problems, Processes, and Approaches. Academic Press, New York.

Madden, J. M. Dornbush, J. N. 1971. Measurement of runoff and runoff carried waste from commercial feedlots. Proc. Intern. Symp. Livestock Wastes. Am. Soc. Agric. Eng., St. Joseph, Mich.

Madhok, M.R. (1934). Agric. Livestock in India 46 70-73.

Magdoff, F., Lanyon, L. and Liebhardt, B. 1997. Nutrietn cycling, transformations, and flows: Implications for a more sustainable agriculture *Advances in Agronomy* 60: 2-73.

Maillard, L.C. (1916). Synthese des matieres humiques par action des acides amines sur les sucres reducteurs. *Ann. Chim. Phys. Vol. No*. 5, 5,258.

Martin, J. P., and Haider, K. (1971). Microbial activity in relation to soil humus formation. *Soil Sci.* 3, 54-63.

Martin, J. P., and Haider, k. (1980). A comparison of the use of phenolase and peroxidase for the synthesis of model humic acid-type polymers. *Soil Sci. Soc. Am. J.* 44, 983-988.

Mathur, S. P., and Morley, H. V. (1978). Incorporation of methoxychlor-^{14}C in model humic acid prepared from hydroquinone. *Bull. Env. Contam. Toxicol.* 20, 268-274.

Mathur, S. P., and Schnitzer, M. (1978). A chemical and spectroscopic characterization of some synthetic analogues of humic acid. *Soil Sci. Soc. Am J.* 42,591-596.

Matsubara, T., and I Wasaki, H. (1972). Occurrence of laccase and tyrosinase in fungi of agaricales and comparative study of laccase from *Russula delica and Russula pseudodelica. Bot. Mag.* (Tokyo) 85, 71-83.

Mayaudon, J. (1968). Stabilization biologique des proteins 14-C dans le sol. In "Isotopes and Radiation in Soil Organic Matter Studies," pp. 177-188. IAEF, Vienna.

McCall, P. J., Vrona, S. A., and Kelly, S. S. (1981). Fate of uniformly carbon-14-ring-labeled 2,4,5-trichlorophenoxyacetic acid and 2,4-dichlorophenoxyacetic acid. *J. Agric. Food Chem.* 29, 100-107.

McCalla, T. M. and Viets, F. G. 1969. Chemical and microbial studies of waste from beef cattle feedlots. Seminar Manage. Beef Cattle Feedlot Wastes, Lincoln, Neb. (mimeo)

McKenzie, R. M. (1989). Manganese oxides and hydroxides. *In* "Minerals in Soil Environments" (J. B. Dixon and S. B. Weed, eds.), SSSA Book Series No. 1, pp. 439 465. Soil Sci. Soc. Am. Madison, WI.

Miller, F.P. and Larson, W.E. 1990. Lower input effects on soil productivity and nutrient cycling. In *Sustainable Agricultural Systems* (Ed. C.A. Edwards), pp. 549-568, SWCS,

Iowa, U.S.A.

Mohan, S., Jayaraj, Purushothaman and Rangrajan, A.V. (1987). Can the use of *Azospirillum* biofertilizer control sorghum shootfly. *Cur. Sci.* 56: 723-725.

Mortland, M. M., and Halloran, L. J. (1976). Polymerization of aromatic molecules on smectite. *Soil Sci. Soc. Am. Proc.* 40, 367-370.

Motsara, M.R. and Bhattacharyya, P. (1992). Status of biofertilizer and constraints in their large scale commercialization. In: National Workshop on commercialization of biotechnologies in agriculture and aquaculture, April 23-24, 1992. Centre for Management in Agriculture IIM Ahmedabad.

Muralikrishna Reddy, V. 1987. Effect of biogas slurry and treated ureas on rice cv. ADT 36. M.Sc. (Ag.) Diss. Annamalai Univ. annamalainager, Tamil Nadu.

Musso, H. (1967). Phenol coupling. *In* "Oxidative Coupling of Phenols" (W. I. Taylor and A. R. Battersby, eds.), pp. 1-94. Marcel Dekker, New York.

Nagar, S. 1975. Biogas plant based on night soil and animal dung. IRCWD News 1446.

Nye, P.H. and Greehland, D.J. 1960. The soil under shifting cultivations. *CBS Technical Communication No. 51*, Harpenden, UK.

Oertli, J.J. 1992. Nutrient management under water and salinity stress. In *Proceedings on the International Symposium on Nutrient Management for Sustained Productivity.* Vol. 1. pp. 138-165. Ludhiana.

Pandey, A. and Kumar, S. (1989). Potential of Azospirilla as biofertilizers for upland agriculture: A review *J. Scientific and Industrial Research,* 43: 134-144.

Parr, J.F., Papendick, R.I., Youngberg, J.G. and Meyer, R.E. 1990. Sustainable agriculture in the United States. In *Sustainable Agricultural Systems* (Ed. C.a. Edwards), pp. 549-568. SWCS, Iowa, U.S.A.

Parris, G. E. (1980). Covalent binding of aromatic amines to hunates. I. Reactions with carbonyls and quinines. *Env. Sci. Technol.* 14, 1099-1106.

Paul, E. A., and Van Veen, J. A. (1978). The use of tracers to determine the dynamic nature of organic matter. *Trans. 11th Intl. Cong. Soil Sci,* 3,61-102.

Perkins, H. F., Parker, M. B. and Walker, M. L. 1964. Chicken manure-its production, composition, and use as a fertilizer. G. Agric. Exp. Stn. Bull. (N.S.) 123.

Piccolo, A., Nardi, S., and Concheri, G. (1996). Macromolecular changes of humic substances induced by interaction with organic acids. *Europ. J. Soli Sci.* 47, 319-328.

Pierce, R. H., Olney, C. E., and Felbeck, G. T. (1974). pp 'DDT adsorption to suspended particulate matter in sea water. *Geochim. Cosmochim. Acta* 38, 1061-1073.

Pieri, Chr. 1995. Technological options for controlling soils organic matter losses in tropical rainfed cropping systems. In *Integrated Plant Nutrition Systems.* FAO Fertilizer and Plant Nutrition Bulletin 12, pp. 181-198.

Pignatello, J. J., and Xing, B. (1996). Mechanism of slow sorption of organic chemicals to natural particles. *Env. Sci. Technol.* 30, 1-11.

Pinnavaia, T. J., Hall, P. L., Cady, S. S., and Mortland, M. M. (1974). Aromatic radical cation formation on the intracystal surfaces of transition metal layer lattice silicates. *J. Phys. Chem.* 78, 994-999.

Prasad, R. and Power, J.F. 1991. Crop residue management. *Advances in Soil Science* 15: 205-251.

Rennie, R.J. (1991). Canadian Legume Inoculants. Evolution of an Industry. In: Expert consultation on legume inoculant production and quality control. FAO Rome, 87-105.

Rewari, R.S. (1984). Summarized results of microbiology trials, All India Coordinated Research Project on Improvement of Pulses, New Delhi.

Rewari, R.S. (1985). Summarized results of microbiology trials, All India Coordinated Research Project on Improvement of Pulses, New Delhi.

Rewari, R.S. (1991). Canadian Legume Inoculants. Evolution of an Industry. In: Expert consultation on legume inoculant production and quality control. FAO Rome, 87-105.

Ross, S. (1989). "Soil Processes. A Systematic Approach." Routledge, London.

Roy, R.N. 1995. Integrated plant nutrition systems. In *Integrated Plant Nutrition Systems.* FAO Fertilizer and Plant Nutrition Bulletin 12, pp. 49-66.

Ryan, J. 1997. Future Directions of applied soil fertility research in West Africa and North Africa. In *Accomplishments and Future Challenges in Dryland Soil Fertility Research in the Mediterranean Area* (Ed. J. Ryan), pp. 335-340.

Safe, S., Ellis, B. E., and Hutzinger, O. (1976). The *in vitro* hydroxylation of 4'-chloro-4-biphenylol by a mushroom tyrosinase preparation. *Can. J. Microbiol.* 22, 104-106.

Sahu, S.K. Nandu, 1988. Direct and residual effect of biogas residue application with nitrogen on rice yield. Int. Rice. Newsl. 13(4): 28.

Sanchez, P.A. 1976. *Properties and Management of Soils in the Tropics.* John Wiley, New York.

Sankaram, A. (1996). Eco-farming Relevance and Rewards. Survey of Indian Agriculture, 1996. The Hindu. p. 35-37.

Sarkar, J. M., Malcolm, R. I., and Bollag, J. M. (1988). Enzymatic coupling of 2,4-dichlorophenol to stream fulvic acid in the presence of oxidoreductases. *Soil Sci. Soc. Am. J.* 52, 688-694.

Sawyer, C. 1960. Chemistry for Sanitary Engineers. McGraw-Hill Book Co., New York.

Scheffer, F., Meyer, B., and Niederbudde, E. A. (1959). Huminstoffbildung unter katalyischer. Einwirkung naturlich vorkommender Eisenverbindungen im Modellversuch. *Z. Pflanzenernaehr. Bodenkd.* 87, 26-44.

Schmid, L. A. and Lipper, R. T. 1969. Swine waste characterizations and anaerobic digeston. Proc. Agric. Waste Manage. Conf., Cornell Univ., Ithaca, N. Y.

Schnitzer, M. (1978). Humic substances: Chemistry and reactions. *In* "Soil Organic Matter" (M. Schnitzer and S. U. Khan, eds.), pp. 1-64. Elsevier, Amsterdam.

Schnitzer, M., and Khan, S. U. (1972). "Humic Substances in the Environment." Marcel Dekker, New York.

Sen, S. and W.C. Bonde, 1962 Effect of time of application of FYM at different levels J. Indian soil Sci. 10:61-63.

Senesi, N., and Testini, C. (1980). Adsorption of some nitrogenated herbicides by soil humic acids. *Soil Sci.* 130, 314-320.

Shanker, R., Kaiser, J. P., and Bollag, J. M. (1991). Microbial transformation of heterocyclic molecules in deep subsurface sediments, *Microb. Ecol.* 22, 305-316.

Sharma, J.P. and Singh, M. (1971). Response of rice to phosphatic and nitrogenous fertilizers with and without phosphobacterin culture. *Indian J. Agron.* 16: 15-18.

Shindo, H., and Huang, P. M. (1982). Role of Mn (IV) oxide in abiotic formation of humic substances in the environment. Nature (London) 298, 363-365.

Shindo, H., and Huang, P. M. (1984a). Significance of Mn (IV) oxide in abiotic formation of organic nitrogen complexes in natural environments. *Nature* (London) 308, 57-58.

Shindo, H., and Huang, P. M. (1984b). Catalytic effects of manganese (IV). Iron (III), aluminum, and silicon oxides on the formation of phenolic polymers. *Soil Sci. Soc. Am. J.* 48, 927-934.

Shindo, H., and Huang, P. M. (1985a). Catalytic polymerization of hydroquinone by primary minerals. *Soil Sci.* 39, 505-511.

Shindo, H., and Huang, P. M. (1985b). The catalytic power of inorganic components in the abiotic synthesis of hydroquinone-derived humic polymers. *Appl. Clay Sci.* 1, 71-81.

Simmons, K. E., Minard, R. D., and Bollag, J. M. (1988). Oxidative coupling and polymerization of guaiacol, a lignin derivative. *Soil Sci. Soc. Am. J.* 52, 1356-1360.

Simmons, K. E., Minard, R. D., and Bollag, J. M. (1989). Oxidative co-oligomerization of guaiacol and 4-chloroaniline. *Env. Sci. Technol.* 23, 115-121.

Singh, J.B., 1989. Biogas slurry manure. Moving Technology. 4:34-36.

Sjoblad, R. D., and Bollag, J. M. (1981). Oxidative coupling of aromatic compounds by enzymes from soil microorganisms. *In* "Soil Biochemistry" (E. A. Paul and J. N. Ladd, eds.), Vol. 5, pp. 113-152. Marcel Dekker, New York.

Smith, A. E., and Muir, D. C. G. (1980). Determination of extractable and nonextractable radioactivity from prairie soils treated with carboxyl – and ring-labeled ^{14}C-2,4-D. *Weed Res.* 20, 123-129.

Smith, S.J. Schepers, J.S. and Porter, L.K. 1990. Assessing and managing agricultural nitrogen losses to the environment. *Advances in Soil Science* 14: 2-43.

Somesegaran, Padama, (1991). Inoculant production with emphasis on choice of carries, methods of production and reliability testing/quality assurance guidelines. In: Expert consultation on legume inoculant production and quality control, F.A.O., 87-105.

Sorokin, A., Seris, J. L., and Meunter, B. (1995). Efficient oxidative dechlorination and aromatic ring cleavage of chlorinated phenols catalyzed by iron sulfophithatocyanine. *Science* 268, 1163-1166.

Steelink, C., and Tollin, G. (1967). Free radicals in soil. *In* "Soil Biochemistry" (A. D. Mclaren and G. M. Peterson, eds.), pp. 147-169. Marcel Dekker, New York.

Stevenson, F. J. (1976). Organic matter reactions involving pesticides in soil. *In* "Bound and Conjugated Pesticide Residues" (D. D. Kaufman, G. G. Still, G. D. Paulson, and S. K. Bandal, eds.), pp. 180-207. ACS Symp. Series 29, Washington, DC.

Stevenson, F. J. (1994). "Humic chemistry: Genesis, Composition, Reactions," 2nd ed. John Wiley & sons, New York.

Stevenson, I. L., and Schnitzer, M. (1982). Transmission electron microscopy of extracted fulvic and humic acids. *Soil Sci.* 133, 179-185.

Steward, B.A. and Robinson, C.A. 1997. Are agroecosystems sustainable in semi-arid regions? *Advances in Agronomy* 60: 191-228.

Stewart, T. A. and Mcilwain, R. 1971. Aerobic treatment of poultry manure using an oxidation ditch. *In* Livestock Waste Management and Pollution Abatement. Am. Soc. Agric. Eng., St. Joseph, Mich. Publ. *Proc*-271.

Stott, D. E., Kassim, G., Jarrell, W. M., Martin, J. P., and Haider, K. (1983). Stabilization and incorporation into biomass of specific plant carbons during biodegradation in soil. *Plant Soil* 70, 15-26.

Subba Rao, N.S. (1982). Biofertilizers. In: Advances in Agricultural Microbiology, Eds: Subba Rao, N.S. Oxford IBH Publishing Co., New Delhi, 219-242.

Subba Rao, N.S. (1988). Biofertilizer in Agriculture. Second edition, Oxford and IBH Publishing Co. Pvt. Ltd. 1-208.

Sultan A Ismail (1997) Vermicology: Biology of earthworm, Orient Longman, Publ. Hyderabed, Inida.

Swaminathan, M.S. (1995). Rational of integrated Systems. Survey of Indian Agriculture, 1995. The Hindu p. 7-15.

Syers, J.K. 1995. Contribution of soil reserves to plant nutrient supply. In *Integrated Plant Nutrition Systems*. FAO Fertilizer and Plant Nutrition Bulletin. Vol. 12, pp. 243-252.

Taiganides, E. P. and Hazen, T. E. 1964. Physical and chemical properties of animal wastes. Am. Soc. Agric. Eng. Paper 64-315. (unpublished mimeograph)

Taiganides, E. P., Hazen, T. E., Baumann, E. R. and Johnson, H. P. 1964. Properties and pumping characteristics of hog wastes. Am. Soc. Agric. Eng. Trans. 7 (2) 123-129.

Talukder, N.M., M.S. Ali and A. Latif. 1988. Effect of biogas effluent on the yield and quality of rice. Int. Rice Res. Newsl. 37:11-16.

Thorne, K. A., Pettigrew, P. J., Goldenberg, W. S., and Weber, E. J. (1996). Covalent binding of aniline to humic substances. 2, 15N-NMR studies of nucleophilic addition reactions. *Env. Sci. Technol.* 30, 2764-2775.

Tilak, K.V.B.R. (1991). Bacterial Fertilizers. Publication and Information Division, ICAR Krishi Anusandhan Bhavan, New Delhi, 110012, 1-16.

Tilak, K.V.B.R. and Subba Rao, N.S. (1978). Carries for legume (*Rhizobium*) inoculants Fertilizer News, 23 (2): 25-28.

Tilak, K.V.B.R. and Subba Rao, N.S. (1987). Association of *Azospirillum brasilense* with pearl millet (*Pennisetum americannum* L. Leeke). *Biology and Fertility of Soils*. 4: 97-102.

Tisdale, S.K., Nelson, W.L. and Beaton, J.D. 1985. *Soil Fertility and Fertilizers*. McMillan, New York.

Townshend, A. R. Reichert, K. A. and Nodwell, J. H. 1969. Status report on water pollution control facilities in the province of Ontario. Proc. Agric. Waste Manage. Conf., Cornell Univ. Ithaca, N. Y.

Vaishya, U.K. and Gajendragadkar, G.R. (1988). Interaction studies between strains of *Rhizobium* and pigeonpea cultivars. *Narendary Deva J. Agric. Res.* 3(1): 34-36.

Vaishya, U.K. and Sanoria, C.L. (1972). Specificity and efficiency of *Rhizobium* cultures of Bengal garm (*Cicer arietinum* L.). *Indian Journal of Microbiology*. 12: 133-141.

Van Alfen, N. K., and Kosuge, T. (1976). Metabolism of the fungicide 2,6-dichloro-4-nitroanline in soil. *J. Agric. Food Chem*. 24, 584-588.

Veeresh G.K; Rajagopal D and Viraktamath. C.A (1991) (Edited) Advances in Management and Conservation of Soil Fauna 1, Oxford and IBH Publ. Co. Pvt. Ltd. New Delhi, India, pp 925.

Venkataraman, G.S. and Shanmugasundaram, S. (1992). Algal Biofertilizer Technology for Rice. DBT Centre for BGA. Biofertilizer, Madurai Kamraj University, Madurai, 625021, T.N. 1-24.

Venkatramani, G. (1995) Ecological farming- A waile option for future Survey of Indian Agriculture, 1995. The Hindu p. 23-31.

Venkatramani, G. (1996). Eco-technology towards intensive farming. Survey of Indian Agriculture, 1996. The Hindu p. 29-33.

Verma, L., Martin, J. P., and Haider, K. (1975). Decomposition of carbon-14-labeled proteins, peptides, and amino acids: Free nd complexed with humic polymers. *Soil. Sci. Soc. Amer. Proc*. 39, 279-284.

Verma, L.N. (1987). Development of integrated nutrient supply system, use of inorganic and biofertilizers and their economics. In: Seminar on Agricultural Technology National Fertilizer Ltd. New Delhi, 23-29, March, 1987. 1-13.

Verma, L.N. and Bhattacharya, P. (1990). Role of Biotechnology in supplying plant nutrients in the nineties. *Fertilizer News*, 35 (12), 87-97.

Verma, L.N. and Bhattacharya, P. (1991). Developmental and use of biofertilizer in India. India. Fertilizer scene, Annual – 1991, 53-62.

Verma, L.N. and Bhattacharya, P. (1992). Production distribution and promotion of biofertilizers. In: Fertilizer, Organic Manures, Recycleable Wastes and Biofertilizers, Ed. H.L.S. Tandon F.D.C.D., New Delhi 132-147.

Verma, L.N. and Martin, J.P. (1976). Decomposition of algal cells and components and their stabilization through complexing with model humic acid-type phenolic polymers. *Soil biology and Biochemistry*. 3(2): 85-90.

Verma, L.N. and Rawat, A.K. (1991). Survival of *Rhizobium japonicum* as affected by different levels of moisture and organic matter in sandy soils *Jantralbl Microbiol*. 146: 545-547.

Wada, K. (1989). Allophane and imogolite. *In* "Minerals in Soil Environments" (J. B. Dixon and S. B. Weed, eds.), SSSA Book Series No. 1, pp. 1051-1087. Soil Sci. Soc. Am., Madison, WI.

Waksman, S. A., and Iyer K. R. N. (1932). Contribution to our knowledge of the chemical natural and origin of humic: I. On the synthesis of the "humus nucleus." *Soil Sci*. 34, 43-69.

Wang T. S. C., Chem. J. H., and Hsiang, W. M. (1985). Catalytic synthesis of humic acid containing various amino acids and dipeptides. *Soil Sci*. 140, 3-10.

Wang, I. S. C., Kao, M. M., and Li, S. W. (1978a). A new Proposed mechanisms of formation of soil humic substance. *In* "Studies and Essays in Commemoration of the Golden Jubilee of Academia Sinica." pp 357-372. Academia Sinica. Taipei, Taiwan.

Wang, M. C. (1987). Catalytic role of selected soil minerals in the abiotic formation of humic substances and the associated reactions. Ph.D. Dissertation, Univ. of Saskatchewan, Saskatoon, Canada.

Wang, M. C., and Huang, P. M. (1986). Humic macromoleule interlayering in nontronite through interaction and phenol monomers. *Nature* (London) 323, 529-531

Wang, M. C., and Huang, P. M. (1987). Polycondensation of pyrogallol and glycine and the associated reactions as catalyzed be birnessite. *Sci. Total Env*. 62, 435-442

Wang, M. C., and Huang, P. M. (1988). Catalytic power of nontronite, kaolinite, and quartz and their reaction sites in the formation of hydroquinone-derived polymers. *Appl. Clay Sci*. 4, 43-57.

Wang, M. C., and Huang, P. M. (1992). Significance of Mn (IV) oxide in the abiotic ring cleavage of polyphenol in natural environments. *Sci. Total Env*. 113, 147-157.

Wang, M. C., and Huang, P. M. (1994). Structural role of polyphenols in influencing the ring cleavage and related chemical reactions as catalyzed by nontronite. *In* "Humic Substances in the Global Environment and Implications on Human Health" (N. Senesi and T. M. Miano, eds.), pp. 173-180. Elsevier, Amsterdam.

Wang, T. S. C., and Li. S. W. (1977). Clay minerals as heterogeneous catalysts in preparation of model humic substances. *Z. Pflanzenernaehr. Bodenkd.* 140, 669-676.

Wang, T. S. C., Kao, M. M., and Huang, P. M. (1980). The effect of pH on the catalytic synthesis of humic substance by illite. *Soil Sci.* 129, 333-338.

Wang, T. S. C., Li, S. W., and Freng, Y. L. (1978b). Catalytic polymerization of phenolic compounds by clay minerals. *Soil Sci.* 126, 15-21.

Wang, T. S. C., Wang, M. C. and Huang, P. M. (1983). Catalytic synthesis of humic substance by using aluminas as catalysts. *Soil Sci.* 136, 226-230.

Wang, T. S. C., Yeh, K. I., Cheng, S. Y., and Yang, T. K. (1971). "Biochemical interactions among plants." pp 113-120. Nat Acad. Sci. Washington, DC.

Wani, S.P. (1992). Role of N2 fixing bacterial inoculants in upland cereals production. In: National Seminar on Organic Farming. Eds. Rai, M.M. and Verma, L.N., Ravi Printing Press, Jabalpur 186-199.

Wani, S.P., Zambre, M.A. and Lee, K.K. (1988). Population dynamics of nitrogen fixing bacteria associated with pearl millet *Pennisetum americanum* (L.) Leeke Plants. In: Biotechnology of Nitrogen Fixation in the Tropics (Eds. Z.H. Shamsuddin, M.W, Othman, M. Marziah and J. Sundaram), University Pertanian, Malaysia, 21-30.

Weber, J. B. (1980). Adsorption off buthidazole, VEL 3510, tebuthiuron, and fluridone by organic matter, montmorillorrite clay, exchange resins, and a sandy loam soil. *Weed Sci.* 28, 478-483.

Weber, W. J., Jr., and Huang, W. (1996). A distributed reactivity model for sorption by soils and sediments. 4. Intraparticle heterogeneity and phase-distribution relationships under monequilibrium conditions. *Env. Sci. Technol.* 30, 881-888.

Weller, J. B. 1970. Building design, Proc. Symp. Farm Wastes, Inst Water Pollu. Control, Newcastle-upon-Tyne, England.

Wershaw, R. I. (1993). Model for humus in soils and sediments. *Env. Sci. Technol.* 27, 814-816.

Wild, A. 1988. *Russell's Soil Conditions and Plant Growth.* Longman, Harlow.

Witzel, S. A. *et al.*, 1966. Physical, chemical and bacteriological properties of farm waste (bovine animals). Proc. Natl. Symp. Anim. Waste Manage, Mich. State Univ., East Lansing.

Wolf, D. C., and Martin, J. P. (1976). Decomposition of fungal mycelia and humic-type polymers containing carbon-14 from ring and side-chain labeled 2,4-D and chloropham. *Soil. Sci. Soc. Am. Proc.* 40, 700-704.

Woomer. P.L and Swift M.J (1994) (Edited) The Biological Management of Tropical Soil Fertility, John Wiley and Sons. UK pp 243

Xing, B., Pignatello, J. J., and Gigliotti, B. (1996). Competitive sorption between atrazine and other organic compounds in soils model sorbents. *Env. Sci. Technol.* 30, 2432-2440.

Yoshida, M. (1883). Chemistry of lacquer (Urushi). *J. Chem. Soc.* 43, 472-486.

You, I. S., Jones. R. A., and Bartha, R. (19820. Evaluation of chemically defined model for the attachment of 3,4-dichloroaniline to humus. *Bull. Env. Contam. Toxicol.* 29 476-482.

Zodhy, L.L., S.M.S. Badr El-Din, M.A. Khalah and H.K. Abd. E.L. MakNud. 1984. Evaluation of digested slurry as a fertilizer. Egypt. J. Microbiol. 19:131-138.

ZSI, Calcutta (1993) Earthworm resources and vermiculture, pp. 128.